Volume 14

LOSS PREVENTION

Prepared by editors of
CHEMICAL ENGINEERING PROGRESS

a **CEP** technical manual

published by AMERICAN INSTITUTE OF CHEMICAL ENGINEERS

LOSS PREVENTION

Volume 14

Prepared by Editors of

CHEMICAL ENGINEERING PROGRESS

a CEP technical manual published by
AMERICAN INSTITUTE OF CHEMICAL ENGINEERS
345 East 47 Street New York, N.Y. 10017

American Institute of Chemical Engineers
345 East 47 street, New York, New York 10017

Library of Congress Card Number LC74-4373616

ISSN No. 0149—3701
ISBN No. 0-8169-0195-3

INTRODUCTION

Safety in the Chemical Industry continues to be of great concern to the public. This, the 14th volume of *Loss Prevention*, is, like its predecessors, dedicated to the active management of safety problems. A free exchange of data and the publication of responsible opinions are essential to the prevention of future accidents.

The symposium is sponsored by the Area 4e Committee on Chemical and Petroleum Plant Safety. Six sessions are presented covering Hazard Analysis, Transportation, Emissions and Waste Control, Case Histories and Plant Retrofit, Safe Design and Operation of Ethylene and Other Petrochemical Plants and Selected Loss Prevention Topics. A total of 26 papers are offered with discussion in the form of questions and answers at the end of each paper. Six of the papers originated in Great Britain, the Netherlands, Norway, and Poland. We are very gratified for this international expression of viewpoints.

Thanks are due to our session chairmen and co-chairmen, as well as to the authors and to those organizations which provided support in the form of facilities, time, and travel expenses for all who participated. Your continued support of these symposia is very much appreciated. Your input is invited so that we may keep the program in tune with current and future conditions.

J.E. Rogerson

AMERICAN INSTITUTE OF CHEMICAL ENGINEERS
14TH LOSS PREVENTION SYMPOSIUM
JUNE 8-12, 1980 PHILADELPHIA, PA

88TH NATIONAL MEETING
2ND CHEMICAL PLANT EQUIPMENT EXPOSITON

J.E. ROGERSON, SYMPOSIUM CHAIRMAN
THE PROCTER & GAMBLE COMPANY, CINCINNATI, OH

ALPHONSE A. CORONA, JR., SYMPOSIUM CO-CHAIRMAN
MOBILE RESEARCH & CEVELOPMENT CORP., PRINCETON, NJ

TABLE OF CONTENTS

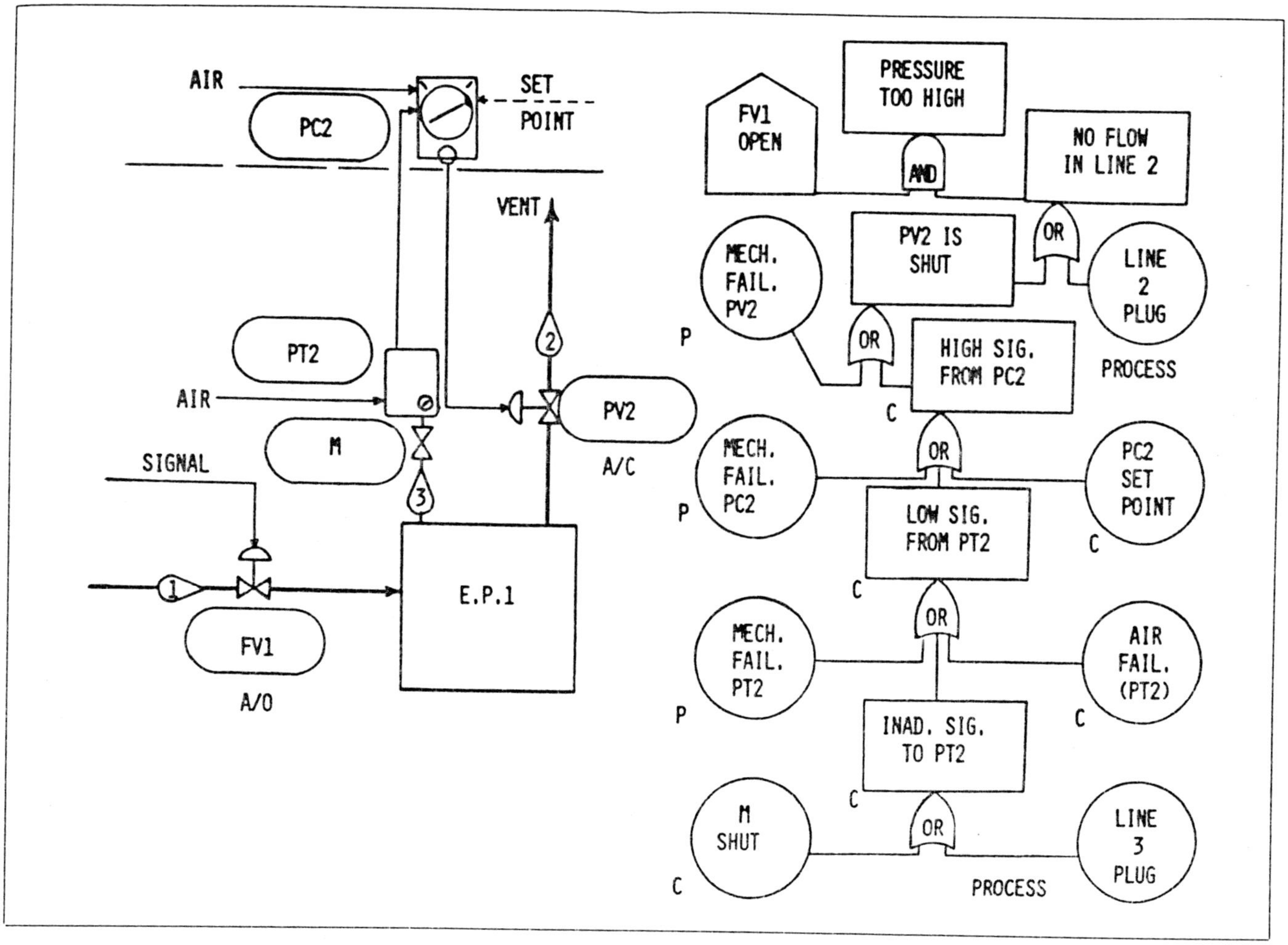

Figure 1. Effect-to-cause control system: primary/command/secondary procedure.

Application of Fault Tree Analysis

One of the advantages of Fault Tree Analysis is that process safety can be improved to any extent desired by restructuring the Tree and recalculating the top event frequency.

R. W. Prugh, Du Port Co., Wilmington, Del.

Recently, there have been considerable discussions on the basic concepts of Fault Tree Analysis and associated computational problems (*1–7*). However, the construction of Fault Trees has been addressed only in a limited manner (*8*). This article presents several standard and novel approaches to Tree construction.

Fault Tree Analysis was first used in the Du Pont Co. in 1966, about five years after its development at Bell Laboratories. This technique could not be immediately applied to chemical process safety analysis because the mathematical evaluation methods had been developed for batch, non-repairable systems. The appropriate equations for continuous processes were devised in the early 1970s, and the use of Fault Tree Analysis has progressed rapidly in the last few years.

It was recognized early on that considerable technical effort would be required to construct and evaluate Fault Trees. Du Pont's Engineering Service Division started to

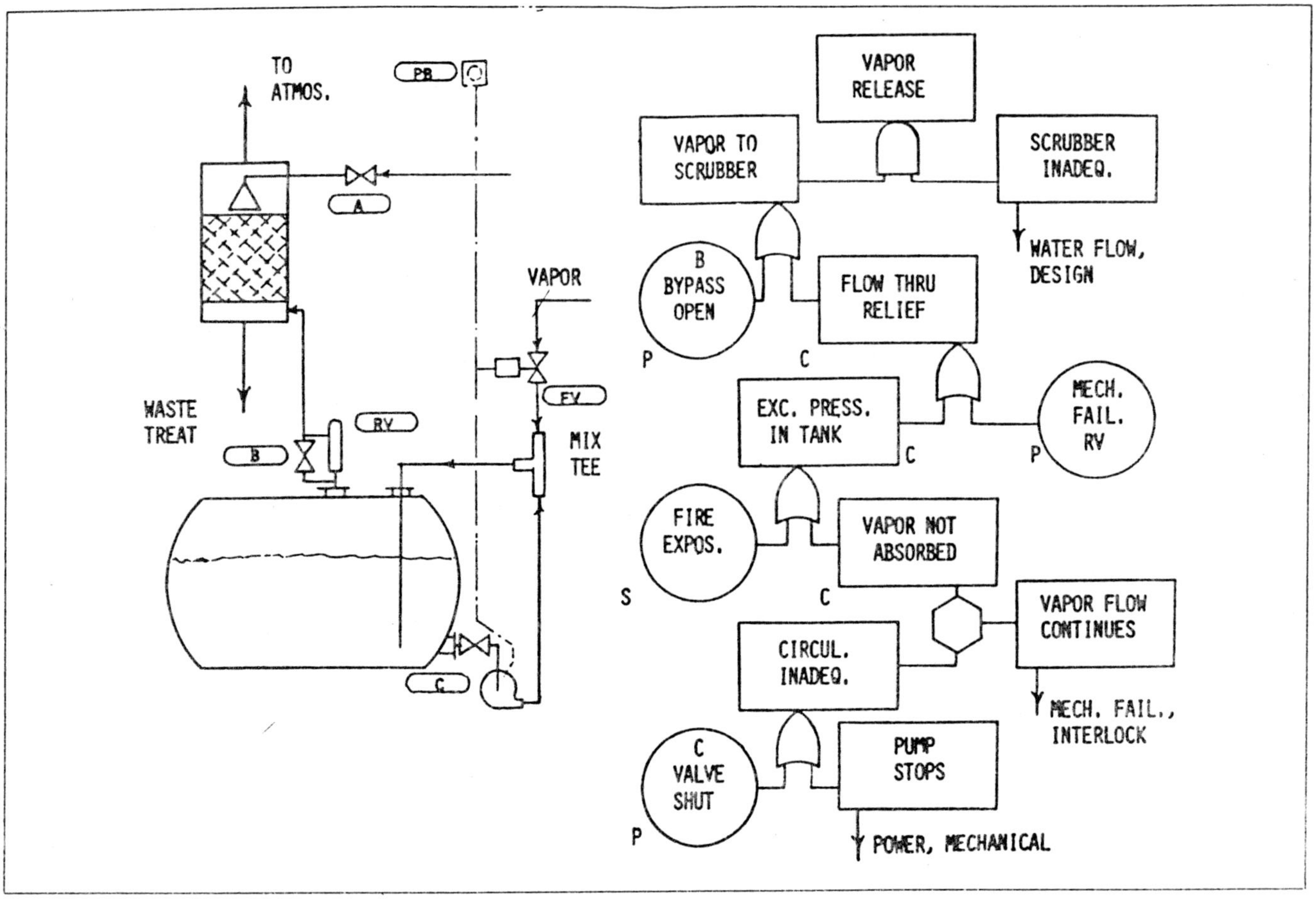

Figure 2. Effect-to-cause process: primary/command/secondary procedure.

develop construction and evaluation procedures which would minimize the effort without reducing the benefits of intensive investigation. Thus, a few guiding principles have been assembled, devised, and disseminated to assist analysts in constructing Trees. The following are several examples.

Effect-to-cause procedure

Figure 1 shows the standard symbols: rectangles, for failure events, including the Top Event at the top of the Tree or branch; circles, for basic causes; *house* symbols for high-probability normal events; and symbols for the AND and OR gates.

Construction of a Fault Tree is a deductive process, and the analyst works backward from effects toward causes. A good example of this method is found in the procedure for constructing branches for control systems. One starts at the component which acts on the process (usually a control valve) and works backward along the signal path toward the source (usually a sensor). Each component along the path has a primary (mechanical or internal) failure mode, and many components also can be commanded to fail by applying an undesirable control signal. For example, the pressure-control valve (PV2) can be *commanded* to close by applying a high-pressure air signal to this air-to-close valve. Similarly, the controller can be commanded to produce an unsafe, high output signal by reducing the signal from the pressure transmitter (PT2). Secondary (external) causes are those which can cause simultaneous failures of several components, as the result of an earthquake, nearby explosion, etc.

Note that the branch below *PV2-is-shut* consists of only OR gates, and this branch would be qualitatively and quantitatively the same with the one OR gate and seven inputs. The reason for proceeding stepwise backward along the signal path is to assure that all of the important causes are included.

The method on backward progression to develop branches can be occasionally applied to processes. This is particularly true if the Top Event is at the end of the process. In Figure 2, the analyst is concerned with venting a toxic material to the atmosphere. He starts backward into the process to develop causes for scrubber failure if the material enters the scrubber and also determines why the material would enter the scrubber. The path leads backward through valves, the storage vessel, the mixing circuit, and into the control systems, to develop the required branches of the Tree.

Vessel rupture is a major concern for safety analysts, because personnel may be injured and property loss may be severe. Figure 3 illustrates many of the common causes of vessel rupture, including mechanical (primary) failure and rupture from excessive internal pressure. At each of the precursor subevents, branches would extend downward to show how the subevent could occur. For example, at *heating/cooling* for a distillation column, there would be a branch showing excessive boil-up without compensating increased condensing capability OR failure of condensing capability without shutoff of boil-up.

It is likely that a relief device provided on the vessel of interest would perform differently for each cause of overpressure. For example, a relief valve may safely vent only 1% of vapor/air explosions, but may safely vent 95% of runaway reactions and 99% of excessive nitrogen-purge rates. Therefore, it may be necessary to construct a Tree which shows the effectiveness of relief for each cause; thus, the Tree would be considerably more complicated than the

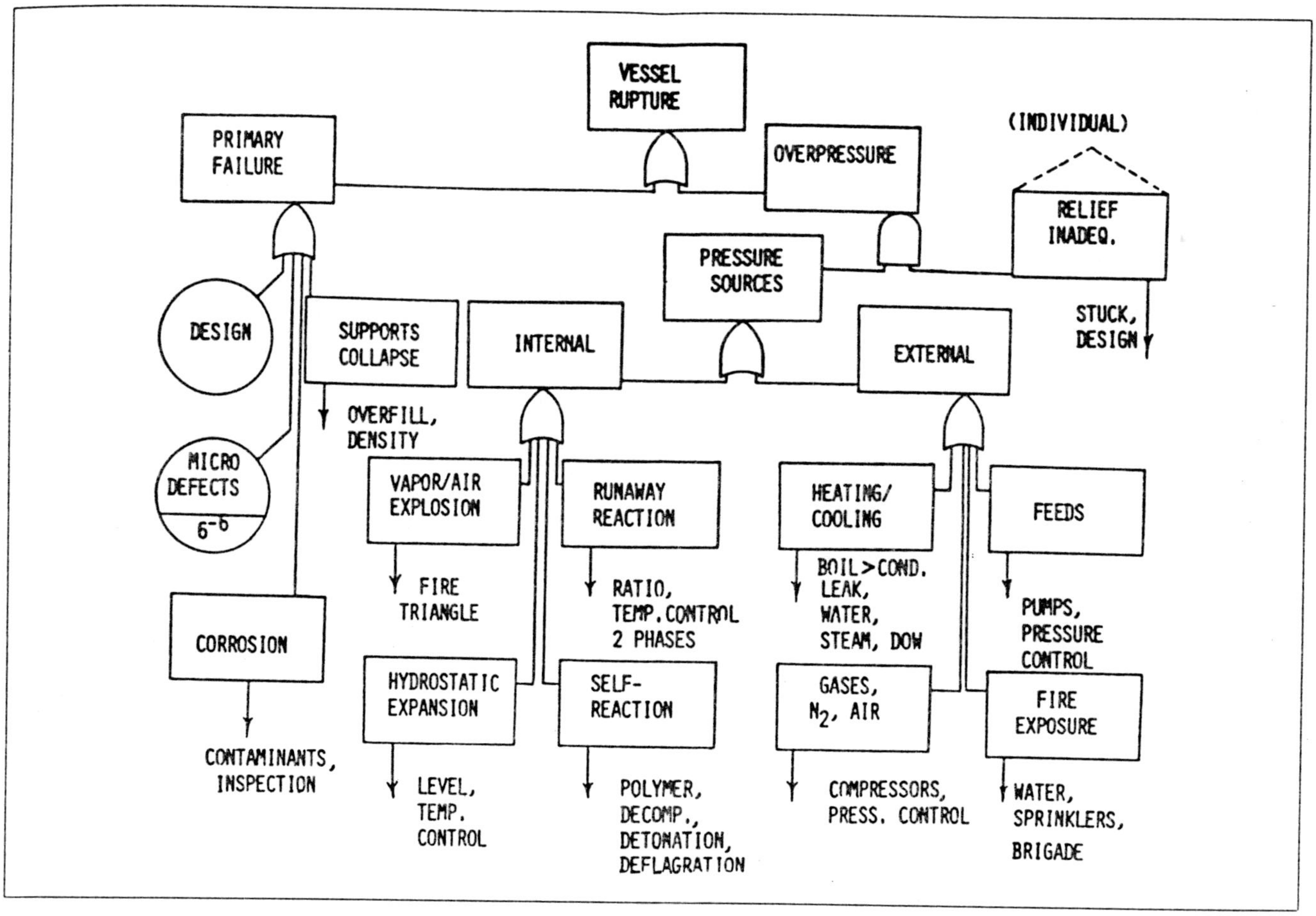

▲ **Figure 3. Source analysis.**

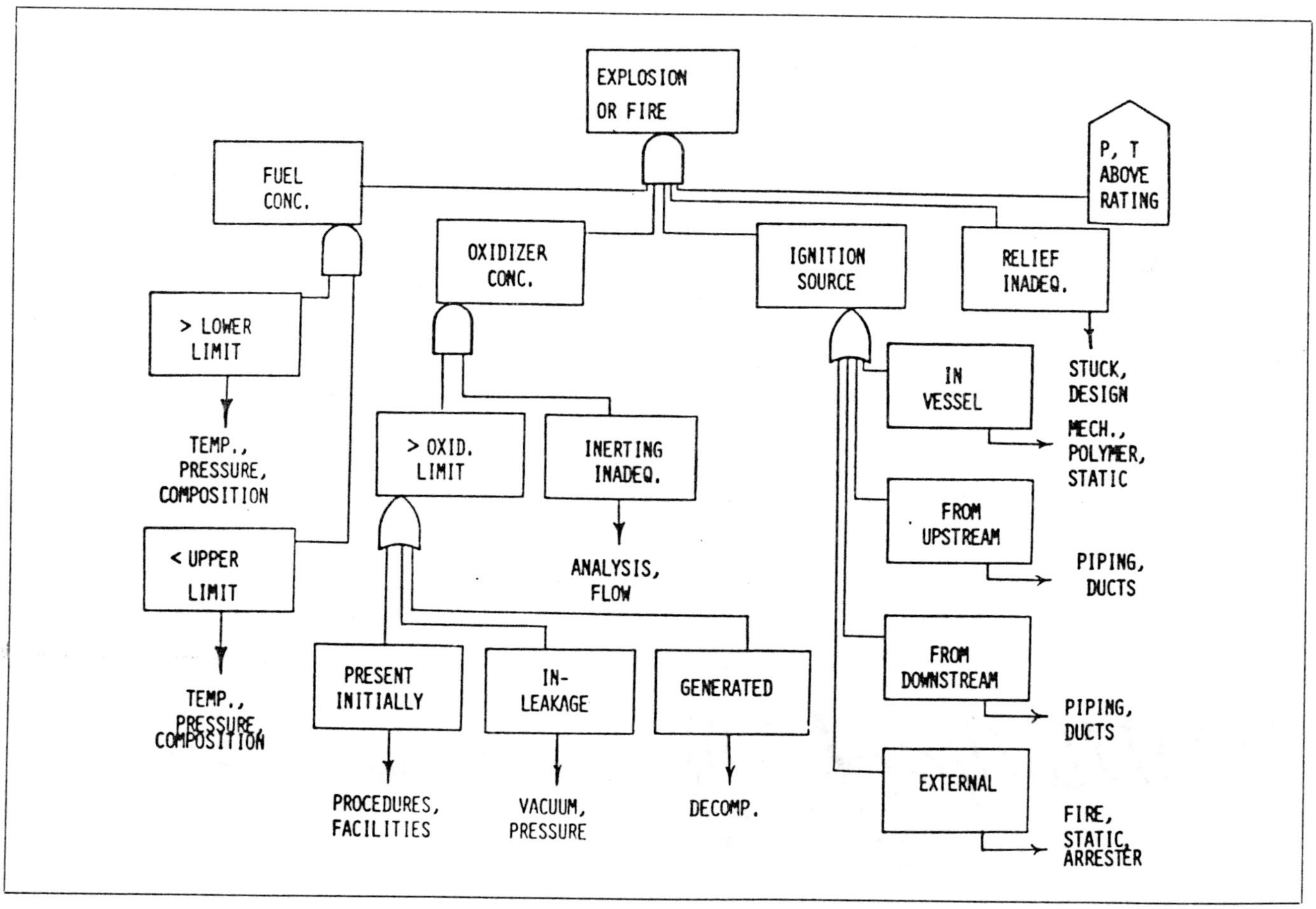

Figure 4. Fire triangle.

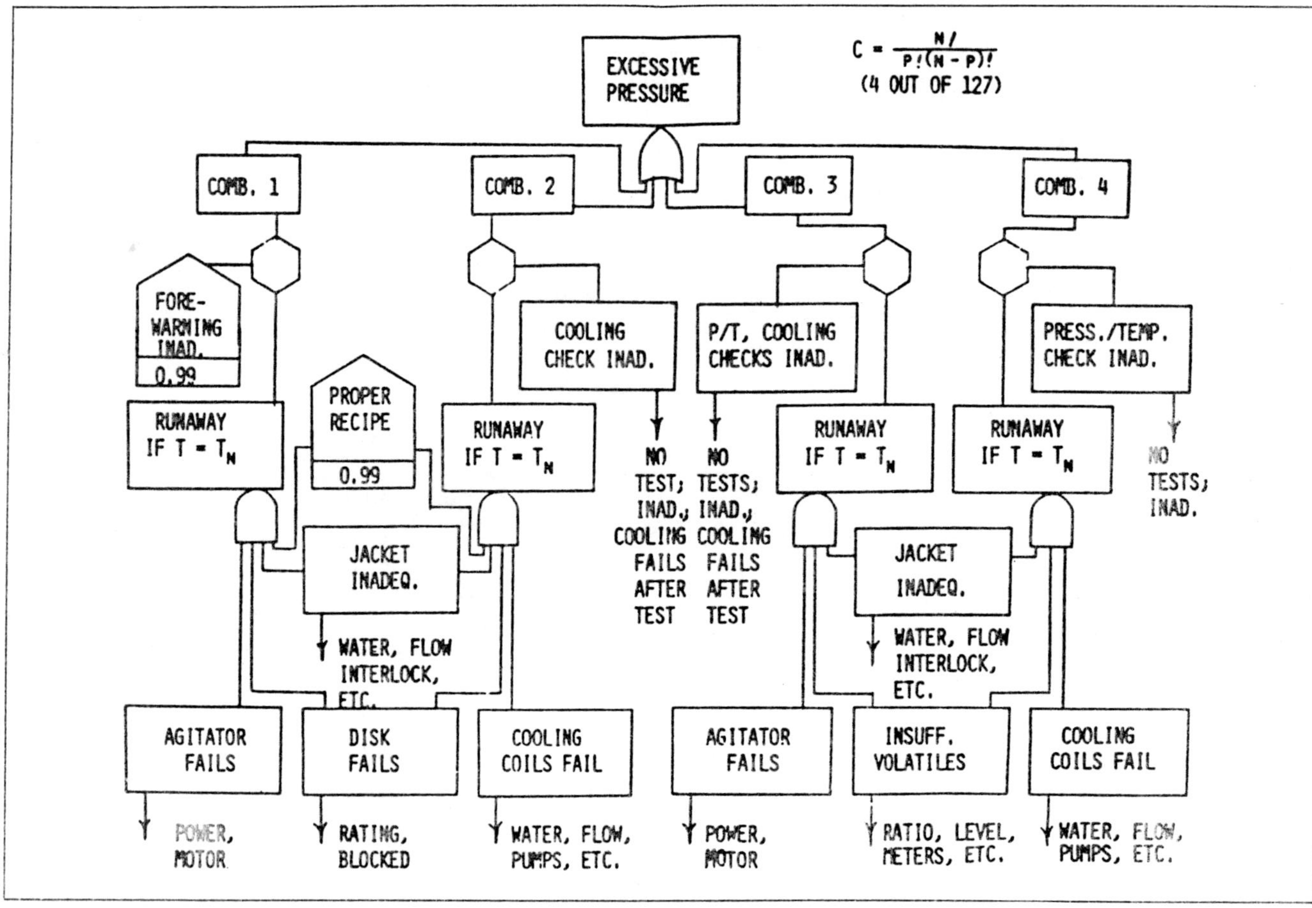

▲ **Figure 5. Batch combinations.**

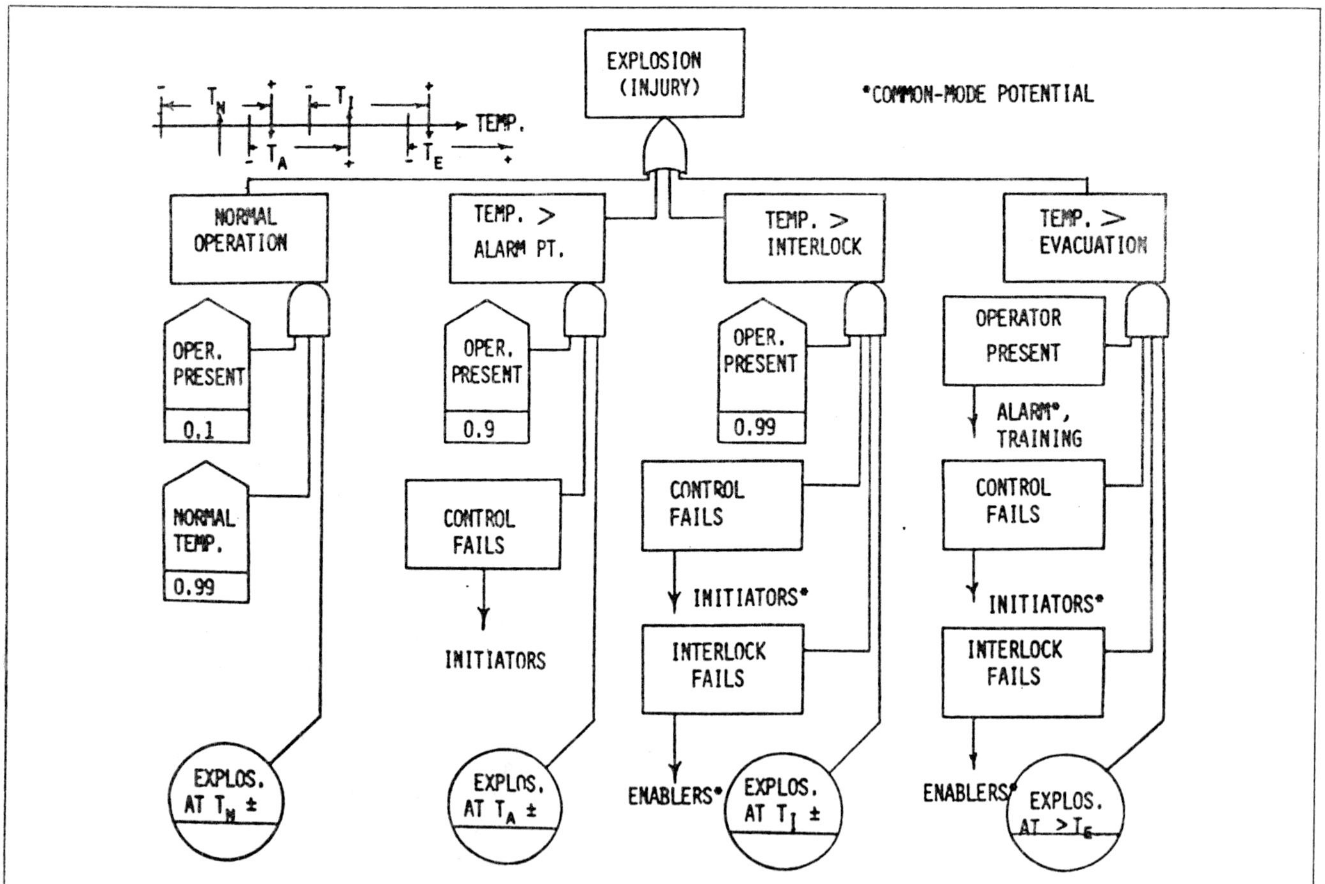

Figure 6. Action points.

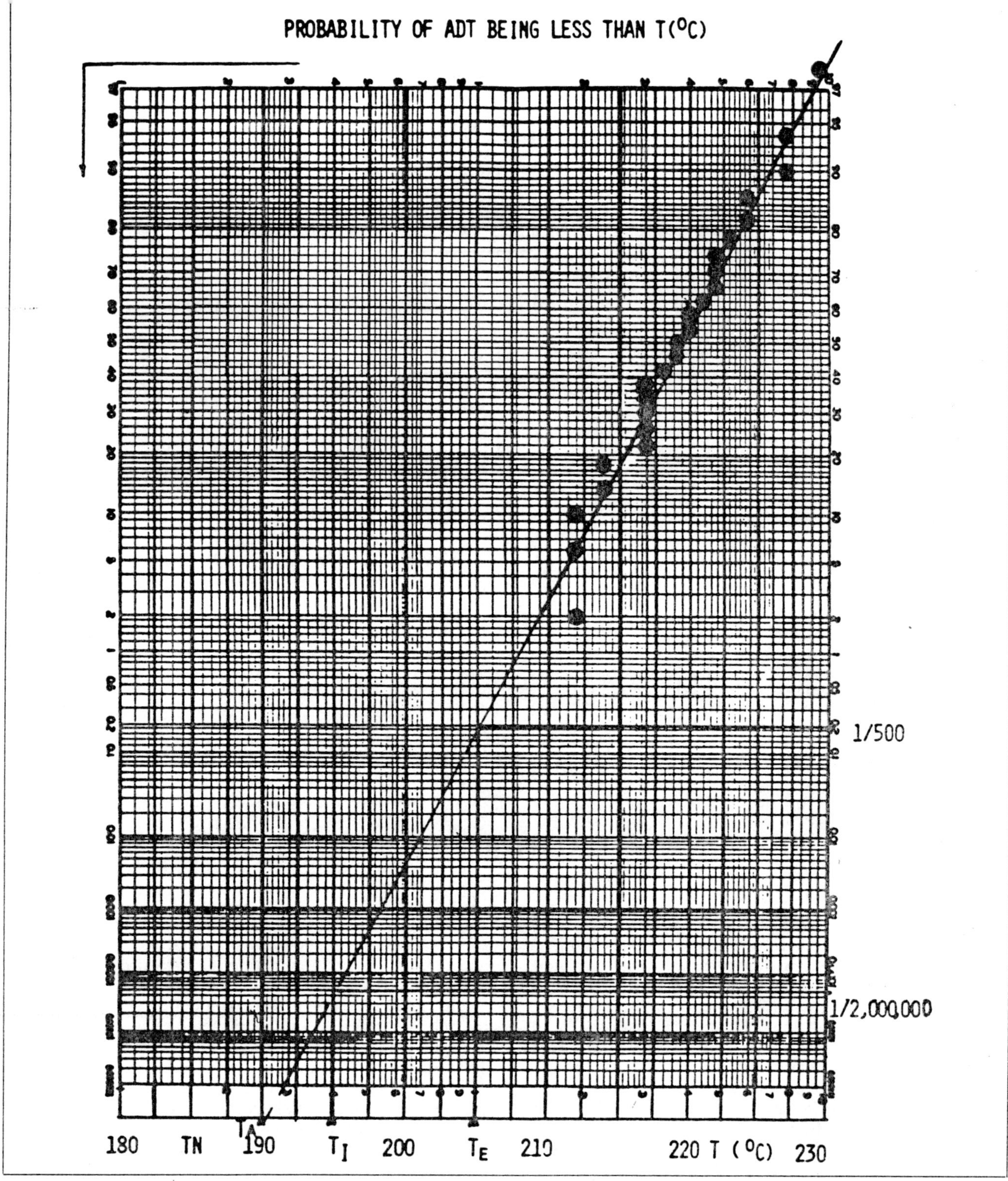

Figure 7. Distribution of auto-decomposition temperatures.

one shown here.

The causes of hazardous fire or explosion, Figure 4, can be described in a Fault Tree similar to the fire triangle. Not only does fuel have to be present, but the fuel concentration must be within the flammable range. Whether or not the concentration is above the lower limit and below the upper limit will depend on temperature, pressure, and composition of the vapor source. Presence of an oxidizer may involve in-leakage of oxidizer and failure of an inerting system. Several types of ignition sources may occur in a process, and the causes for each type should be determined and placed on the Tree. An explosion may not be hazardous, if it can be safely contained or relieved. Thus, the probability of the vessel withstanding an explosion (or fire) and the causes of overpressure-relief (or fire-protection) failure should be developed.

In many well-designed processes, there are several safeguards to prevent hazardous loss of control, Figure 5. These include temperature, pressure, and ratio controls, alarms, overrides or excursion limits, interlocks, emergency-cooling facilities, automatic vent valves, relief valves, and rupture disks. Failure of all of them most probably would cause a serious incident, while failure of only one would be inconsequential. However, failure of three or four in various combi-

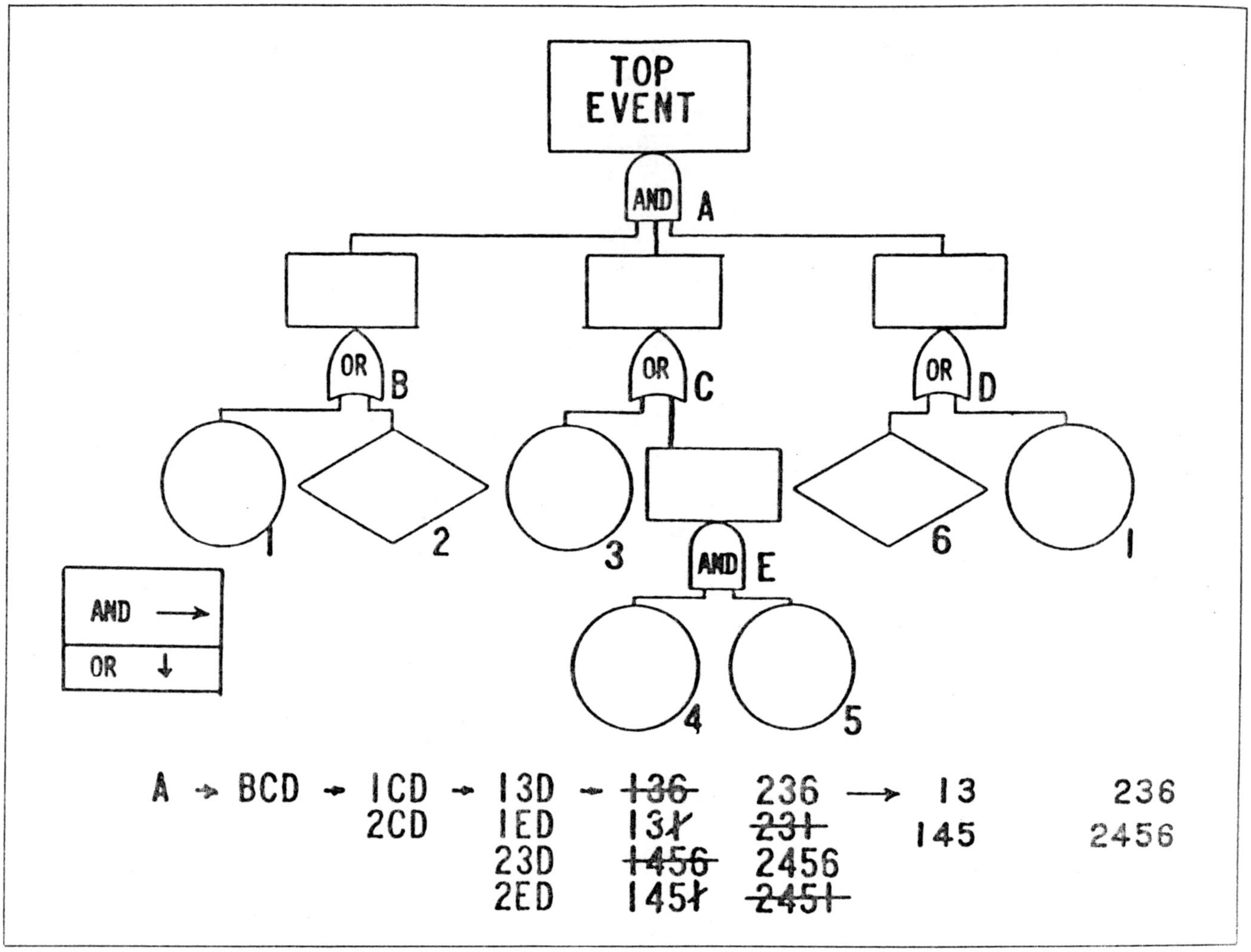

Figure 8. Failure and resolution.

nations may also lead to an undesired incident. The Tree structure shown covers a batch process where one combination of three failures, two combinations of four failures, and one combination of five failures would cause the undesired excessive pressure. In this case, seven safety features were identified, and from the equation for combinations, there were 127 combinations which had to be evaluated to arrive at the four critical combinations shown in the Tree.

One approach to constructing a Fault Tree for a process is to determine the temperature or pressure levels at which automatic systems or the operator should act to control a process. In Figure 6, four important temperatures were identified: the normal operating temperature (185°C), the temperature alarm point (190°C), the temperature at which the interlock system should shut down the process (195°C), and the temperature at which a building-evacuation alarm sounds (205°C). To estimate the probability of explosion and injury, it would be necessary to evaluate the probability of the operator (or other persons) being present, the frequencies of control, interlock, and evacuation-alarm failures, and the probability of the unstable process material exploding at each temperature of interest.

To estimate the probability of explosion, it would be necessary to test the material and determine the temperature at which it would auto-decompose and explode. As is probably typical, there is not one auto-decomposition temperature but, instead, a range of temperatures. The data for tests on 25 batches of the starting ingredient plotted on a probability graph, Figure 7, indicate that there is a normal distribution of auto-decomposition temperatures. If this normal behavior can be assumed to extend to temperatures corresponding to normal operation, and the alarm, interlock, and evacuation temperatures, it is possible to estimate the probabilities needed for evaluation of the Fault Tree. For example, the graph indicates that one batch in each two million could be expected to explode after the high-temperature alarm sounds but before the interlock system could actuate to shut down the process.

This example also introduces the problem caused by common-mode failure. For example, the failure of a temperature sensor may cause the temperature control system to fail and also cause the interlock circuit and evacuation alarm to fail. Common-mode failures are identified when the same basic cause appears in two or more branches which meet at an AND gate. The Tree must be restructured to show the common-mode failure properly, or the subsequent qualitative and quantitative evaluations will be seriously incorrect and in an unsafe direction. A simple procedure for common-mode failure resolution, Figure 8, involves a stepwise replacement of events by inputs to their gates, following the rule: AND gates—inputs placed horizontally in the table; OR gates—inputs inserted vertically in the table.

An incorrect ratio of reactants can lead to runaway reactions or other incidents in many processes. The Tree structure in Figure 9 involves three situations which should be evaluated. Situation 1 is where the flow of one reactant (A) is too high and the flow of the other reactant (B) fails to increase proportionately to maintain the required safe ratio. Situation 2 is the reverse of the first, where the flow of one reactant (B) is too low and the flow of the other (A) fails to

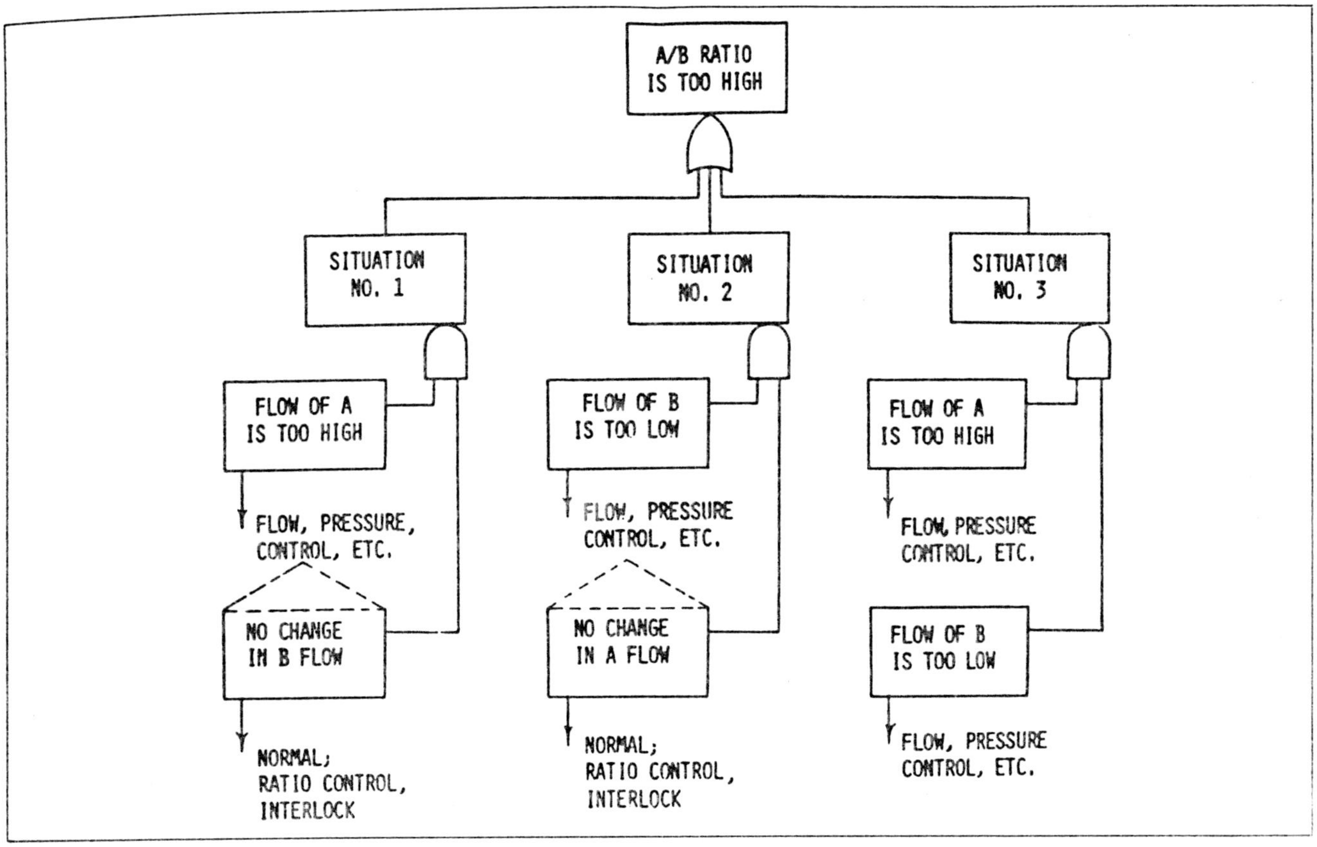

▲ Figure 9. Reactant ratios.

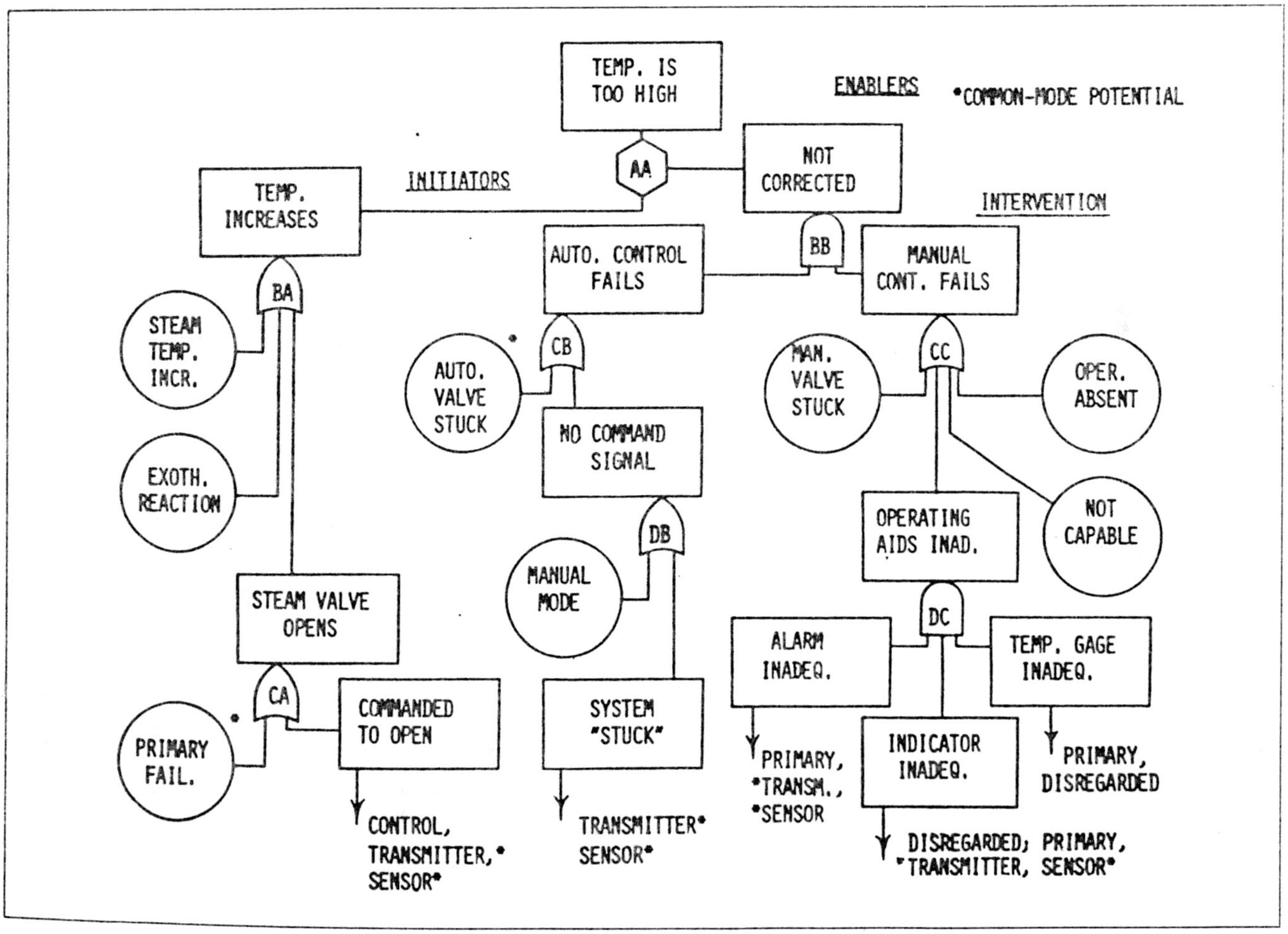

Figure 10. Initiators/enablers/intervention.

decrease proportionately—or stops altogether—to maintain a safe ratio. The third situation is where the reactant flows deviate unsafely in opposite directions at the same time; in most cases, the probability of this occurring is much less than either of the first two situations and does not need to be shown on the Tree.

Initiators, enablers, and intervention

The concept of initiating failure events and enabling failure events can aid in constructing a Fault Tree, Figure 10. Initiating events cause unsafe changes in a process; enabling events allow these initiating events to become more serious. These events allow upward progression of failure events in a Fault Tree. Enabling failures (such as stuck relief valves) are not usually detected when the failure occurs, but only during deliberate tests or as the result of a *demand* by the process. Thus, long failure durations (detection plus repair times) are associated with some enabling failures. On the other hand, the duration of an initiating event (such as failure of a steam pressure regulator or a lightning stroke) usually is short enough to be inconsequential in the fault tree calculations.

For most processes, the operator is an important element of process control. Indicators, alarms, and emergency-shutdown devices are provided by the process designer to enable the operator to intervene in case the automatic controls fail and then bring the process to a safe condition. The important aspects of human intervention (presence, capability, and operating aids) are shown on the Tree for later quantitative evaluation.

"Daisy Chain Manner." Some hazard incidents occur in a stepwise manner, Figure 11. Thus, there is a sequential relationship between the causes. This applies particularly to vehicle accidents and probably to many types of batch processes. As shown in a Fault Tree for a ship accident, the analyst determines how frequently a collision course occurs (while entering a harbor, for example). He then estimates the probability that the ship would fail to change course to avoid collision. In this combination of failures, a collision would occur unless the ship, which is about to be struck, does not move out of the way. Other aspects of accidents are evaluated similarly to estimate the frequency of cargo release.

In many processes, the hazards involved in starting up or shutting down a process are more serious than those which could occur during the continuous phase of operation. This situation might arise, if it is necessary to bypass interlock systems during start-up or to pass through a hazardous ratio of reactants during some types of shutdowns. To combine these different phases of operation in a Fault Tree, it is necessary to combine the batch type operations (the start-ups and shutdowns) with the continuous operation. This is accomplished by determining the per-year frequencies of occurrence and then adding them, as shown in Figure 12. The reciprocal of the sum is the average interval of Top Event occurrences or the probability of the Top Event occurring in terms of chances per year. This example also shows the mathematical method for combining frequencies and durations at AND, INHIBIT, and OR gates. These equations are presented in Table 1.

In summary

These Fault Tree structures do not cover all aspects of all processes, but may serve as models for Fault Tree construction. As Fault Tree Analysis continues to be applied to chemical processes, other useful models will be developed. For example, the Primary/Command/Secondary concept may be applicable to the chemistry of a process: primary (internal) failures may involve instability of ingredients or reactions, command failures may involve mixing, catalysis, etc., and secondary (external) failures would involve fire exposure, loss of utilities, etc.

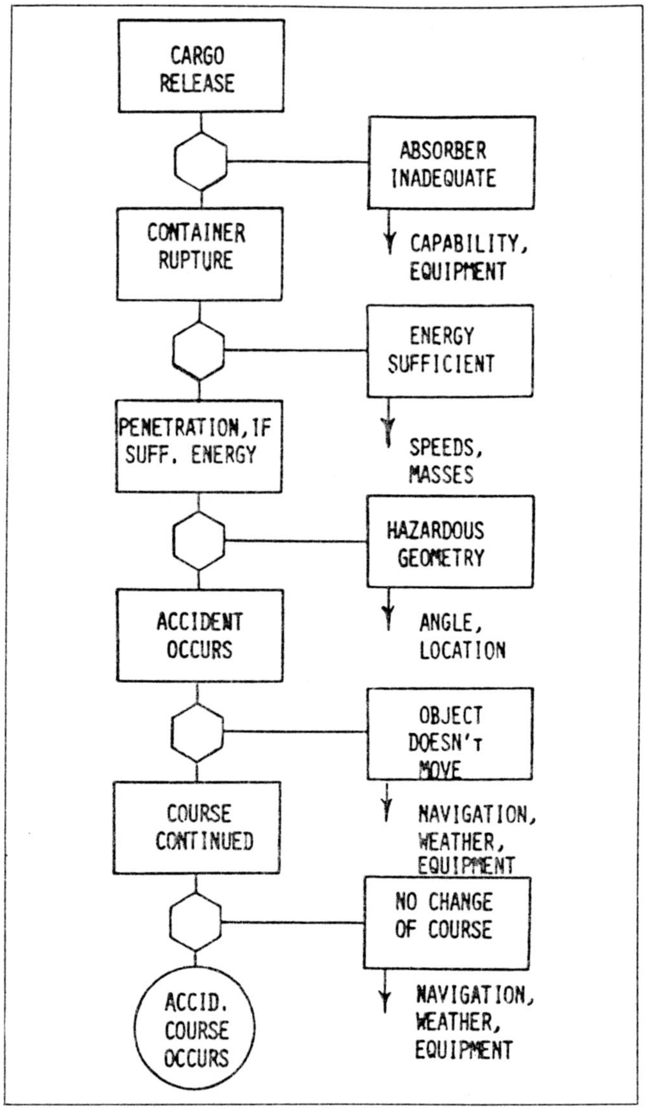

Figure 11. Daisy-chain (cause-to-effect).

One of the important results of Fault Tree Analysis is the calculated frequency of Top Event occurrence. For example, a Fault Tree may indicate that the process would explode at average intervals of 10 years. One of the advantages of Fault Tree Analysis is that the process can be improved by the analyst with subsequent restructuring of the Tree and recalculation of the Top Event frequency to any desired extent. The frequency of process explosion could be reduced to one chance in 1,000 per year (an average interval of 1,000 years) by adding an interlock system to the process controls.

Several questions arise. Would the original 10 years have been acceptable? If not, is the 1,000-year interval obtained with the interlock system acceptable? At what points (frequency or interval) would proposed process improvements be considered required or "gold-plating?" Other considerations include evaluating the exposure of operators, other personnel, and the public to process hazards (in relation to other day-to-day exposures to accidents and Act-of-God hazards), and the extent of property damage and business-interruption losses. Gibson (*9*) discusses some of the considerations and criteria recently taken by some companies for acceptability of process risks. #

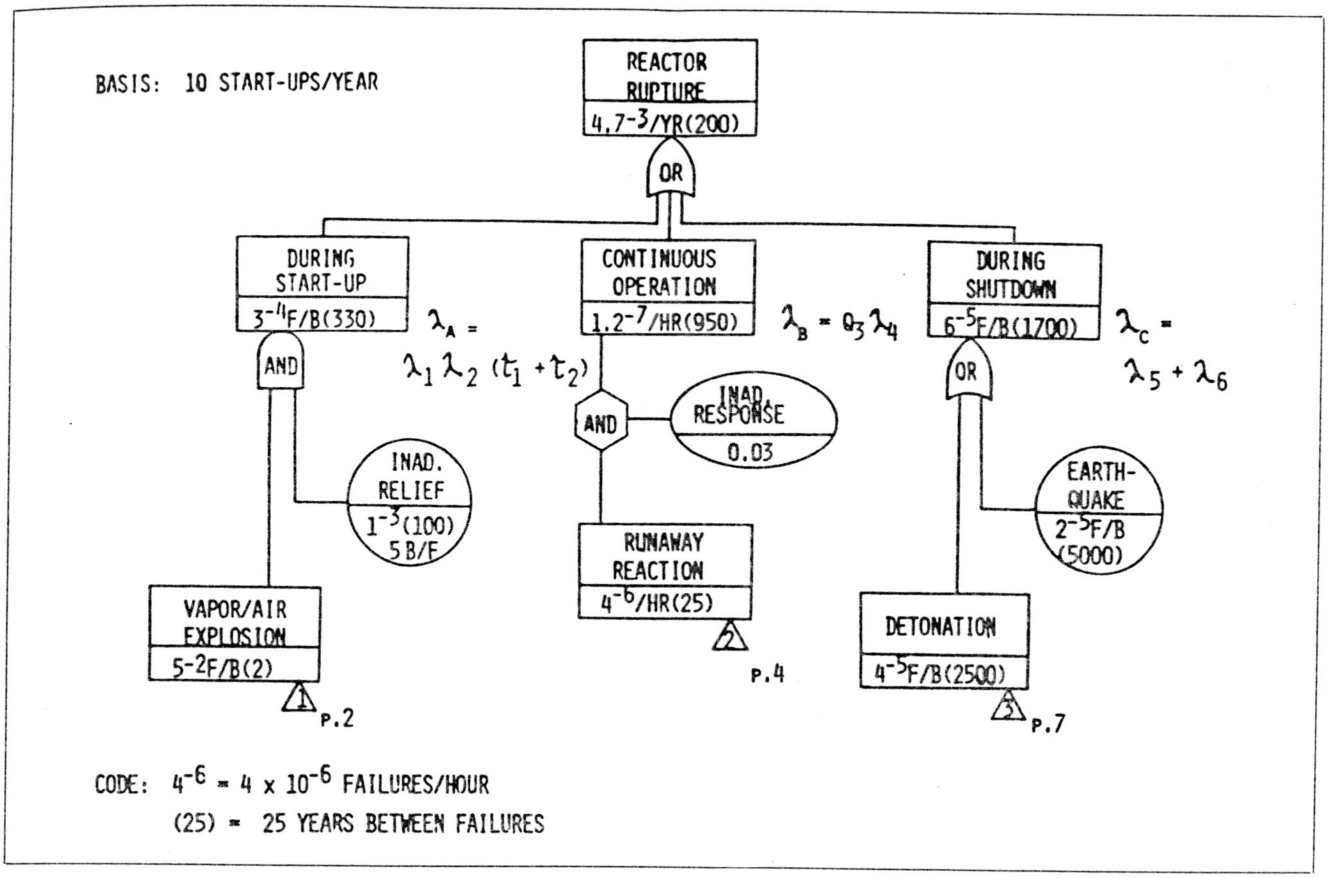

Figure 12. Combining batch and continuous operations.

Table 1. Fault tree equations.

		Logic Gate Formulas		
		2 Inputs	3 Inputs	i Inputs
AND	λ	$\lambda_1\lambda_2(\tau_1 + \tau_2)$	$\lambda_1\lambda_2\lambda_3(\tau_2\tau_3 + \tau_1\tau_3 + \tau_1\tau_2)$	$\Sigma\,(1/\tau_i) \times \Pi\, Q_i$
	τ	$\dfrac{\tau_1\tau_2}{\tau_1 + \tau_2}$	$\dfrac{\tau_1\tau_2\tau_3}{\tau_2\tau_3 + \tau_1\tau_3 + \tau_1\tau_2}$	$\dfrac{1}{\Sigma\,(1/\tau_i)}$
OR	λ	$\lambda_1 + \lambda_2$	$\lambda_1 + \lambda_2 + \lambda_3$	$\Sigma\,\lambda_i$
	τ	$\dfrac{\lambda_1\tau_1 + \lambda_2\tau_2}{\lambda_1 + \lambda_2}$	$\dfrac{\lambda_1\tau_1 + \lambda_2\tau_2 + \lambda_3\tau_3}{\lambda_1 + \lambda_2 + \lambda_3}$	$\dfrac{\Sigma\, Q_i}{\Sigma\,\lambda_i}$
INHIBIT	λ	$\lambda_1 Q_2$	$\lambda_1 Q_2 Q_3$	—
	τ	τ_1	τ_1	—

The above relationships apply only where the product of λ and τ (which equals Q) is substantially less than 1.0.

Literature cited

1. Brown, D. B., "Systems Analysis and Design for Safety," Chapter 5, p. 152, Prentice-Hall (1976).
2. Fussell, J. B., "Fault Tree Analysis—Concepts and Techniques," p. 133, in "Generic Techniques in Systems Reliability Assessment," by E. J. Henley and J. W. Lynn, Noordhoff International Publishing (1976).
3. Gibson, S. B., "The Design of New Chemical Plants Using Hazard Analysis," p. 135, in "Process Industry Hazards," AIChE Symposium Series No. 47 (1976).
4. Malasky, S. W., "System Safety," Chapter 5, p. 142, Spartan Books (1974).
5. Rasmussen, N., *et al.*, "Reactor Safety Study," Appendix II, USAEC; WASH-1400 (August, 1974).
6. Yellman, T. W., "Comments on Fault Trees—A State of the Art Discussion," *IEEE Transactions on Reliability, R-24* (**5**), p. 344 (December, 1975).
7. Young, J., "Using the Fault Tree Analysis Technique," p. 827, in "Reliability and Fault Tree Analysis," by R. E. Barlow, *et al.*, Soc. for Ind. & Appl. Math. (1975).
8. Powers, G. J., and F. C. Tompkins, "Fault Tree Synthesis for Chemical Processes," *AIChE J.*, **20**, p. 376 (March, 1974).
9. Gibson, S. B., "Hazard Analysis and Numerical Risk Criteria," AIChE 14th Annual Loss Prevention Symposium (1980).

R. W. Prugh, who has had 20 years of experience in Du Pont's explosives, organic chemicals and safety and fire protection units, is now in the Hazards Engineering Dept. of the firm. He is a registered engineer in Delaware, Pennsylvania, and New Jersey.

DISCUSSION

RUSSEL HILL, Oxirane International: I think you've done a great job of giving us an outline of how to attack problems that don't show up in the textbook very much. I would like to ask a question about the validity of the data that you said was available on the reliability of components. Do you consider the data valid, too conservative, too optimistic? Does the data take into account the differences of operating conditions?

RICHARD PRUGH, Dupont: Some of the data are very good, particularly those of Anyakora, Lees, and Farmer (*1-5*). They presented in their well-documented publications the number of equipment items they looked at in the field and they've gone through maintenance records to find out how many failures there were.

Unfortunately, there are also a lot of consensus data around; that is where operations on one side says "look at all the failures we're having," and maintenance over here says "look how many you have in service." So when these two people get together, they essentially are arriving at a failure rate; that's consensus. That's certainly not "hard" data, and that certainly is not as good as actually counting failures and units in service. I think you have to determine how accurate the data are, from the documentation provided.

There has been some discussion about that in one publication (*1*) and a supervisor in one of our Gulf coast plants tried to assess the different environmental conditions of high humidity, temperature, and vibration, etc. (*6*); that's all the data we have right now.

TREVOR KLETZ, ICI: You mentioned availability of data. The best source that I know of is the data bank operated by the United Kingdom Atomic Energy Authority to which any company can subscribe for about 3,000 per year. This gives you access to all the data that the Atomic Authority themselves and other companies have put in.

Their data on instruments and electrical equipment are very good; their data on mechanical equipment are not so good, but they are the best data available. AIChE is considering setting up its own data bank. I understand that Mr. Edwards of Fluor is taking the initiative on this and any one interested should get in touch with him. He is interested in finding out the level of support for a data bank in this country operated by the AIChE.

DAVID ALLEN, Foster Wheeler: Do you have any experience with or thoughts about computerized techniques for fault tree synthesis? We, in Foster Wheeler, have written computer programs for fault tree synthesis and analysis and have used them to construct and analyze trees containing 900 gates. I believe that Powers has synthesized trees of over 100 gates using a similar program. We believe that these programs greatly facilitate fault tree analysis, and perhaps allow process engineers rather than fault tree experts to perform such analyses. These are techniques that people should look into.

PRUGH: Yes, Gary Powers of Design Sciences, Inc. and the Carnegie-Mellon are trying to develop a computer program that would actually construct Fault Trees; that is, do the logic that's involved in constructing trees. Several programs have been available for drawing a Tree once you have the logic developed and there are several programs available for calculating event frequencies based on the inputs, to eventually obtain the Top Event frequency (*7*). But as far as I know, there is nothing yet available that would actually construct the Tree.

Eventually, I think we're going to have a computer program where you just describe a process to the computer from a Process and Instrument diagram, with "truth table" input/output models for each component, and the computer will develop the logic and construct the Fault Tree. I think that's several years away. We stilll have to use a "library" of Tree structures or digraphs, right now.

REFERENCES

1. Anyakora, S.N., et al., *Chem. Eng.*, 396 (Nov., 1971).
2. Lees, F.P., *Chem. Eng.*, 418 (Sept., 1973).
3. Lees, F.P., *I. Chem. Eng. Symp. Series*, 47:73 (1976).
4. Engel, J.R., *Nuclear Safety*, **15** (4), 387 (Aug., 1974).
5. Farmer, F.R., *I. Chem. Eng. Symp. Series*, **34**, 82 (1971).
6. Barbin, F.H., *Instrumentation Tech.*, **20**(11), 44 (Nov., 1973).
7. Powers, G.J., and S.A. Lapp, *Chem. Eng. Prog.* **72**(4), 89 (April, 1976).

Hazard Analysis and Risk Criteria

Management's duty is not to eliminate hazards, because that is impossible, but to reduce them to a socially acceptable level.

S. B. Gibson, ICI Americas, Inc., Wilmington, Del. 19897

Bad hazard analysis has been justifiably criticized, but thorough, unbiased hazard analysis is a valuable tool in the decision process. The analyst and decision maker must publicize their methods and objectives to win the confidence of society and to demonstrate that they are in the best interests of society. Consistent and effective decisions for acceptable risk require such criteria as the risks to employees and to the environment.

The following statement from the Canadian Atomic Energy Control Board's bulletin on "Risk of Energy Production" illustrates the objectives of a good hazard analysis.

"Risk to human health does not start when an electricity generating plant or solar plant puts out energy. It starts when an engineer first puts pencil to paper, continues as materials are acquired, fabricated and installed, and terminates only when the installation is dismantled at the end of its useful life. . . . In other words, all risks must be included for a fair evaluation, not just the risk which is most obvious."

Most of the criticisms levelled at hazard analysis are actually intended for bad hazard analysis.

Exhaust all relevant data

A simple definition of hazard analysis is the collection of *all* relevant information on a subject, quantifying as much of it as is reasonably practicable in terms of its benefits and disbenefits and making this information available to the decision maker. Far from being the subject of criticism, hazard analysis should be welcomed, as it is totally irresponsible for any decision maker to make up his mind without all the facts.

It must be stressed that the decision process is not part of hazard analysis. Hazard analysis simply presents as much information as possible objectively in an attempt to reduce (not necessarily eliminate) the subjectivity of many decisions. The decision maker may be a politician, trades union member, work force, or manager; if one wishes to exclude the scientists and technocrats from the decision process, it is relatively easy to limit their contribution to the role best suited to their abilities, presenting information in a quantitative manner.

It could still be argued that an analysis will be biased in favor of the assessor. But, in that case, it would be an example of bad hazard analysis. In any case, a decision maker worth his salt should discuss the assumptions made and query data used and should identify undue bias. Incidentally, it is disturbingly noteworthy that a biased analysis is interpreted by the media as deceit if the analyst is in industry and as righteous concern for society's interests if

0149-3701/81/3782 $02.00 © 1981 AIChE

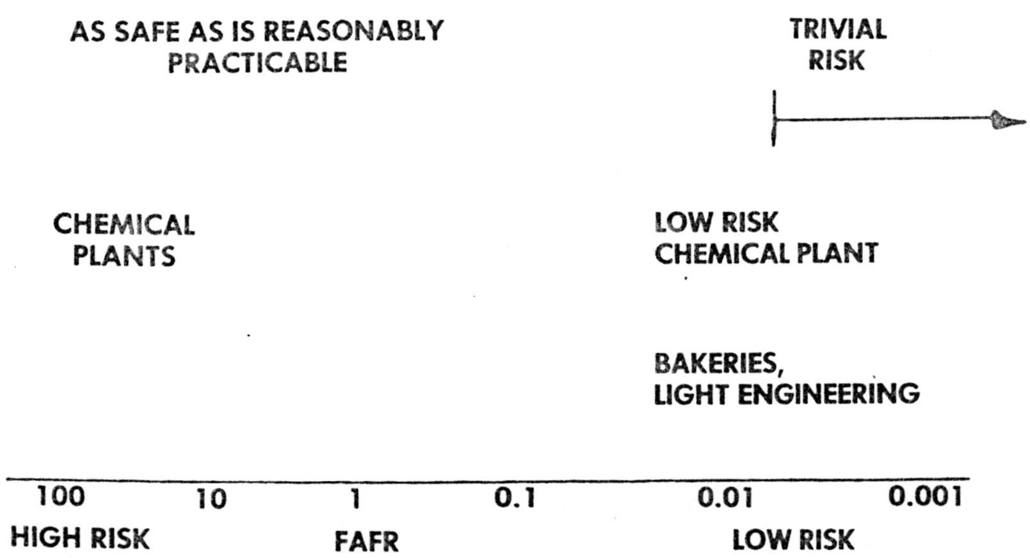

Figure 1. Scale of risk.

the analyst is anyone else.

Good hazard analysis is in everyone's interest, and we must not be afraid to publicize our efforts in this area as attempts by a responsible industry to use our resources consistently and efficiently. The rigorous quantitative approach focuses attention on the most critical areas (which often prove not to be the obvious or the politically sensitive ones) and frequently identify important factors which otherwise would not come to light.

One of the particular merits of the quantitative approach to decision making is that it enables the adversaries to argue over the figures. They may agree on a compromise figure or they may agree to disagree. In many cases, the rigorous analysis may show that the answer is not sensitive to the subject of the disagreement. Whatever the outcome of the argument, it is much more meaningful to discuss figures, however inaccurate they may be, than purely subjective factors.

What's an acceptable risk?

The problem is very complex when one is judging risk. One seldom can eliminate completely the risk in an activity. So, when does the risk become acceptable? One basis for judgment used in British law is that one should be as safe as is reasonably practicable. Reasonably practicable is defined as implying "that a computation must be made in which the quantum of risk is placed on one scale and the sacrifice involved in the measures necessary for averting the risk (whether in money, time, or trouble) is placed on the other, and that, if it be shown that there is a gross disproportion between them—the risk being insignificant in relation to the sacrifice—the defendants discharge the onus upon them."

This is apparently clear guidance when there is a "gross disproportion." But, how big is "gross," and are there not situations where a simple disproportion would justify a risk? Clearly, this is an area where quantitative analysis could lead to more consistent decisions.

Another interesting aspect of quantitative analysis is highlighted by this quotation. It implies that there is a risk which may be considered insignificant *per se* so that any sacrifice is grossly disproportionate; i.e., a risk so low that we shouldn't bother about it. It is obvious that people's attitudes are important in deciding what is an acceptable risk, and it is appropriate to consider here the role of attitude surveys.

The attitudes and reactions of the public/society must be taken into account in any analysis, but the formation of attitudes is a very complex process. The media strongly influence the public opinions and to emphasize the sensational, the tendency often prevails over objective balanced reporting. People's attitudes may well be formed on unbalanced information; attitudes will vary widely.

How then do we use this information? Do we base our analysis on the average attitude? If so, how can this be calculated? Do we base it on the favorable or unfavorable half or should it be a quarter or the extreme 10%?

Attitude surveys are only of limited help in the decision-making process, but they provide valuable research in the communications field. They can help us to direct the information into the best place so that society will be fully and properly informed and can judge whether the decision is made responsibly. Until that time, the decision makers must devote some of their efforts to convincing society that they can be trusted to take the decisions in the best interests of society. This leads on to the $64 question: what are the best interests of society?

Table 1. FARs for industrial activities (UK).

Activities	Year	FAFR
Manufacturing Industry	1971–74	2
Chemical Industry	1971–74	4
Construction Industry	1971–74	9
Mining and Quarrying	1971–73	10
Air Crew	1964–73	200
Professional Boxers	1963–74	20,000

Table 2. FARs for nonindustrial activities (UK).

Activities	Year	FAFR
Travelling by Train	1963–72	3
Staying at Home	1972	4
Travelling by Bus	1963–73	4
Travelling by Car	1963–73	48
Travelling by Air	1964–73	190
Canoeing	1962–72	670
Gliding	1964–73	1,000
Motor Cycling	1963–73	1,040

Table 3. FARs for industrial activities (US).

Activities	Year	FAFR
Chemical Industry	1977	3
Manufacturing Industry	1977	5
Trucking	1977	11
Agriculture	1977	27
Construction	1977	30
Mining & Quarrying	1977	31

Table 4. FARs for nonindustrial activities (US).

Activities	Year	FAFR
Travelling by Bus	1977	2
Travelling by Train	1977	2
Staying at Home	1977	3
Travelling by Air	1977	20
Travelling by Car	1977	100
Travelling by Motor Cycle	1977	680

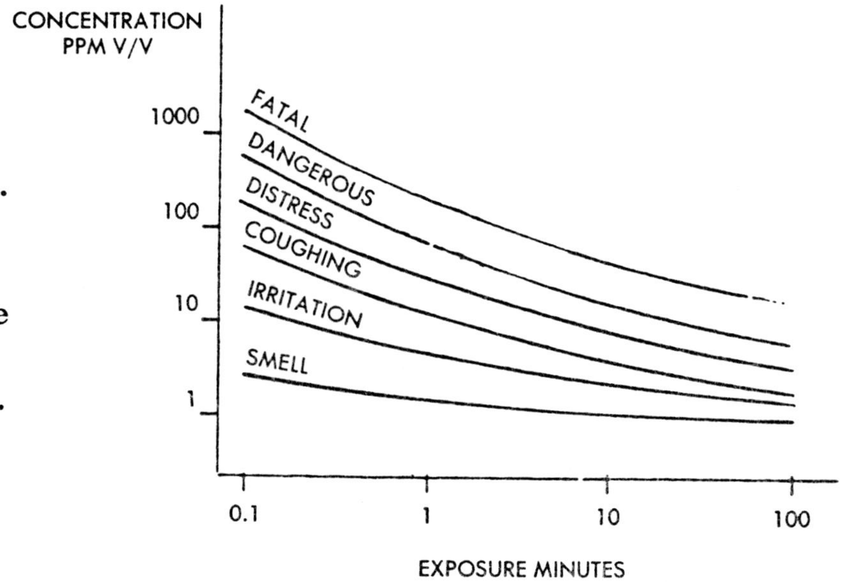

Figure 2. Effects of exposure to chlorine.

Best interests of society

The whole subject of criteria for risk depends on the answers to the following three basic questions.

1. *What is the objective of society?*
- To eliminate risk completely?
- Return to nature?
- To gradually improve our quality of life and standard of living by a steady process of technological development?

These together with many other answers, practical and impractical, represent beliefs held by one group or another in society. Since many of them are mutually exclusive, any choice would offend some groups.

2. *What is the society?* Is it the employees and local community surrounding a plant; our company alone; the chemical industry; the state; the United States; or, does it go even wider?

3. *What is the time period of concern?*
- The immediate present?
- The current tax year?
- Until the next presidential election?
- The next two generations?
- The next 1,000 years?

The answers to these three questions may change with time or the particular problem in hand. However, a socially and morally responsible company would be aiming to develop a steady improvement in the quality of life and standard of living of the community it serves and affects by a process of technological improvement.

If this is generally accepted as a goal, society will either have to collect and weigh *all* the facts about the decisions being made on their behalf, or trust the decision makers (whether they are in government, commerce, industry, or elsewhere) to be socially and morally responsible people with similar objectives.

ICI in Britain employs hazard analysis as a management tool which helps to achieve an acceptable safety performance in the widest sense. This analysis also helps to balance further use of resources in reducing risk against the other benefits that those resources could produce for society (in the form of customers, employees, shareholders, etc.) by enhancing the profitability and long-term prosperity of the company. Considerable publicity is given to the use of this tool so that more people can contribute to the decision process.

Criteria for acceptable risk

To illustrate practically how it has been used, we trace the development of two criteria for use in hazard analysis. We shall start from the premise that all human activity is fraught with risk. The duty of management, as a responsible part of society, is not to eliminate hazards, because that is impossible, but to reduce them to a socially acceptable level.

Let's assume that we can measure risk along an axis as shown in Figure 1. Intuitively, one assumes that chemical plants will have high risk activities. There are many other high risk activities which have been highlighted by poor performance in the past and are marked by the authorities as worthy of close scrutiny. At the low risk end, one might expect to have things like bakeries, light engineering works, and other activities which are not considered to be particularly hazardous. In fact, the risks in such activities are considered to be trivial and are not the subject of public or official concern.

The size of the risk does not really feature on the scale of Figure 1, because one might well consider a jumbo-jet flying over a crowded area to be a low risk despite the fact that if it crashed, it would cause hundreds of fatalities and injuries. Similarly, when dams are built, it is considered to be a low risk, even though the collapse of a dam could cause even higher casualty rates. However, the public and the authorities tend to assume chemical plants are high risk and to place them at the high end of the scale.

If we now decide we can measure risks, risks comparable to those at the lower end should be categorized as trivial; even risks which are up at the high end are very rarely expected to be completely eliminated, since this would almost certainly require the activity to be completely abandoned. They are expected to be reduced as far as is reasonably practicable.

If we now assume that we can measure or predict risks for any activity, there will be some level of risk down at the right hand end of the scale which we can deduce from measuring or estimating the risks of those activities already seen to be down at the right end of the scale, such that we can consider that risk to be trivial. There is no reason why every chemical plant must cause risks at the high end of the scale. We now have a better understanding of the risks involved. The use of improved safety techniques, such as hazard and operability studies (*1*) and hazard analysis (*2*), enables us to identify the risks more confidently. Modern

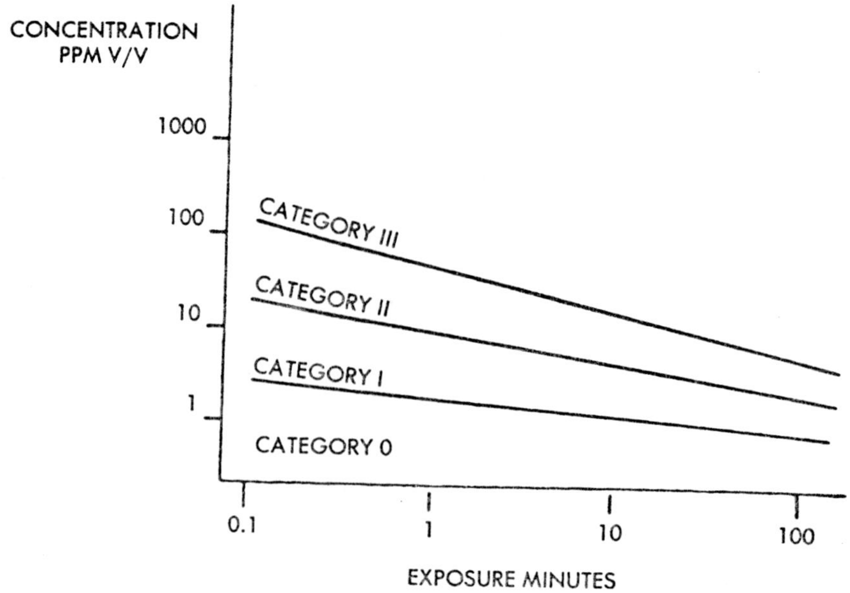

Figure 3. Categories of chlorine emission hazard.

control and protective systems provide the means to anticipate the upset conditions and take appropriate action.

Using all the knowledge at our disposal, it is indeed our intention to design all our new chemical plants such that the predicted risk for them is equivalent to those activities down at the right hand end of the scale. If we can predict with reasonable confidence that this is indeed the situation, there is no reason why anyone should expect the risks of such a chemical plant to be reduced even further. By tradition, chemical plants are seen to be dangerous and automatically expected to be at the left hand end of the scale. Therefore, the authorities and the public often require further more stringent precautions.

Chemical plants can be designed with a risk factor which is trivial. Consequently, we ought to be able to convince society that what is trivial for other situations is also trivial for chemical plants and that the industry should not be automatically expected to improve whatever is produced by a factor of ten, or hundred, or whatever takes their fancy. Therefore, we could adopt the idea that there is a scale of risk with trivial risks at the right hand end and a large region where risks are as low as is reasonably practicable, Figure 1.

There may also be a level far beyond the left hand end where the risk is totally unacceptable, but this is an area we shall not explore in this article. Taking this idea a little further, we must quantify the scale. Although not the complete answer, one suggestion would be to graduate it in terms of the Fatal Accident Frequency Rate, which will be explained below. Then, the problem resolves itself into defining: 1. what is a trivial level of risk and 2. so far as is reasonably practicable in a consistent manner.

Chemical industry: low risk area

We shall use the Fatal Accident Rate (FAFR) as a parameter for measuring risk. The fatal accident rate is defined as the number of deaths one would expect in 10^8 exposed hours. To get a feel for what 10^8 exposed hours means, consider a typical chemical factory employing about 1,000 people. If we consider those 1,000 people working an average lifetime, say 40 years, and each employee with a little overtime works about 2,500 hours a year, their lifetime work will be 1,000 × 40 × 2,500 hours: that is, one hundred million worked hours. So, the FAFR is the number of deaths that we could expect from a group of 1,000 people

Table 5. Exposure to chlorine in order of increasing severity.

Category	I	II	III
Effect	Smell nuisance	Vegetation damage Mild casualties	Possibly serious effects
Public Reaction	Minor reaction	Much adverse comment	Public outcry
	Some complaint	Some compensation	Serious consequences
Acceptable Frequency	≃2 per year	1 in 10 years	1 in 100 years

during their working lifes.

Tables 1 and 3 show some Fatal Accident Rates for different industrial occupations. For comparison, Tables 2 and 4 show the fatal accident rates which can be expected in leisure time pursuits and everyday situations.

Tables 1 to 4 indicate that risks up to about an FAR of 5 are rather trivial, the sort of risks we are exposed to every day of our lives and to which we give little thought. Whereas, risks from 10 upwards are sometimes cause for considerable concern. It is reasonable to say, therefore, that the risks from the chemical industry are at the low end of the spectrum of life's risks. ICI (UK) is reputed to have a good safety record and is not being constantly castigated by the British Health and Safety authorities for its poor performance. We can take this record as some indication that society considers the risks in the company to have some measure of acceptability.

The statistics for ICI employees in Britain show that the average risk for the years 1945 to 1970 was an FAR of about 6. This figure is higher than expected, because our statistics were concerned only with the hourly paid workers and did not include the salaried staff, who being often remote from operating plants are on average at much lower risk. Now, a target should show some improvement over the average and we noted over a 10-year period from 1955 to 1964, we had in fact achieved an FAFR of 4. We also noted that only about half of the fatal accidents were caused by hazardous chemicals or processes, and the remaining half caused by accidents such as falls or being struck by moving or falling objects.

We concluded, therefore, that these latter risks must remain the province of the traditional safety approach, but that by close attention to design and operation, we should aim to ensure that on all new plants, the sum of specific process risks as opposed to general background risk, should be not greater than an FAR of 2. This FAR was reached subjectively by arguing that our past record was good, that risks in ICI were at the low end of the spectrum of life's risks, and that this figure was an improvement over our past performance. It is also axiomatic that our past average is made up of high and low risks, and if in future we reduce the high risks to our target level, we shall inevitably improve our overall performance.

It is our view that these arguments stand up well on moral grounds: the question of cost does not enter into them. We would suggest that this target FAFR of 2 can be used to define a trivial level of risk for our employees. This level is such that risks less than this can be considered effectively zero and risks in excess of this should be reduced as far as is reasonably practicable or until they are reduced below the FAFR of 2.

Reasonably practicable criterion

The other example defining so far as is reasonably practicable can be illustrated by the toxic gas emission criterion developed for accidental toxic releases affecting the public. It was developed as a result of an incident at a chlorine-producing works in England when chlorine escaped from the factory and affected a group of school children. The school children were not seriously injured, but naturally, the events caused considerable public concern, and a team was set up to study the problem. The team recognized that "this shall never happen again" and "escapes should be absolutely eliminated." These meaningless subjective criteria which could be achieved only by shutting down the factory. One of the first tasks of the team was to establish socially acceptable risks associated with chlorine emissions.

The physiological effects of chlorine emissions on the public fall into three categories. First, there is a chance that a chlorine emission may be smelled, thereby causing a nuisance to the public outside the plant. Although highly undesirable, such an event is unlikely to cause any harm, and will therefore be socially acceptable if it happens only occasionally. A more severe escape could cause some degree of distress to people or damage to vegetation which could cause a nuisance and which could lead to claims for compensation. Thirdly, a major emission of chlorine could lead to personal injury or risk to life outside the plant.

These three categories of risk were identified in order of increasing severity as Categories I, II, and III. Category 0 was defined as no nuisance to the public. By collecting published and private data on the toxicity of chlorine gas, Figure 2 was produced indicating the effects of exposure to chlorine. Three hazard categories were then defined more precisely in terms of their effect by indicating the range of chlorine concentration and exposure times covered by each, Figure 3. Thus, three categories of chlorine emission hazard were defined in relation to their effect on the public.

However, we have not yet gotten a usable criterion. We still have to decide what sort of frequencies would be acceptable for each of these emissions. In the minds of those deciding on the criterion were two main factors. One, if emissions happened too frequently, the reaction of the public would be such as to cause the plant to be closed. Secondly, if too stringent targets were set, the penalties on operation and the cost of preventing the incidents would be intolerable burdens on the business.

With these two factors in mind, target frequencies for Category I of about two per year, for Category II of once every ten years, and for Category III of once every hundred years were suggested, Table 5.

A quantitative assessment of the size and probability of chlorine escapes from all the plants on the works and an evaluation of the modifications required to reduce them to the target frequencies showed that the criterion was a reasonably practicable solution to the problem. A similar criterion has now been used for all ICI (UK) chlorine-producing works, as well as adopted by a number of other companies in Europe and the U.S.A., to whom our toxic gas emission assessment techniques have been licensed.

In conclusion

Hazard analysis properly executed is a valuable aid to the decision maker. Its advantages are most obvious when used for comparing alternate routes to the same goal. There are other occasions when the decision maker needs a criterion to decide whether a situation is acceptable or not, and we have demonstrated that such quantitative criteria can be developed even in very emotive areas.

Acknowledgments

Acknowledgments are due to ICI (UK) where the criteria were developed, and ICI Americas where the techniques are being introduced, for permission to present this article. #

Literature cited

1. "A Guide to Hazard & Operability Studies," Chemical Industries Assoc., London.
2. Kletz, "Practical Applications of Hazard Analysis," *Loss Prevention*, Vol. 12, AIChE, New York (1978).

S. B. Gibson, who earned his degree in math and physics at Keele Univ., Great Britain, is with ICI Americas where he is responsible for hazard analysis.

DISCUSSION

STAN SKINNER, Monsanto: I do a fair amount of travelling. I read from your table that air travel is ten times as safe in the U.S. as in the U.K. I find this a bit difficult to believe. I will soon have to fly home across the Atlantic and wonder what is the risk for this crossing. Would this be the arithmetical mean of the U.S. and U.K. values, or the logarithmic mean?

S. BARRY GIBSON, ICI: There was a table prepared not very long ago which gave accident rates for most major airlines. There was a tremendous variation in the figures. I think the Australian airline, Quantas, had the best records, but Air Lebanon (some orders of magnitude poorer) was to be avoided at all costs.

RUSSELL HILL, Oxirane International: Some years ago, either you or another division of ICI published a similar study, in connection with the review of the design of oxygen mixing stations in ethylene oxide plants, in which you concluded that fatal accident frequency of one per 33,000 years in specific installations was acceptable. How does that correlate with the numbers that you present today?

GIBSON: This was another division of ICI. The one in 33,000 years corresponds to a fatal accident rate to an individual on the plant of 0.35; the difference there was they were not only considering risks to individuals.

They were also concerned that the ethylene oxide plant could cause a major hazard that might well kill a lot of people. As I mentioned earlier, one becomes involved in a more philosophical argument when accidents involve lots of people rather than just one or two. For example, is it better to kill 100 people one at a time over 100 years; or is it much worse to kill 100 people altogether once every two hundred years? That was part of the background in settling up criteria for the ethylene oxide plant. They decided that an acceptable risk to an individual was 0.35 FAFR (which was 10% of the average FAFR for the chemical industry at the time) and this is equivalent to a major hazard rate of 3×10^{-5} per year or in other words one major hazard every 33,000 years.

Howard Fawcett, Geo. Met Technology: I think I understand these words, fatalities, on both sides of the ocean. Something that bothers me and has for a long time are those injuries which are not fatal but nevertheless cause permanent incapacitation. In my opinion, and I am speakig now as an American, the ANSI code plus the OSHA code on top of it in more recent times have completely mixed up these numbers, so frankly I have very little faith in them. I wonder if you looked at these numbers in terms not just of fatalities, I understand what a fatality is, but in terms of those injuries which are really incapacitating or which in fact have resulted in a serious loss of life expectancy to the person who happens to be involved.

GIBSON: The fatal accident frequency rate is a convenient term to use as a measure for serious accidents. Obviously, when we are designing a new plant we can't examine the line diagram in sufficient detail to identifv everv place where an employee could cut his finger or splash his arm or eye or sprain his back. So, what we are trying to do is to identify accidents which could be serious, and the fatal accident frequency rate is a convenient term for measuring the risk of that accident.

The accident may well cause a distribution of injuries from death to a cut finger, but the fact that a fatality is possible fits it into the class which can be categorized by the FAFR metric. To some extent you answered your own question, when you said you understand what is meant by a fatality; a serious injury has many meanings so the FAFR has a less ambiguous meaning.

FAWCETT: Thank you.

W.B. HOWARD, Monsanto: I would like to ask a further question relative to something you alluded to a moment ago in answering Mr. Russell Hill's question, and this is acceptability of large accidents. We might have one major airplane to fall in 100 years, and the average loss of life is pretty low. Nevertheless, that one airplane fall accident is totally unacceptable as indicated by the public reaction to it; not what we may want but what the public thinks regarding a DC-10 falling in Chicago and, say, 300 people losing lives or something like this.

Have you worked up any numbers relative to what are acceptable for single major accidents apart from long-term averages of many small accidents? I believe this is an area to which we haven't given enough public attention to in treating FAFRs. We have a general rule in this country relative to newspaper publicity. When a fatal accident in a chemical plant results in the death of five people, that gets total countrywide newspaper coverage; if it is less than five, let's say two or three, it would get normally statewide coverage; maybe two 200, 300, 500 lives. What treatment can we give to these?

These can be real also in chemical industries. This is indicated by a major vapor cloud explosion in Brazil which has not been talked about much; it didn't get into the newspapers nearly as much as Flixboro, but this one in Brazil cost something like 40 lives.

GIBSON: Because this is a very difficult subject we ought to speak about the sort of risks that are involved in all sorts of activities. Unfortunately, the media, which strongly influences public opinion, and the environmental groups do not address risk; they address perceptions of risk. I think the more education the people receive, the more perceptions of risk become equivalent to risks.

Now, there is no number that really measures risk. Someone gave me a very good analogy once: "Measuring risk is like measuring largeness; largeness has all sorts of connotations." So, there is no one number we can use in measuring risk. The difference between people's perceptions of risk and the risks themselves can be illustrated by the dear old lady whose knees are positively knocking as she climbs up the airplane steps; she is terrified at flying but it is just as dangerous or just as safe for her as a businessman who climbs up the steps and flies with equanimity.

Now, you say what have we done towards establishing a target for major hazards. The ethylene oxide plant we talked about earlier had a target of once every 33,000 years; that was a big risk and that was the sort of number used. In an attempt to develop a number with a better logical basis, we considered a statement in a government advisory committee report for the United Kingdom Health and Safety Executive. The statement was: if assessments "indicated with reasonable confidence that in a particular plant a serious accident was unlikely to occur more often than once in 10,000 years, this might perhaps be regarded as just on the borderline of acceptability bearing in mind the known background of risk faced every day by the general public."

From such guarded wording it seems that once every 10,000 years is somewhere near the target. Again, if you look at that figure "once every 10,000 years" and assume the people involved in the incident are killed, it works out to an FAFR of 1.1 which is inside our FAFR target.

Another aspect we considered were the major accidents that have occured of which we have records in the chemical industry, the big ones that killed more than about five people. If you list them all and make certain crude assessments of the number of chemical plants in the world of the type that would have such accidents, one can estimate the frequency of a major hazard to be about once every 1000 plant years. So, if we aim to improve by a factor of 10 over our present performance, we again come to a target of 1 every 10,000 years for a major hazard. So, for whatever it's worth, there is a suggested figure for a big risk.

In actual fact, the number of people involved in an incident occurs on the denominator and the numerator of the calculation. So, they cancel out. That may sound like a rather callous, but that is a problem we were considering in the previous question. However, generally the accidents only involve one or two people, and you determine the number of people involved in the same way as you determine what the accident is going to be by stringent, systematic analysis of the process. When you look at the process and identify a way in which the accident will occur, you also have to look and see how many people are likely to be involved. It is a pity that David Lewis' paper in session 67 is not to be presented, because that has some very good data on fatality radii for explosions; this information also is very difficult to get hold of.

SIDNEY BROOKS, Dupont: Has ICI done any work in quantifying the dollars involved to get to the fatal accident frequency rating? How responsive is your management to design changes?

GIBSON: This is one of the reasons for having a numerical target. There are risk-averse managers and risk-prone managers. You, thus, will get one group of management which is willing to spend any amount of money on safety and another which has opposite views. This is a very unbalanced situation and a totally inefficient way of spending your safety dollars.

Having a numerical target is a good way of getting consistent spending. But, my answer to your question is that when we are designing new chemical plants, it is not difficult to change the design, to change the scope at an early stage, and to assure oneself that one is going to meet the target. So far, we have not had any difficulties in changing the design pressures, temperatures, making reactors smaller, putting in extra protective systems and things like that. We've not had any difficulty in achieving our target of an FAFR of two for new chemical plants without significantly affecting project viability.

The problem comes when you do an assessment on an existing plant and find it has a high FAFR. To reduce it requires a shutdown and physical changes (rather than just changing lines on a drawing) and this is where the cost escalates enormously. Then, we do have problems, and one has to do cost benefit and risk benefit analysis. This involves using money values for loss of life, loss of goodwill, public reaction; difficult but not impossible areas. As for the tendency to go a little bit better, having the numbers available enables me to say, if you do that you are wasting money there, why don't you spend it over here where the numbers are not so good and I was going to say good common sense usually prevails, but that's my opinion. Well, thank you very much.

System Reliability Engineering Methodology for Industrial Application

Some of the available computer programs that can be used to provide input to the engineering effort are listed here, since most industrial systems reliability enginering frequently involves too much information for manual techniques. Also discussed are several theoretical problems that can result in erroneous conclusions and recommendations.

J.S. Arendt,
JBF Associates, Inc., Knoxville, TN
J.B. Fussell
University of Tennessee, Knoxville, TN

System reliability engineering is that part of system design which is concerned with improving system performance with regard to system malfunctions. Decreasing the expected number of these malfunctions, decreasing the expected time the malfunctions exist or some combination of these two improves system reliability. During the past several years the analytical tools available to the reliability engineer have been expanded and a number of problems with the methodology have been identified. The purpose of this article is to present the state of reliability engineering methodology and some unresolved problems. Risk assessors commonly use techniques of system reliability analysis; however, risk assessment is not the subject of this article.

We will assume that the system malfunctions of interest are identifiable. We will also assume that if all the system components function, the system is capable of carrying out the desired mission. Therefore, we will only be concerned with system reliability engineering techniques that are based on changes in states of the system constituents. Furthermore, we will assume that the system reliability engineering analysis is to be based on an abstract model of the system. That is to say, actual observation of system malfunctions with subsequent changes to the system design, as a method of system reliability engineering, is not considered in this article.

0149-3701/81/3795 $02.00 © 1981 AIChE

Reliability analysis of systems commonly encountered in practice requires the use of a computer because of the enormous amount of information that must be processed. Of the methods presently available, the use of fault tree analysis is most prevalent because of the availability of computer software for analyzing fault trees and because of its widely understood format. Many system reliability theoreticians believe that present production methods such as fault tree analysis are out of date or soon will be. However, many reliability engineers use fault tree analysis methodology when a state-of-the-art technique is required. Certainly a large gap exists between the state-of-the-theory and the state-of-the-practice of reliability engineering methodology. In this article, we are concerned with methods used for reliability engineering which have been applied to problems and have been reported in various documents; however, some theoretical problems that have an impact on engineering effort are presented. Presently available production methods suffer from several of these theoretical limitations.

ANALYSIS STEPS AND EXISTING CAPABILITY

System reliability analysis can be divided into six steps: (1) system definition, (2) logic model construction, (3) failure mode determination, (4) quantitative evaluation, (5) uncertainty analysis, and (6) formulation of conclusions and recommendations. These steps are described briefly below. There are computer programs that can be used as an aid in several of these steps. It is not the purpose of this article to provide detailed information concerning the capabilities of all of these computer programs; however, a partial list of the available programs is provided in Table 1 following the article.

System Definition

System definition encompasses (1) the acquisition of detailed information about the functional and physical interconnections of the system components and (2) the establishing of system physical and analytical bounds on the system model. Analytical bounds must be established for all the system malfunctions of interest, called the TOP events, each of which defines an analysis problem. The analytical bounds include the following information for each analysis problem.

1. Initial conditions that describe the initial configuration of all components which have more than one operating state. For instance, a relay or valve can be open or closed. This state, as defined by the correctly operating system, must be noted so that the proper component failure modes will be used in the analysis.
2. The limit of resolution of the analysis. This defines what level of system entity is considered basic to the analysis.
3. Events assumed to have zero or unity existence probability. For example, sabatoge might be neglected.
4. Other necessary assumptions.

All the formal published work, of which we are aware, concerning the requirements for defining a problem for the purposes of system reliability analysis has been done by those interested in computerized logic model construction Therefore, the available literature [1-4] on the practicalities of system definition is scarce and somewhat academic.

Logic Model Construction

Logic model construction is the effort to bridge the gap between the system and the analysis effort. The model itself is some representation of the Boolean failure or success logic associated with a particular analysis problem. Each system malfunction to be studied requires an individual logic model. Examples of types of logic models are fault trees [5-8], cause-consequence diagrams [9-13], block diagrams [14] and a listing of the system failure modes.

Failure Mode Determination

Failure mode determination is the listing of the combinations of basic system component malfunctions that will result in the TOP event. Such a collection of basic events is called a cut set; and if each event is necessary to the set in order to assure the TOP events occurrence by the set, the set is a minimal cut set. [15-17] A major problem in reliability analysis of complex systems is the astronomical number of minimal cut sets that are often encountered. Failure modes and effects analyses [6,18] can be considered as a method for determining minimal cut sets directly from the system for a host of TOP events; however, usually only one-event minimal cut sets are determined.

All component failures represented in a logic model are statistically dependent. When one of these dependencies manifests itself as the simultaneous malfunctioning of several components in a system, the set of these malfunctions is called a common-cause failure or common-mode failure. Identification of potential common-cause failures is often not an important factor in analyzing chemical systems because system malfunctions are usually dominated by independent hardware failures. Usually, the most important common-cause failures with respect to a specified TOP event are those that cause the occurrence of all the events in at least one minimal cut set for that TOP event. If the mechanism of a common-cause of failure is recognized, it can be added to the list of minimal cut sets for the TOP event as a one event cut set. That is to say, if the cut set list that is obtained from the logic model is used as a basis for the search for common-cause failure potential, this list can be expanded to include the important dependencies among the component failures. We generally do not recommend that any attempt be made to include common-cause considerations directly in the system logic model. Doing so compounds the problem of obtaining the minimal cut sets. Computer programs exist that can aid in common-cause failure identification by "brute force" searches for generic common-cause failure potential among all the events in a minimal cut set. None of these programs has the capability of identifying potential for propagation of failures, e.g. failure of a component because of increased stress resulting from another component failure.

Quantitative Evaluation

Quantitative evaluation of system reliability is concerned with synthesizing reliability characteristics for the TOP event from basic component characteristics. Most published methods for quantitative evaluations [19-20] require minimal cut sets as input to represent the system model. These cut sets may include common-cause events.

Three major assumptions of all production quantitative methodologies are:

1. The minimal cut sets adequately represent the system for the purpose of the analysis.
2. The common-cause failure analysis has resulted in a final cut set list such that the basic events can be treated in the quantitative analysis as statistically independent without introducing significant error.
3. The failure model is also the repair model, if repair is considered. This assumption is discussed in more detail later.

From these minimal cut sets and the statistical distributions of time to failure and repair of each event in these cut sets, several quantitative reliability characteristics can be calculated (sometimes only approximately) for a specified TOP event of interest. Sometimes more than one of these characteristics is applicable to decision-making efforts. Several of the more frequently used event characteristics are defined below.

Availability and Unavailability. The availability is the probability that a specified fault event does not exist at some specified time in the future. The unavailability is the probability that the event does exist and numerically is equal to unity minus the availability. The general assumption used in the availability calculation is that no system component failures exist at time t = 0.

Availability is usually associated with repairable systems. In which case, the availability at time t contains no information about system failures before time t. The availability is a useful characteristic if the occurrence of the specified system failure is tolerable at least some expected fraction of the time. The average unavailability during a time interval is the expected fraction of the time the system failure exists and is a useful characteristic when system downtime has economic implications.

Reliability and Unreliability. The reliability is the probability that a specified fault event has not occurred during a specified time interval. The unreliability is the probability that the event has occured one or more times in the time interval and is the numerical complement of the reliability. Repairable system reliability calculations account for the possibility of a redundant component failing and being repaired without system failure necessarily occurrring. No analytical method exists for calculating repairable system reliability under a consistent set of assumptions for an arbitrary logic structure. The system reliability can be determined directly when the system failure rate is known. Vesely's [19] exact method for calculating the system expected number of failures results in a failure intensity such that, if this failure intensity is divided by system availability, an approximation to the system failure rate, and hence the system reliability, is obtained. No one has shown that this approximation to system reliability is always "conservative." However, the expected number of failures is always an upper bound on the unreliability and is a reasonable approximation of the unreliability when it is not large (e.g., $\leq$ 0.1).

When the expected number of failures is known, the authors have never encountered a problem where conclusions and recommendations depended on also knowing the reliability. Therefore, while not being able to calculate the system reliability exactly is recognized as a theoretical problem, we do not consider it a problem in practice.

System reliability is sometimes an interesting system characteristic and is the pertinent factor of merit when the system failure is considered intolerable. For catastrophic system failures, such as explosions, system reliability is reported; system availability has limited engineering meaning for such failures.

Expected Number of Failures. The expected number of failures is the statistically expected number of occurrences of a specified fault event over a specified time interval. The expected number of failures is in a sense a more global system characteristic than either availability or reliability because it contains information about the system failure and repair performance during a time interval. If the cost of each failure is known, the expected number of failures multiplied by the cost per failure gives the expected cost during the time interval resulting from that failure.

Importance of Basic Events. Several probabilistic methods [21] can be used to compute the importance of basic events in fault-tree analysis. All the methods assess the importance of basic events by a numerical ranking although the interpretation describing the relation of the occurrence of a basic event to the occurrence of the TOP event is different in each case. The calculation of relative component importance is less effected by weak failure data than are the calculated reliability characteristics of the TOP event.

Measuring the relative importance of system constituents may (1) identify components toward which system upgrade efforts should be directed and (2) suggest the most efficient way to diagnose system failure through a repair checklist.

Uncertainty Analysis

The purpose of an uncertainty analysis is to determine the uncertainty in the quantitative results which is attributed to uncertainty in the component data input to the analysis. No method exists to determine uncertainty resulting from a less than adequate logic model, inappropriate assumptions, or breach of the theoretical capability of the quantitative methodology.

Uncertainty in component data results because component failure and repair characteristics (1) are not known perfectly from any application (i.e. less than an infinite amount of data is available), (2) actually vary among systems, and (3) are out of date because of the historical nature of data collection. Therefore, no data bank can ever contain specific values of failure and repair rates that are applicable to the analysis of a specific system. Data from a data bank should, at best, be viewed as input to be used by the analyst as an aid in deciding on appropriate statistical descriptions for propagation input data uncertainty through system reliability computation.

Formulation of Conclusions and Recommendations

Appropriate conclusions and recommendations are not generally determinable from quantitative results alone. However, valuable input to these conclusions can be realized by investigating (1) the cause of any anomolous quantitative results, (2) the importance of the individual component malfunctions to the main system failure, (3) the uncertainty in the results that arises from uncertainty in the input data, and (4) the relation of the results to the results of other comparable studies. The quantitative characteristics that are calculated in many systems evaluations should be viewed as describing the best performance the system can possibly achieve. Since all cut set lists are incomplete, the modeling is not generally guaranteed to be conservative and, at present, no method for quantifying this lack of conservatism has been demonstrated. Therefore, quantitative system reliability results usually cannot be used to prove that a system is "good enough". Even though systems evaluations are not complete, the potential exists in these techniques to confidently demonstrate that a system's reliability performance is better than alternatives.

Small values for expected frequency and probability often result from the quantitative studies. Interpreting these numbers is difficult when only one system or a small number of systems is involved. In this case we can interpret these small numbers by stating that we feel more confident about the system's performance if the failure characteristic of merit is 10^{-5} than if the number is 10^{-3}.

Formulation of conclusions and recommendations, then, is the engineering effort that uses as input all the analysis assumptions and results. The assumptions are as important as the results. A potential problem is that the analyst is seldom aware of all the assumptions he had made.

ANALYSIS PROBLEMS

Presented below are five problem areas that we believe have substantial impact on the ability to treat problems encountered in practice. All five can lead to poor conclusions and recommendations if they are not recognized. Presently available computer aids will accept situations subject to these problems and produce erroneous results. The problems presented here do not represent an exhaustive list.

Repair Models

Most quantitative methods almost always assume that the failure logic model is also the repair logic model. Although this is a valid assumption in some cases, it causes errors serious enough to result in erroneous conclusions and recommendations in many cases encountered in practice. Some effort has been made to overcome this modeling problem when using cause-consequence diagrams [22]; however, no production quantitative methods are available for cause-consequence analysis.

Any fault tree that shows secondary failure development is not a valid repair model. A secondary failure is an event that causes component or subsystem failure for which it is not held accountable. Secondary failure can dominate the causes of system failure. For example, consider the malfunction of a computer system. A failure-prone air conditioning system could dominate the causes of failure of the computer. However, repair of the air conditioner does not constitute repair of the overall system since the computer must still be repaired in any case. In this example, we have "OR" failure logic and "AND" repair logic.

Existing computer programs and theoretical methodologies for the repairable case encourage the use of the failure logic as the repair logic. It is easy to generate simple, practical examples where this problem introduces factors of 1000 error in the calculated results. Since, in our experience, few analysts are aware of this problem, we believe this to be a potential major source of erroneous information input to the decision-making process from system reliability engineering analysis and risk assessment.

Dependencies in Event Trees and Cause-Consequence Diagrams

Event trees [13,22] and cause-consequence diagrams [9-13] are used to organize complex problems for probabilistic evaluations and are frequently a tool used during risk assessment. These models are particularly useful when describing accident progressions when systems and procedures are available to cope with an accident initiating event. These systems and procedures are sometimes called Lines of Assurance (LOAs). Conventionally, failure logic for the LOAs is developed using fault trees. Although cause-consequence diagrams use more symbols and contain more information, they are logically equivalent to event trees. Both models begin with an initiating event and branch successively, reflecting the success or failure of the LOAs, to the possible consequences of the initiating event, as shown in Figure 1, where F is the failure and S is the success of the appropriate LOA.

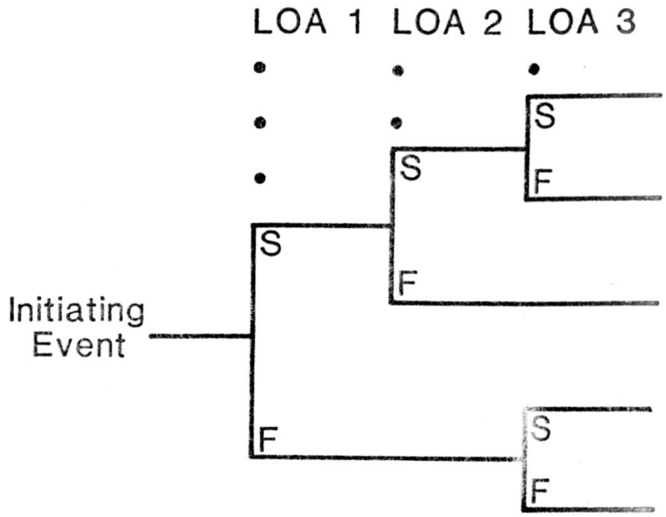

Figure 1. Sample event tree.

Usually only one fault tree is developed for a given LOA, but sometimes more than one fault tree is required. In the example shown in Figure 1, if LOA 1 succeeds, the fault tree for LOA 3 could be different than for the case when LOA 1 fails. Cause-consequence diagrams allow such fault trees to be recognized on the diagram when higher order LOAs are dependent on the success or failure of lower order LOAs. However, the reverse is not true. Cause-consequence diagrams do not have the capability for recognizing the dependence of lower order LOAs on the success or failure of the higher order LOAs and therefore compromise the modeling effort. When using event trees or casue-consequence diagrams, the analyst should lay the diagram out so that the dependence of the lower order LOAs on the higher order LOAs is minimized.

This problem can be treated correctly by developing fault trees for each accident chain rather than for a specific LOA. However, a major feature of event trees and cause-consequence diagrams is that a problem can be logically developed in segments and one fault tree can be constructed to fit into several segments. A theoretical price is paid, however, since the fault tree developed for a LOA is generally not correct for all the accident sequences in which it appears.

Maintainability Calculations

Very little theoretical work has been done concerning the calculation of complex system maintainability characteristics. The time the repair activity begins is stocastically determined by the component failure distribution; therefore, techniques for treatment of components with other than constant repair rates are tedious or theoretically unknown. In most cases methods for repairable system evaluations assume components have constant repair rates.

In our opinion, assuming the component repair rates are constant, does not usually introduce serious limitations. The asymptotic unavailability of repairable a component is dependent only on the mean time to repair and the mean time to failure and is nearly reached after about three mean times to repair in most cases of practical interest. In addition, all system reliability characteristics of interest in practice only depend on the repair distribution through their dependence on the component unavailabilities. Therefore, assuming constant component repair rates only affects system results during the early transient, and even this effect is slight in most cases based on our unpublished studies.

Production methods for system reliability analysis assume statistically independent component repairs. This assumption can be more serious than assuming independent failures. The assumption is often not valid because of repair policy, unannounced failures, and a limited number of available repair crews in the event of several, dependent or independent, component failures. While considerable effort is being made to deal with dependent (common cause) failures, we know of no effort to remove the requirement of assuming independent component repairs.

At present analysts cannot determine the system repair rate. This also precludes quantitative treatment of delay situations in a logic model. For example, the fault tree delay gate is used when a fault event will result in further system faults if not repaired in a specified delay time; however, no quantitative treatment is possible in existing computer programs.

Binary Failures

While any system constituent may have many failure modes or states, for systems evaluations, the analyst must assume that a given failure mode or state exists or it does not exist. Sometimes a system fault is a continuous function of the output of many components. For example, an overheat condition may occur if the sum of the heat generation of a number of components exceeds a given value. Since the binary assumption is made in production methods that a component is overheating or it is not overheating, modeling such situations is at best an approximation.

Phased Missions

Time-dependent failure logic is difficult, and often impossible, to model correctly with presently available methodology. Phased mission analysis is directed toward system level changes in failure logic.[23-26] Examples of systems that undergo phased missions are plentiful: emergency core cooling systems in nuclear power plants and fire fighting systems in chemical plants are two such examples. As the mission proceeds the components remain the same while the system changes. Presently developed theoretical methods for phased mission analysis are restricted by the following additional assumptions: (1) the times that the system failure logic changes are known and hence are not dependent on stochastic events, and (2) nothing in the systems is repairable.

Treating a phased mission as a series of conventional reliability problems can lead to substantial error in the results because of reasons that include (1) the component failure characteristics are dependent on the time the mission begins, not the time a phase begins, and (2) the system can experience a failure at the time the failure logic changes (at the phase boundary) because a cut set with all events failed is present in the latter logic but the cut set was not failed in the former logic. Present theoretical methods for analyzing phased missions are limited. The need to be able to treat repairable systems undergoing a phased mission is a pressing problem. Montague[27] has recently developed an exact method for calculating the expected number of failures for a system undergoing a phased mission.

SOME COMMENTS ON DATA

Lack of data has resulted in the largest criticism of system reliability engineering and risk assessment methodology as discussed in this article. Sufficient data, as well as standardized methodology, are important if system reliability engineering methodology is to be widely accepted. Nevertheless, for many purposes of system reliability engineering, a set of data that is rigorously defendable on a statistical basis is not required.

Present data sources contain only an

"averaged" value for failure rates rather than detailed time-dependent values. Our studies have shown that, in most cases for system assessments, an average failure rate is adequate and conservative except at times near system start-up.[28] Also sensitivity studies can be effective even when data is "soft."

In quantitative evaluations, data serves an often overlooked, important function. During the system modeling effort the analyst is continually making decisions concerning events to include in and events to be omitted from his model. Events omitted from the model are assumed to have a negligible rate of occurrence. To ensure a model that adequately represents the system for the purposes of the analysis, the analyst must have access to knowledge equivalent to a data store.

Reliability data is unlike other data such as nuclear cross sections. The basis of cross section data is a finite number of never changing nuclei all with identical properties. The basis of reliability data is stochastic, involving ever-changing components and environmental conditions. In addition, reliability data has the additional dimension of being the description of a statistical distribution rather than a fixed but unknown value. In other words, even under perfect measurement conditions, reliability data for a single event cannot be determined by a single test, but rather must be determined from years of observation of equipment in various environments during which time the equipment designs can change. In addition, it is frequently easy to determine how many components of a certain type failed in a certain way during a known time period. Unfortunately, it is often not known how many identical components did not fail during this time interval; that is, the population of components in use is not known. Collecting and evaluating reliability data is very frustrating.

CONCLUSIONS

During the past ten years, system reliability methodology has advanced from the academic stages to a formal engineering tool. We believe the assumptions made during an analysis are as important as the results. Unfortunately some of the assumptions used in reliability analyses, particularly in computer-aided analyses, are not realized by analysts. Although lack of data has resulted in a major criticism in the use of system reliability engineering techniques, we feel that the objectives of many analyses can be realized using a set of data that is not statistically defendable. While several methodology problems exist, the major limitations to the advance of formal application of system reliability engineering techniques is the shortage of trained personnel and the lack of standardized methodology.

LITERATURE CITED

1. J. B. Fussell, Synthetic Tree Model-A Formal Methodology for Fault Tree Construction, ANCR-1098, Aerojet Nuclear Co., Idaho Falls, ID, March 1973.

2. J. R. Taylor and E. Hollo,"Experience with Algorithms for Automatic Failure Analysis," Nuclear Systems Reliability Engineering and Risk Assessment, SIAM, 1977, pp. 759-777.

3. G. J. Powers, "Computer Aided Synthesis of Fault Trees," NATO Advanced Study Institute on Generic Techniques of System Reliability Assessment, Nordhoff Publishing Co., 1974.

4. G. E. Apostolakis, et al., "The Use of Decision Tables in the Systematic Construction of Fault Trees," Nuclear Systems Reliability Engineering and Risk Assessment, SIAM, 1977.

5. D. F. Haasl, "Advanced Concepts in Fault Tree Analysis," System Safety Symposium, Seattle: The Boeing Co., June 8-9, 1965.

6. H. E. Lambert, Systems Safety Analysis and Fault Tree Analysis, UCID-16238, Lawrence Livermore Laboratory, University of California, May 1973.

7. J. B. Fussell and D. P. Wagner, "Fault Tree Analysis as a Part of Mechanical System Design," NBS Special Publication 487, 25th Meeting of the Mechanical Failures Prevention Group, August 1977, pp. 289-308.

8. P. A. Crosetti and R. A. Bruce, "Commercial Application of Fault Tree Analysis," Ninth Reliability and Maintainability Conference, Annals of Reliability and Maintainability-1970, Vol. 9, p. 230.

9. D. S. Nielsen, The Cause-Consequence Diagram Method as a Basis for Quantitative Accident Analysis, RISØ-M-1374, Danish Atomic Energy Commission, 1971.

10. D. S. Nielsen, "Cause-Consequence Diagrams," NARS Publication 2, Nordic Working Group on Reactor Safety, December, 1972.

11. D. S. Nielsen, Use of the Cause-Consequence Charts in Practical Systems Analysis, Conference on Reliability and Fault Tree Analysis, Operations Research Center, Berkeley, California, September 3-7, 1974.

12. R. A. Evans, "Fault Trees and Cause-Consequence Charts," IEEE Transactions on Reliability, Vol. R-23, April 1974, p.1.

13. J.B. Fussell, et al., A Collection of Methods for Reliability and Safety Engineering, ANCR-1273, Aerojet Nuclear Company, Idaho Falls, ID, April 1976.

14. B. J. Garrick and W. C. Gekler,"Reliability Analysis of Engineered Safeguards," Nuclear Safety, Vol. 8, September-October 1967.

15. J. B. Fussell, "Fault Tree Analysis-Concepts and Techniques," Generic Techniques in Systems Reliability Assessment, Nordhoff, Leyden, 1976, pp. 133-162.

16. R. E. Barlow and H. E. Lambert, "Introduction to Fault Tree Analysis," Reliability and Fault Tree Analysis, SIAM, 1975, pp.7-35.

17. P. Chatterjee, Fault Tree Analysis:Minimal Cut Set Algorithms, ORC74-2, Operations Research Center, University of California, Berkeley, CA., 1974.

18. W. E. Jordan, "Failure Modes, Effects, and Criticality Analysis," Proceedings of Annual Reliability and Maintainability Symposium, San Francisco, 1972, IEEE,New York, 1972, p. 30.

19. W. E. Vesely, "A Time Dependent Methodology for Fault Tree Evaluation," Nuclear Engineering and Design, 13, 1970, pp. 337-360.

20. J. B. Fussell, "How to Hand-Calculate System Reliability and Safety Characteristics," IEEE Transactions on Reliability, R-24, August 1975, pp. 169-174.

21. H. E. Lambert, "Measures of Importance of Events and Cut Sets in Fault Trees," Reliability and Fault Tree Analysis, SIAM, 1975, pp. 77-100.

22. D. S. Nielsen and B. Runge, "Unreliability of a Standby System with Repair and Imperfect Switching," IEEE Transactions on Reliability, Vol. R-23, No. 1, April 1974, pp. 17-28.

23. H. Ziehms, Reliability Analysis of Phased Missions, Thesis, Naval Postgraduate School, Monterey, CA., December 1974.

24. J. D. Esary and H. Ziehms, "Reliability Analysis of Phased Missions," Reliability and Fault Tree Analysis, University of California, Berkeley, September 1974.

25. G. R. Burdick, et al., "Phased Mission Analysis: A review of New Developments and an Application," IEEE Transactions on Reliability, Vol. R-26, April 1977.

26. D. J. Campbell, A Procedure for Determining the Importance of Basic Events to Systems Undergoing Phased Mission, Master's Thesis, NERS-78-07, Department of Nuclear Engineering, University of Tennessee, Knoxville, December 1978.

27. D. F. Montague, Nuclear Science and Engineering, 74, 199-209, (1980).

28. G. A. Foster and J. B. Fussell, On the Effect of Assuming Exponentially Distributed Component Failure and Repair Times During System Reliability Assessments, NERS-78-02, Department of Nuclear Engineering, University of Tennessee, Knoxville, March 1978.

29. W. E. Vesely and R. E. Narum, PREP and KITT: Computer Codes for the Automatic Evaluation of a Fault Tree, IN-1349, Idaho, August 1970.

30. J. B. Fussell, E. B. Henry and N. H. Marshall, MOCUS- A Computer Program to Obtain Minimal Sets from Fault Trees, ANCR-1156, Aerojet Nuclear Company, Idaho, March 1971.

31. R. B. Worrell, Set Equation Transformation System (SETS), SLA-73-0028-A, Sandia Laboratories, Albuquerque, New Mexico, May 1974.

32. D. M. Rasmuson and N. H. Marshall, "FATRAM - A Core Efficient Cut Set Algorithm," IEEE Transactions on Reliability, Vol. R-27, No. 4, October 1978, pp. 250-253.

33. S. M. Semanderes, "ELRAFT - A Computer Program for the Efficient Logic Reduction

Analysis of Fault Trees," IEEE Transactions on Nuclear Science, Vol. NS-18, February 1971, pp. 481-487.

34. P. K. Pande, M. E. Spector, and P. Chatterjee, Computerized Fault Tree Analysis, TREEL and MICSUP, ORC 75-3, Operations Research Center, University of California, Berkely, California, April 1975.

35. D. B. Wheeler, et al., "Fault Tree Analysis Using Bit Manipulation," IEEE Transactions on Reliability, Vol. R-26, No. 2, June 1977, pp. 95-99.

36. W. J. Van Slyke and D. W. Griffing, ALL-CUTS - A Fast Comprehensive Fault Tree Analysis Code, ARH-ST-112, July 1975.

37. R. C. Erdmann, et al., A Method of Quantifying Logic Models for Safety Analysis," Nuclear Systems Reliability Engineering and Risk Assessment, SIAM, 1977, pp. 732-754.

38. P. J. Pelto and W. L. Purcell, MFAULT - A Computer Program for Analyzing Fault Trees, BNWL-2145, Battelle Northwest Laboratories, Richland, Washington, November, 1977.

39. J. Olmos and L. Wolf, A Modular Approach to Fault Tree and Reliability Analysis, MITN-209, Master's Thesis, Department of Nuclear Engineering, MIT, August 1977.

40. G. R. Burdick, N. H. Marshall and J. R. Wilson, COMCAN - A Computer Program for Common-Cause Analysis, ANCR-1314, Aerojet Nuclear Company, Idaho Falls, Idaho, May 1976.

41. C. L. Cate and J. B. Fussell, BACFIRE - A Computer Program for Common Cause Failure Analysis, NERS-77-02, Nuclear Engineering Department, University of Tennessee, Knoxville, May 1977.

42. D. M. Rasmuson, et al., COMCAN II - A Computer Program for Common Cause Failure Analysis, TREE-1289, EG&G Idaho, Inc., Idaho Falls, Idaho, September 1978.

43. O. Platz and J. V. Olsen, FAUNET - A Program Package for Evaluation of Fault Trees and Networks, RISØ-348, Denmark, September 1976.

44. W. Y. Gateley, D. W. Stoddard and R. L. Williams, GO - A Computer Program for Reliability Analysis of Complex Systems, KN-67-704, Kaman Sciences Corp., Colorado Springs, Colorado, April 1968.

45. J. B. Fussell, et al., SUPERPOCUS - A Computer Program for Calculating System Probabilistic Reliability and Safety Characteristics, NERS-77-01, Nuclear Engineering Department, University of Tennessee, Knoxville, February 1977.

46. W. E. Vesely, F. F. Goldberg, FRANTIC - A Computer Code for Time-Dependent Unavailability Analysis, NUREG-0193, United States Nuclear Regulatory Commission, March 1977.

47. H. E. Kongso, RELY4: A Monte Carlo Computer Program for System Reliability Analysis, RISØ-M-1500, Denmark, June 1970.

48. "Reactor Safety Study," WASH-1400 (NUREG-75-/014), United States Nuclear Regulatory Commission, October 1975.

49. S. D. Matthews, MOCARS: A Monte Carlo Code for Determining the Distribution and Simulation Limits, TREE-1138, EG&G Idaho, Inc., Idaho Falls, Idaho, July 1977.

50. D. M. Rasmuson, et al., User's Guide for the Reliability Analysis System (RAS), TREE-1168, EG&G Idaho, Inc. Idaho Falls, Idaho, September 1977.

51. J. S. Arendt, et al., EXCON - A Computer Program for System Risk Assessment, JBFA-128-78, JBF Associates, Inc., Knoxville, Tennessee, October 1978.

52. L. S. Baker, et al., TREDRA - A Computer Program to Draft Fault Trees, JBFA-128-78, JBF Associates, Inc., Knoxville, Tennessee, October 1978.

53. H. E. Lambert and F. M. Gilman, The IMPORTANCE Computer Code, UCRL-79269, Lawrence Livermore Laboratory, University of California, 1977.

Table 1 Some Computer Programs Presently Used in System Reliability Engineering

Purpose	Program Name	Comments
Determine minimal cut sets from fault trees	PREP[29]	Uses combinatorial testing in a Boolean equation
	MOCUS[30]	Uses a top-down Boolean substitution method
	SETS[31]	Uses Boolean manipulation to determine prime implicants
	FATRAM[32]	Top-down algorithm similar to MOCUS with refinements for efficiency
	ELRAFT[33]	Uses the unique factorization property of prime numbers
	MICSUP[34]	A bottom-up algorithm using Boolean substitution
	FAULTRAN[35]	Technique is based on binary coding of events and bit manipulation for tree reduction
	ALLCUTS[36]	Almost identical to MOCUS
	WAM-CUT[37]	Uses a Boolean manipulation and minimization algorithm
	MFAULT[38]	Uses a bottom-up algorithm and calculates approximate values of TOP event existence probabilities
	PL-MOD[39]	Uses modular decomposition and calculates TOP event quantitative characteristics
Common-cause failure identification aid	COMCAN[40]	Determines possible common-cause failure candidates using "brute force" searching algorithm, matching component susceptibilities with locations or special conditions
	BACFIRE[41]	Similar to COMCAN.
	COMCAN II[42]	Uses a reduced fault-tree approach to determine minimal cut sets having failure potential due to common causes
Quantitative evaluations	FAUNET[43]	Written for minicomputer; uses a top-down pruning and bottom-up modularization, Boolean substitution algorithm; also calculates TOP event existence probabilities
	GO[44]	Uses a Boolean manipulation algorithm; also calculates point estimates of TOP event characteristics
	KITT 1 & 2[29]	Calculates time-dependent availability, reliability, and expected number of failures from component failure rates and repair times; also calculates component importance
	SUPERPOCUS[45]	Calculates same time-dependent characteristics as KITT using bounding approximation methods
	WAM-BAM[37]	Calculates the TOP event existence probabilities using Boolean manipulation

Table 1 Continued

Purpose	Program Name	Comments
	FRANTIC [46]	Calculates effect on average unavailability and time-dependent instantaneous unavailability due to periodic system testing; testing characteristics include the test interval, test duration time, repair time, and human-caused failure potential associated with the testing
	RELY4 [47]	Uses Monte Carlo importance sampling to determine minimal cut sets
Uncertainty analysis	SAMPLE [48]	Monte Carlo simulation of failure rate and repair time distributions of basic events in order to determine uncertainty of TOP event reliability characteristics
	MOCARS [48]	Monte Carlo code for determining distribution and simulation limits of TOP event reliability characteristics
	RAS [50]	Integrated package of reliability analysis computer programs which includes MOCUS, FATRAM, POCUS, KITT-1, and COMCAN
	EXCON [51]	Determines system risk using cause-consequence diagrams
	TREDRA [52]	Draws fault trees from MOCUS-type input on a plotter
	IMPORTANCE [53]	Calculates various measures of component importance and other quantitative information

DISCUSSION

ED NAIDUS: Does any of your work lead you to a method of treating a single event probability so that redundancy isn't the only logical answer? I refer to a single event that is a single failure which is so serious that the only answer is redundancy. Redundancy ordinarily is an improvisation based on pure opinion or perhaps experience. It is a very common problem to have a single failure without statistical evidence with regard to component behavior that would lead to an unacceptable result. So, redundancy usually is the system chosen. But, is there any rational way of tackling this problem?

ARENDT: Most system reliability analysis techniques are directed towards identifying the important system failure modes with respect to a given main system failure of interest. Usually, in chemical systems, these important system failure modes are the failures of single pieces of equipment. One way of improving system performance is to reduce or eliminate the number of single failure points in the system. This can sometimes be accomplished by adding redundant components.

Adding redundancy is only one option for improving system performance and one that I believe is too often automatically accepted as the proper choice. Many other factors influencing the choice of methods of improving system reliability have to be considered such as improved maintainability and component overdesign.

One gentleman asked a question about automated logic model construction.

Jerry Fussell, my co-author, was one of the first researchers in the area of automated logic model construction. I have a fundamental concern about automated logic construction. I feel that the use of automatic logic model construction will not alleviate the lack of qualified personnel who are able to formulate logic models through the conventional manner. I feel it's an important area of research, but I don't believe we want to reach the point where we think we can supply a "black box" with a system schematic and expect that the resulting logic model is logically pure and correct. I believe automated logic model construction should take its place as an aid for analysts to use with discretion and caution.

One other comment: most of the measures of risk that have been defined today are measures of average risk. I prefer to think of risk as the doublet of the frequency of the event paired with the consequence of the event. This way we have more information. It's more beneficial to know that an event with a consequence of 10^{-2} per year than to know that its average risk is one. Compare this to a situation where the consequence 10^{8} and the expected frequency is 10^{-8} per year. I believe that it is much better to report coupled frequency and consequence measures of risk rather than the product.

Ignition of Flammable Vapors by Human Electrostatic Discharges

The minimum ignition energy for an acetone/air mixture was determined to be three times greater for a human electrostatic discharge than for a purely capacitive spark, determined by direct comparison in the same experimental apparatus. Extrapolation of the results allows application of published minimum ignition energies.

R. W. Johnson
Hercules Inc., Cumberland, MD.

"Hazard analysis" has found wide applicability in the Chemical Processing Industries as a quantitative method for reducing risks of personal injury and death in hazardous processes, and for providing management with information which will aid in optimizing the effectiveness of the dollars spent on plant safety. Although a relatively recent technique in the area of loss prevention, hazard analysis has developed into a comprehensive methodology, bring together such diverse tools as fault tree analysis, material response testing, statistics and probability, and financial risk control.

At the heart of hazard analysis is the determination of the probabilities of occurrence of identified hazards. Engineering experience and tools such as fault trees can be used successfully to identify potential hazards, but only upon the assigning of realistic probabilities to these hazards can decisions be made as to whether a standard of safety has been met and which potential hazards contribute most to the probability of occurrence of the undesired event.

0149-3701/81/3806 $02.00

Hercules' hazards analysts, in applying the HERC® technique to the quantitative evaluation of process hazards,(1-4) must frequently resolve questions dealing with electrostatics. One significant area of uncertainty has been a question as to how relevant purely capacitive electrostatic initiation data are to the prediction of the initiation probability of flammable vapors by human spark discharge.

The objectives of this work were to investigate in detail the initiation of flammable vapors by human electrostatic discharge, and to verify the probability of initiation of an acetone/air mixture due to electrostatic discharge from a human operator.

The human spark ignition of flammable vapor requires several factors to coincide in the same place at the same time:

1. Presence of a vapor/air mixture within a fairly narrow concentration range.

2. Sufficient electrostatic charge generated on an operator by rising from or sliding across a seat, removing an outer garment, induction from a nearby charged object, separating sheets of plastic or rubber, miscellaneous physical activity, etc.

3. Storage of the charge on the operator for a sufficient length of time. This requires a low absolute humidity to preclude charge leakage.

4. Spark discharge to a large or grounded object in the same place where the vapor/air mixture exists.

This study assumes the presence of both a flammable vapor and a human spark discharge, and focuses on the probability of ignition of the vapor given the above conditions.

For electrostatic discharge initiation of solids, liquids, vapors, and dusts, the material response characteristics are usually determined in a apparatus which provides a purely capacitive electrostatic discharge; i.e., with very low resistance in the discharge path between where the charge is stored and where the spark occurs. However, in cases where the charge is not stored on a conductor, or resistances exist in the discharge circuit (as is the case with a human electrostatic discharge, the skin acting as the major resistance), it is questionable whether the purely capacitive material response data can be applied for a quantitative assessment of probability. Consequently, in this study the actual in-process situation involved in human electrostatic discharge and ignition of flammable vapors was simulated as closely as possible to verify the application of purely capacitive material response data to human spark discharges.

DOES A HUMAN SPARK HAZARD REALLY EXIST?

The in-process energy related to electrostatic discharges is the energy actually imparted to the ignitable material by the spark or sparks passed between two objects. This energy is closely related to, but not necessarily equivalent to, the quantity of electrostatic energy stored on the ungrounded or semiconductive item under study. In electrical terms, the electrostatic energy E (joules) stored on an object with capacitance C (farads) and voltage V (volts) is given by

$$E = 1/2\ CV^2$$

A human operator insulated from ground has a capacitance in the range of 150 to 300 pF,(5) and potentials of at least 10 kV can be generated on an operator when quickly removing an outer garment or rising from a chair.(6) Thus, an energy of $\frac{1}{2} \times (300 \times 10^{-12}) \times (10^4)^2 = 1.5 \times 10^{-2}$ joules (15 mJ) can easily be generated on a human operator. Since the minimum energies required to ignite flammable vapors by purely capacitive electrostatic discharges are in the range of 0.02 to 2.4 mJ,(7) it was tentatively concluded by the Bureau of Mines in 1953 that "precautions should be taken to prevent accumulation of electrostatic charges on individuals around operations involving handling of explosives that may ignite with electrostatic energies of 0.015 J or less."(8) In a later Information Circular,(9) the bolder conclusion was drawn, based on the same calculation, that "almost any flammable mixture can be spark ignited by commonly occurring electrostatic discharges."

Contrary to these conclusions, studies at the Shirley Institute(10,11) (Manchester, England) have indicated that an energy perhaps 60 to 100 times greater than the minimum spark ignition energy is required to ignite a flammable vapor/air mixture by a spark discharge from a human operator.(10,11) These studies, which will be discussed in a later section, point out the need for direct material testing which simulates as closely as possible the actual in-process situation. This is in contrast to relying solely on comparing calculated in-process energies with material response testing performed in an apparatus which might be quite different from the process configuration.

Furthermore, if the above-mentioned studies are correct, recommendations based on the Bureau of Mines criteria would be unnecessarily conservative, thus causing an adverse cost impact on processing operations.

COMPARISON USING SAME APPARATUS

The basic approach used in this study was to compare the minimum energies necessary to ignite the same vapor/air mixture in the same apparatus by means of sparks originating from static electricity stored in two different ways: first, stored in a purely capacitive (very low resistance) circuit; second, stored on a human operator insulated from ground. The former gives the published "minimum ignition energies;" whereas, the latter gives the direct in-process initiation probabilities of interest.

An acetone/air mixture was used throughout

this study for reasons of availability and wide applicability of the results in the Chemical Processing Industries. The published minimum ignition energy of a stoichiometric acetone/air mixture at atmospheric pressure is given as 1.15 mJ.(7,12) Therefore, if an extrapolation based on the 60-100 factor suggested by the workers at the Shirley Institute is made, then the minimum human spark ignition energy of the acetone/air mixture could be between 70-120 mJ. This is above the range of normally-occurring human spark discharges, in which case one would not expect ignition to occur.

APPARATUS AND PROCEDURE

The apparatus used in this study is shown diagramatically in Figures 1 and 2. The spark chamber was a transparent cellulose acetate cylinder, approximately one liter in volume, with the bottom fitted with a funnel filled with glass wool to serve as a flame arrestor. The top of the chamber was loosely covered with a piece of filter paper for the purpose of preventing surrounding air from freely entering the chamber. The electrodes, either 13 mm-diameter steel spheres or 2 mm-diameter steel rods with hemispherical tips, were mounted on shafts which passed through opposite sides of the chamber. The high voltage electrode was fixed in place, and the grounded electrode was attached to a dial-type micrometer and could be moved in and out to adjust the spark gap. The high voltage was supplied by a Van de Graaf generator, manually controlled by a foot switch. A multirange electrostatic voltmeter was used to measure the static voltage.

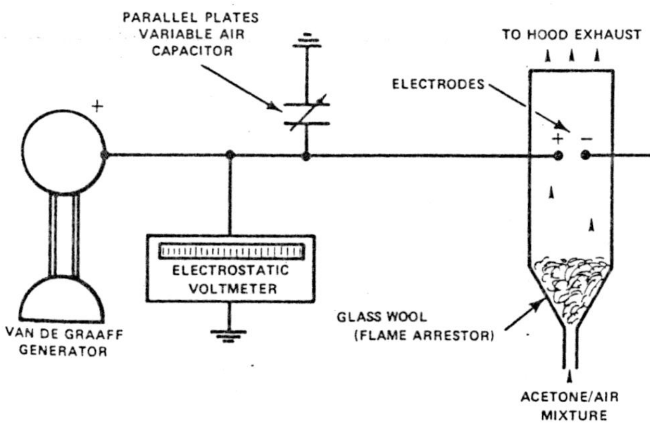

Figure 1. Capacitive spark discharge circuit.

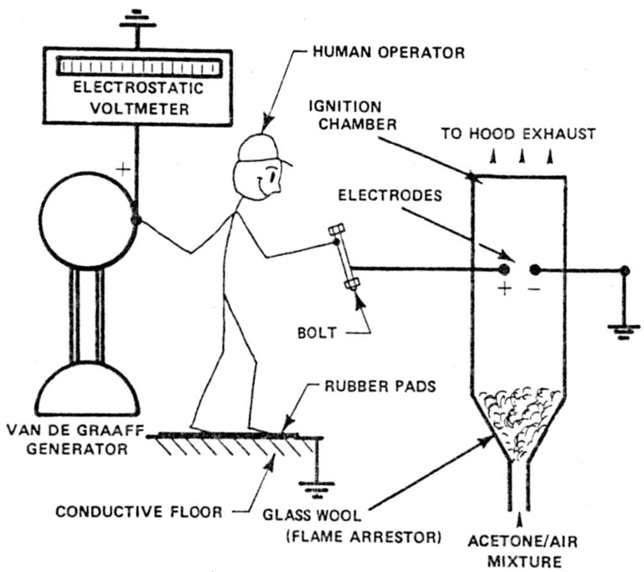

Figure 2. Human spark discharge circuit.

On the human spark trials, the operator had one hand on the Van de Graaf generator, while gripping a bolt wired to the fixed electrode with the other hand. The operator was insulated from a conducting floor by rubber mats and, in some trials, a thick polypropylene block, in order to vary the capacitance of the operator.

After adjusting the spark gap and acetone/air ratio, the capacitor or human operator was slowly charged until a spark occurred between the electrodes. The voltage immediately before sparking, and whether or not the mixture ignited, were then recorded. The criteria for positive ignition were the witnessing of a flame, loud noise, and blowing the filter paper from the top of the ignition chamber. Time was allowed between trials for the system to regain thermal eqilibrium (at least two minutes after spark with no ignition, and at least five minutes after a spark with ignition). For each set of runs, the capacitance of the entire high-voltage circuit was measured with an impedance bridge.

MINIMUM IGNITION ENERGY OBTAINED

Initial tests were performed to compare a minimum ignition energy obtained with the present apparatus with the value reported in the literature. The resulting minimum for a stoichiometric acetone/mixture was found to be 1.40 mJ, which compares favorably with the published value of 1.15 mJ. It should be noted that "minimum ignition energy" is here taken to mean the lowest energy for which one ignition in ten trials is observed.

Preliminary tests were also conducted to determine the optimum acetone/air ratio (see Figure 3). It was found that the optimum ratio was slightly on the lean side of stoichiometric. In all subsequent tests, both purely capacitive and human spark, an acetone/air ratio of 0.90 times stoichiometric was used (equivalence ratio = 0.90), corresponding to a concentration of approximately 4.5% acetone in air.

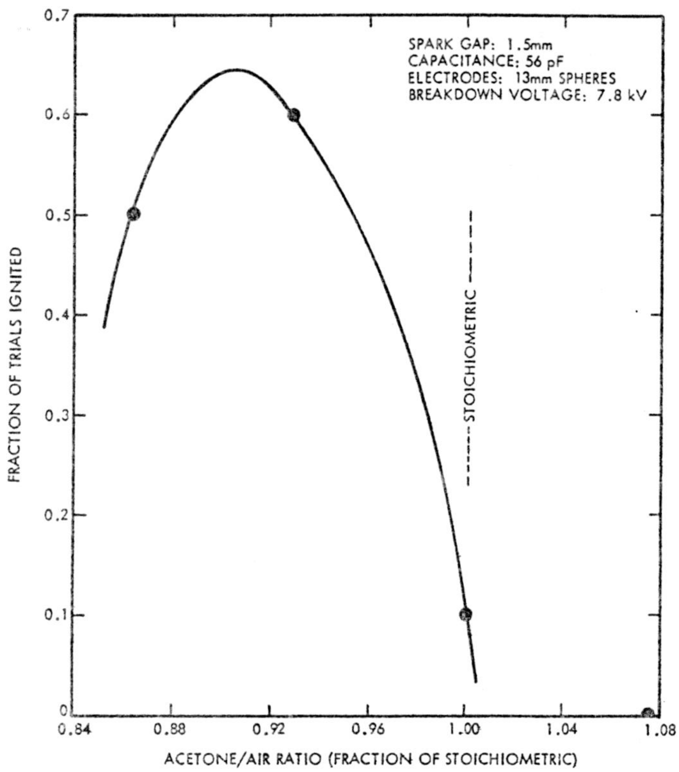

Figure 3. Ignition vs. acetone/air ratio.

ONLY A SMALL DIFFERENCE BETWEEN HUMAN SPARK AND CAPACITIVE SPARK

The results of the capacitive and human spark ignition trials are given in Table I. These results are also given in the form of probit curves in Figure 4.

Taking one ignition in ten trials to be the definition of minimum ignition energy, it was found that for a purely capacitive discharge the minimum ignition energy for a mixture of 4.5% acetone in air is 1.27 mJ; whereas for a human spark discharge, the minimum ignition energy is 3 mJ. Thus, the human spark discharge needed to be only 2.4 times as energetic as the purely capacitive spark to ignite the same acetone/air mixture. This result contrasts sharply with the conclusion reported in the literature that the human spark must be 60 to 100 times more energetic than the purely capacitive spark to ignite the same vapor/air mixture. Experimental differences which might account for this discrepancy are as follows:

(1) In the previous investigations, natural gas/air and coal gas/air mixtures were used to compare minimum ignition energies.

(2) In the present work, the electrodes were fixed; and, once the circuit attained breakdown voltage, time was required to wait for a spark to cross the gap. For the Shirley Institute trials, the finger of the charged operator was brought up to a grounded steel ball inside a box containing the vapor/air mixture.

(3) The discharge for the Shirley Institute trials was directly from the operator's finger; whereas in the present work, the operator held a bolt in his hand which was electrically connected to a steel electrode.

(4) In the apparatus setup used in the human spark trials of this study, a small but perhaps significant capacitance existed between the human operator and the electrode. The insulated wire between the electrode and the bolt held in the hand of the operator had a capacitance of approximately 18 pF. This corresponds to 10.5% and 15.2%, respectively, of the 136 pF and 171 pF human spark circuits, and only has 33% of the required capacity to ignite the acetone/air mixture by itself (i.e., 33% of the 55 pF in the purely capacitive spark circuit, Figure 2). Taking this numerically into account, it would be expected that a human spark about three times (instead of 2.4 times) more energetic than a purely capacitive spark is needed to ignite the same vapor/air mixture. However, it is not known whether the 18 pF of capacitance between the operator and the electrode alters the minimum human spark energy in any way other than

merely adding capacitance to the circuit.

TABLE I

RESULTS OF SPARK IGNITION TRIALS*

DISCHARGE	CAPACITANCE (pF)	BREAKDOWN VOLTAGE (kV)	SPARK ENERGY (mJ)	IGNITION/ TRIALS
CAPACITIVE	55	6.64	1.21	0/10
	55	6.80	1.27	1/10
HUMAN	136	6.48	2.85	0/10
	136	6.65	3.00	1/10
	136	6.70	3.05	2/10
	136	6.81	3.15	6/10
HUMAN	171	6.59	3.34	0/10
	171	6.69	3.42	1/10
	171	6.72	3.45	2/10

*4.5% Acetone in air at 25°C; 2 mm diameter rod electrodes

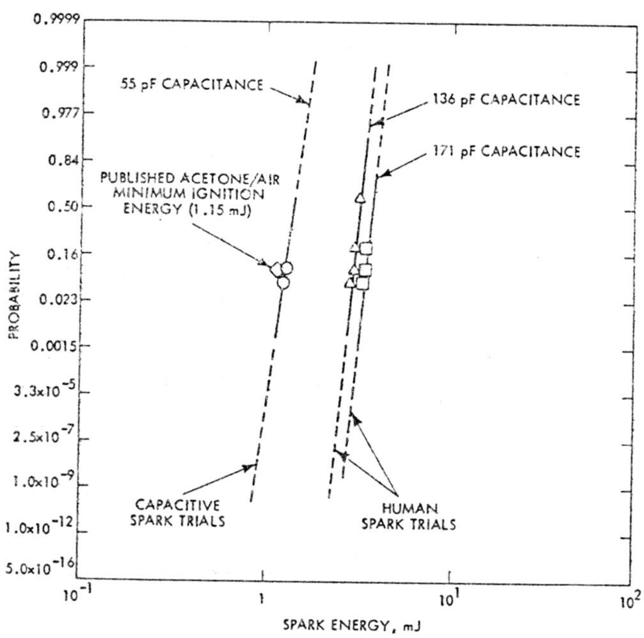

Figure 4. Probit curves for acetone/air spark ignition.

SUMMARY

In testing an acetone/air mixture, it was found that the minimum ignition energy for a human spark discharge was three times greater than for a purely capacitive discharge, measured in the same experimental apparatus. Extrapolating these results, it is anticipated that any vapor/air mixtures with minimum ignition energy of 5 mJ or less can be ignited by sparks drawn from a human operator. This would include nearly all hydrocarbon/air mixtures, since the range of minimum ignition energies for such mixtures is about 0.02 to 2.4 mJ. Consequently, precautions must be taken to prevent electrostatic generation, storage, and discharge in areas where flammable vapors are present. Typical precautions and control measures include inerting, the use of conductive plastics, the use of conductive shoes and floors, increasing the humidity in the process area, and grounding and bonding of all metallic processing equipment.

The probit curves in Figure 4 can be used to find the probability of ignition of an acetone/air mixture by a human electrostatic discharge, given the spark energy calculated from the capacitance of the operator and the static voltage generated (or induced) and stored on the operator. It can be seen that at the above-mentioned spark energy of 15 mJ, the probability of igniting an acetone/air mixture is essentially unity (100%).

Care should be exercised in applying the results of this study to different process situations and different flammable vapors. Nevertheless, it is felt that the human spark test apparatus used in this study produced results that would be more conservative than most configurations found in actual processes. Further study in this area should be directed toward the testing of other flammable vapors, such as methanol and propane, as well as dust clouds of various finely-divided combustible materials.

LITERATURE CITED

1. Richardson, R. H., et al., "Hazards Analysis Through Quantitative Interpretation of Sensitivity Testing," Hercules Incorporated, Annals of New York Academy of Sciences, 152, Article 1, 269-282 (October 28, 1968).

2. Hoffman, P. D., "Hazard Evaluation and Risk Control," Loss Prevention, 8, 80 (1974).

3. Albaugh, L. R., et al., "How Hazards Analysis Can Help," Loss Prevention, 12, 30 (1978).

4. Katz, M. J., "Hazard and Risk Evaluation," Loss Prevention, 10, 127 (1976).

5. Cleves, A. C. and J. F. Sumner, Ministry of Aviation, Explosives Research and Development Establishment, Report No. 18/R/62 (17 August 1962).

6. Wilson, N., J. Electrostatics, 4, 67-84 (1977/1978).

7. Calcote, H. F., C. A. Gregory, Jr., C. M. Barnett, and R. B. Gilmore, Ind. Eng. Chem., 44, 2659 (1952).

8. Brown, F. W., D. J. Kusler and F. C. Gibson, Bureau of Mines, U.S. Dept. of Interior, Report of Investigations No. 5002 (September 1953).

9. Van Dolah, R. W., M. C. Zabetakis, D. S. Burgess, and G. S. Scott, Bureau of Mines, U.S. Dept. of Interior, Information Circular No. 8137 (1963).

10. Henry, P. S. H., Static Electrification, 1971, London, Inst. of Phys. Conference, Serial No. 11, p. 212.

11. Wilson, N. Shirley Institute Report 28/286/13 (November 1973).

12. Fenn, J. B., Ind. & Eng. Chem., 43, No. 12, pp. 2865-2869, (1951).

JOHNSON, R.W.

DISCUSSION

TREVOR KLETZ, ICI: I would like to relate an incident which illustrates what you've been saying.

Two or three years ago during one of the rare hot summers that we have in the United Kingdom, a colleague of mine was on holiday and he stopped to fill up his car with petrol, I beg your pardon, to fill up his automobile with gasoline. He got out of his car, took the filler cap off the end of the gasoline tank and held it in his hand. Since it was a warm day, he used the opportunity to take off his pullover and threw it into the car; this caused a separation of charge and charge was left on him. When he was about to put the gasoline filler cap back on the car, a spark jumped from the filler cap to the car and a flame appeared on the end of the filler pipe. It couldn't travel back into the gasoline tank, because the vapor in the tank was too rich. There was only a lazy flame on the end of the pipe which he put out with a rag.

A particular point of interest is that he was holding a metal object in his hand. However, these circumstances are quite exceptional, not usual. Our first though when dealing with a leak on the plant is not to take off the our clothing, there are other things to do first. I have always taken the view that there is no need to make sure that people earth themselves by wearing conducting footwear. Do you agree?

ROBERT JOHNSON, Hercules, Inc.: I would agree with your comment...

ANONYMOUS: I am not sure I understand why a human spark has to be three times the energy of a classic spark to ignite an inflammable mixture in the first place. Those molecules don't know where the spark is coming from; neither does the spark.

JOHNSON: I would agree with your comment that it would only be under exceptional circumstances that these types of conditions would be found. We were trying to attain a conservative end to see what kind of bounds there were. We work within an explosive manufacturing facility, manufacturing solid rocket propellant.

Thus, nearly all of our operating areas have lead floors and operators are required to wear conductive shoes. We must take into account the severity of the consequences of a fire or explosion, and the probability that a vapor/air mixture will be present.

There are other means by which static electricity can be controlled other than conductive footwear, most notable of which would be via humidity control. As a general rule of thumb, if an enclosed operating facility is at normal operating temperatures around 70° to 80°F, it should be kept above 60° relative humidity in order to effectively control static electricity. At high humidity levels, you would have charge leakage to the ground by means of increased surface conductivity.

This was a more pragmatic work. There has been work done dealing with the theoretical aspects. One of the suggested mechanisms is that the resistance in the circuit due to the skin of the operator slows down the charge mobility of the charge discharging into the circuit, and this somehow decreases the immediate spark energy available to the flammable vapor. Another explanation is that multiple sparks occur from a discharge by a human operator. Thus, only the first, most energetic spark would be the one that would ignite the flammable vapor/air mixture.

Full-Scale Explosion Study of Relief Vents Suitable for Protecting Large Structures

Experimental study indicates that substantial economies may be achieved by estimating explosion venting requirements based on "partial volume" explosions without significant sacrifice of asset protection.

E.S. Naidus
APC Corp., Hawthorne, NJ

In 1975 an extensive series of controlled and instrumented building explosions were carried out at the Fenwal Explosion Research Laboratories in Hollister, Massachusetts. The details and analysis of the results were reported at the Loss Prevention Symposium, A.I.Ch.E., Kansas City (1976) (1). A number of questions raised by the earlier experimental work have been addressed in the work reported here.

The governing hypothesis underlying the work in 1975 and 1979 is that a large percentage of industrial explosions (occurring external to process equipment as a result of dust or vapor deflagrations) involve a "cloud" of combustible material that is a small (1/6 to 1/10) fraction of the total enclosing structure.

As early as 1951, P.E. Cotton (2) published his estimates of the fraction of the enclosure volume actually deflagrating for use in explosion pressure calculating. Others (3, 4, 5, 6, 7) have treated the same subject but relatively few experiments have been reported that permit quantitative treatment of the problem.

It should be recalled that most reported data on controlled experimental explosions have been carried out on dust or vapor clouds that filled the test volume completely with stoichiometric or near-stoichimetric mixtures of air with combustible material. The practical problems of designing usable structures to withstand or to survive explosion pressure loadings (impulses) require a better understanding of the forces on such structures from typical interior explosions.

DISCUSSION:

Our experimental approach was to use the well-documented propane-air combustible mixture in a fixed (but adjustable) volume (fuel chamber) within a substantially larger light-construction (plywood bolted to light steel angle) building, equipped for explosion relief venting by means of varying size openings and types of vents.

The specific questions that were attacked were:

1. The effect of varying the explosion volume with a fixed building volume.
2. The effect of varying the explosion relief vent area.
3. The response of alternate vent designs that would lead to more widely applicable and lower cost relief vents.

One fundamental aspect of the dynamics of interior explosions is the evaluation of the "impulse" (pressure x time) as it affects the

0149-3701/81/3893 $02.00

structure and as it acts on the explosion relief vent. It has been conventional for insurance and code officials to call for explosion venting in terms of "pounds per square foot". This can lead to the bizarre situation of complying with a 20 lb./sq. ft. specification by using a one-half inch thick steel plate. To avoid this startling possibility, most guidelines caution the vent user about "inertia" and suggest that a lightweight panel be used with various releasable attachments. Even then, the question of the response time of fasteners is rarely discussed. (8) (9). The use of the term "impulse" is a sound technical way to get to the physical reality of the cause of damage by an explosion. A structure rated at a nominal 488 kg/sq. m. (100 lb/sq. ft.) will not sustain such a load for much over 0.5 seconds delivered over a substantial area. To obtain protection, a relief vent must dissipate such impulses by opening within 50-60 milliseconds from a force not exceeding 244 kg/sq. m. (50 lb./ sq.ft.) (or some equally low value of the impulse). A second practical consideration is the need for early dissipation of low-energy dust explosions - so-called "puffs" - especially in structures where dust can collect and be re-dispersed by the energy of the "puff" and the resultant vibration of the structure. The quick opening of vents allows removal of some dust, some heat, and some pressure and the vents will remain open to dissipate a secondary explosion.

In bucket elevators (or long, narrow structures), where the ratio of length of the structure to crosssection area is 3 or more, more complex dynamics may dominate, but the basic concept will remain valid. The initial explosion in a bucket elevator maintains a relatively high pressure and flame temperature in either direction (in an unvented condition); as this condition prevails, the explosion becomes even more intense due to the accelerating effect of the expansion of combustion products along the enclosed space. (It should be noted that the separation by venting, of the effects of the pressure wave and the flame front, is possible due to the much higher velocity (300 to 400 m/sec.) of the pressure wave than that of the flame front (3 to 4 m/sec.)). By early venting of the initial explosion and by spacing vents along the elevator and related structures, pressures and temperatures are lowered by dissipation through the vents, thus reducing the potential damage in the elevator and minimizing the transport of excessive pressures and temperatures to other connected structures. (10) (13)

EXPERIMENTAL CONDITIONS:

The plywood and light steel frame structure was identical to that used in previous experiments. Provision was made for mounting an explosion relief vent (4' x 4', 4' x 6', 4' x 8') on one wall. The explosion cell was constructed as a nearly cubical box with thin polyethylene blow-out film adhered to make a gas-tight enclosure. The propane: air mixture (5:95) was introduced to the chamber by adding a weighed quantity of propane, turning on a small fan for two minutes, then maintaining 30 seconds without fan before ignition. The ignition source was 2 1/8" diameter electrodes, 1/16" spacing, 15,000 v. source, 22 watts spark. Data was obtained at 1.8 cu. m. (64 cu. ft.), 1.36 cu. m. (48 cu. ft.), and 1.02 cu. m. (36 cu. ft.) explosion cell volumes by moving the back wall of the chamber forward. The vents were standard 4' x 8', 4 x 6' and 4' x 4' models with the tubing and pin-clip fasteners used in 1975; also similar units using aircraft cable with collapsible washers as an alternate system. (See Figures 1, 2)

Explosion tests were also run with completely "free openings" (4' x 8', 4' x 6' or 4' x 4'). The pressure sensors were C.E.C. #4-402 (0-15 psig) connected to a C.E.C. 5-124A high speed recording oscillograph. As a secondary reference, high speed movies were taken of some of the tests showing the behavior of the vents at the critical period of pre-release, release, and "fly-away"; vents that were provided with aircraft cable tethers for the releasable members were also tested and found to be practical for retaining the releasable elements.

RESULTS:

The data obtained is summarized in Tables 1, 2, 3. The experimental configurations are shown in Figures I and II.

Comparing the averaged data for three different explosion chamber volumes with two sizes of vent openings, we find:

TABLE 1

FUEL SIZE	VENT SIZE	P_{max} (Average)	R_{max} (Average)
1.8 cu. m (64 cu. ft.)	1.2 m x 2.4 m (4' x 8')	91.4 kg/m (0.13 si.)	253.1 kg/m/sec. (3.6 psi/sec.)
1.36 cu. m (48 cu. ft.)	1.2 m x 2.4 m (4' x 8')	154.7 (0.22)	583.5 (8.3)
1.02 cu. m (36 cu. ft.)	1.2 m x 2.4 m (4' x 8')	63.3 (0.09)	295.3 (4.2)
1.8 cu. m (64 cu. ft.)	1.2 m x 1.2 m (4' x 4')	147.6 kg/m (0.21 psi)	442.9 kg/m/sec. (6.3 psi/sec.)
1.36 cu. m (48 cu. ft.)	1.2 m x 1.2 m (4' x 4')	246.1 (0.35)	660.9 (9.4)
1.02 cu. m (36 cu. ft.)	1.2 m x 1.2 m (4' x 4')	56.2 (0.08)	442.9 (6.3)

The inherent variability of the explosion conditions of propane: air content is due to the uncertainty of complete mixing and filling, as well as to uncontrolled turbulence effects. The trend of increasing P_{max} with chamber volume (increased mass of combustibles) seems to be as expected but for the anomaly of higher P_{max} values at the 1.26 cu. m. (48 cu. ft.) volume.

These results have led us to explore a possible "resonant" effect due to accidental acoustic reinforcement for specific chamber dimensions and shape or due to specific pressure sensor locations. (See Figure 2)

Regression analyses* on two main sets of data (obtained several years apart) are plotted on Figure III.

Correlation coefficients are greater for the data obtained with actual vents than that for "free opening" but taken together, the performance of vents is very close to that of "free opening".

*Mr. Richard Schwab (Allied Chemical Corporation) contributed his expertise to carry out independent calculations on our data for which we are very appreciative.

EFFECT OF LOCATION OF P_{Max}

The usual position of the pressure sensor was in the plane of the vent opening, close to one edge. All the data in Table 1 were taken in such an arrangement.

Figure II shows the location of identical sensors in three locations whose signals were simultaneously fed to three identical high speed oscillographs. Table 2 gives the data obtained in this series: (each value given is an average of three).

It is of interest to note that a low value of P_{max} at the vent (2) sensor is obtained only when the venting is through a free opening, while the P_{max} data with commercial vents in place show fairly close values, regardless of location. It can be seen as well that all P_{max} values decrease with increasing vent area (whether venting is free opening or commercial vent).

The variation of P_{max} and R_{max} values is taken to be dependent primarily on the variability of mixing in the fuel cell.

By theory, a constant volume (quantity) of a constant ratio or reactants should give identical values for P_{max} and R_{max}. The linear plot of R_{max} vs. P_{max} tends to sub-

TABLE 2

RUN #	OPENING	TYPE		BOX (1)	VENT (2)	CORNER (3)
15	1.2 m x 2.4 m (4'x 8')	Open	P_{max}	70.3 kg/m (0.10 psi)	35.1 kg/m (0.05 psi)	105.4 kg/m (0.15 psi)
			R_{max}	393.7 kg/m/s (5.6 psi)	372.6 kg/m/s (5.3 psi/s)	499.1 kg/m/s (7.1 psi)
-	1.2 m x 2.4 m (4' x 8')	XRW Vent	P_{max}	91.4 (0.13)	91.4 (0.13)	105.4 (0.15)
			R_{max}	450.0 (6.4)	351.5 (5.0)	393.7 (5.6)
16	1.2 m x 2.4 m (4' x 8')	XRV VENT	P_{max}	119.5 (0.17)	105.4 (0.15)	147.6 (0.21)
			R_{max}	583.5 (8.3)	541.3 (7.7)	639.8 (9.1)
17	1.2 m x 1.2 m (4' x 4')	Open	P_{max}	210.9 (0.30)	105.4 (0.15)	210.8 (0.30)
			R_{max}	914.0 13.0	583.5 8.3	984.3 14.0
18	1.2 m x 1.2 m (4' x 4')	XRW VENT	P_{max}	351.5 (0.50)	351.5 (0.50)	386.7 (0.55)
			R_{max}	1546.8 (22.0)	1546.8 (22.0)	1757.7 (25.0)

stantiate this hypothesis; the failure of the R_{max}/P_{max} lines to have zero intercepts suggests that there may be another systematic influence on the data. (See Figure III)

The two general types of explosion relief vents are shown in reproduction of standard product sketches. The XRV model is identical with the XRV tested in 1975. The two XRW models are substantially modified, lower cost units for vertical mounting on large structures.

All types were adequately restrained to within 0.5 m (18") of the opening by an aircraft cable type of tether. (See Appendix)

A relationship between R_{max} and P_{max} and V (volume of the explosive mixture) has been proposed by W. G. Chappell (9) as follows:

$$R_{max} = \frac{3(K)^{1/3}\ (P)^{2/3}}{(V)^{1/3}} \qquad (R_{max})^3 = \frac{27(K)\ P_{max})^2}{V}$$

When V is constant

(as in 1975 data and B series, 1979):

$$(R_{max})^3 = K^1\ (P_{max})^2$$

A plot of R_{max} vs. $(P_{max})^{2/3}$ showed no significant improvement in linearity of the plot over the R_{max} vs. P_{max} graph.

The deviation of the R_{max} vs. P_{max} slopes from a zero intercept suggests that there may be a "damping effect" of the air volume surrounding the fuel chamber as well as shape and turbulence effects that will require further study.

EXPLOSION VENTING DATA

TABLE 3

	P_{Max} (psi)	R_{max} (psi)
WITH VENT	to convert (psi) values shown: multiply by 703.07	to convert (psi/s) values shown: multiply by 703.07
A Series	.23, .20, .20	9.6, 6.7, 9.1
	.13, .13, .13, .13	3.5, 4.2, 4.4, 3.2
	.11, .09	3.2, 3.6
	.20, .25	4.7, 7.6
	.27, .36, .38	6.3, 10.0, 12.0
B Series (Plotted in Figure III)	.13, .16, .15	6.4, 6.3, 6.7
	.13, .15, .15	5.0, 7.7, 9.1
	.15, .20, .16	5.6, 10.0, 6.3
	.19, .17, .16	8.3, 8.3, 8.3
	.18, .15, .15	14.0, 7.7, 8.3
	.21, .21, .18	12.0, 9.1, 8.3
	.50, .50, .55	22.0, 22.0, 25.0
WITH OPENING		
A Series	.15, .23, .20	8.5, 7.1, 6.3
	.20, .10, .10	4.2, 4.9, 3.4
	.09, .09, .08	4.2, 2.2, 6.3
	.25, .20, .20	12.0, 8.3, 6.3
B Series (Plotted in Figure III)	.10, .10, .10	3.3, 3.8, 5.6
	.05, .07, .06	3.3, 3.7, 5.3
	.15, .12, .12	4.2, 7.1, 7.1
	.30, .30, .30	13.0, 9.0, 23.0
	.15, .15, .20	8.3, 5.0, 15.0
	.30, .35, .40	14.0, 11.0, 20.0

CONCLUSIONS:

1. The commercial explosion-relief vents can be evaluated for large scale design purposes by the methods described.

2. The commercial vents in this study perform close to the theoretical limit (a free opening) in response to low-order, but hazardous explosion forces (impulse on a building).

3. Vent models offering effective and economical pressure-relief along with replaceable and tethered panels have been described and tested.

All vent models showed the typical high speed response shown in oscillograph tracings and high-speed motion picture film, estimated at less than 50 milliseconds from ignition to release.

Each series of tests run with "free openings" and vents gave linear plots of R_{max} vs. P_{max} with virtually no distinction between the values and the slope of the lines for "free openings" and the vents. (See Figure III). Such results suggest that the vents offer little more resistance to the pressure impulses (in the range under study) than the column of air to be moved through the corresponding "free opening".

As a matter of interest, the data obtained in 1975 (7) was replotted so as to compare with the present data (See Figure III). The linearity and close parallelism reinforces the conclusion that valid measurements of venting behavior have been obtained. The regression analyses for both the 1975 and 1979 work is presented in Table 4 below. (Also see Figure III).

	EQUATION OF LINES	CORRELATION COEFFICIENTS
1979		
Vents	$R_{max} = 36.5\ (P_{max}) - 3.0$	0.91
Openings	$R_{max} = 33.1\ (P_{max}) - 1.8$	0.80
1975		
Vents	$R_{max} = 42.8\ (P_{max}) + 0.8$	0.96
Openings	$R_{max} = 39.2\ (P_{max}) + 1.4$	0.73

Experimental work confirms limited field experience with actual building explosions in which such units released in the expected manner and served to protect the buildings and contents from serious damage. (See Appendix)

It now remains to tackle the special problems confronted in agricultural and other combustible dust and vapor hazards such as grain elevators, storage enclosures, conveying enclosures, blending and mixing structures and other large volume enclosures in which combustible ducts or vapors may arise from the accidental release from process equipment. On a practical level we are participating in a scale model grain elevator study of venting and we have furnished several full scale, but small, grain elevator installations. (13) Information obtained in such work will be correlated with our experimental activity already reported.

Special recognition should be given to Parker Peterson, John Call, and Warner Dalzell of the Fenwal Company whose efforts made this presentation possible.

APPENDIX 1.

Field experience with explosions and vents are not often witnessed and recorded. The occasional documentation of "Real Life" incidents can, therefore, be of great value.

1. In early 1977, an explosion occurred in a "highly protected" resin-solvent ink plant. The building was a so-called pre-engineered metal building (50' x 80' x 15'), equipped with 32 roof-mounted explosion vents (4' x 8') and with wall panels secured with "approved" explosion release fasteners.

An overheated stirrer drive led to overheating of the kettle contents; a small fire ensued that was controlled with a hand extinguisher. However, fumes were strong enough to drive personnel out; as they were leaving, a low frequency rumble "woosh" was noted, followed by an overhead fire ball that extended the length of the building.

The explosion caused 10 of the XRV vents to release and the heat from the fire ball caused 14 additional units to drop out. None of the wall panels equipped with "approved" explosion relief fasteners were released.

The damage was limited to the contents of the kettle, the stirrer, and some scorched surfaces. Operations resumed within several hours.

Assessment by fire and insurance engineers was that the building, contents and some personnel would have been lost if the explosion had not been vented.

2. In 1980, an explosion ocurred in a solid waste disposal building. The building housed heavy equipment used to crush solid waste for a major city in Wisconsin. Previous experience indicated that explosions of varying intensity would occur regularly. Explosion vents (XRV-P) were installed on the roof, at a vent ratio of 1 sq. ft./50 cu. ft. of building volume. Over a one-year period, the vents were released and replaced about once a month. A recent explosion was sufficiently violent to blow all the vents out and damaged the panels severely, requiring new panels. No damage to any of the facility or injury to personnel occurred.

LITERATURE CITED

1. Cotton, P.E., N.F.P.A. Quarterly, 45, (October, 1951).

2. Hartman and Nagy, Ind. Eng. Chem., 49, (October, 1957).

3. Brasie and Simpson, Loss Prevention, 2, CEP Technical Manual (1968).

4. Allen and Athens, Loss Prevention, 2, CEP Technical Manual (1968).

5. Runes, E., Loss Prevention, 6, CEP Technical Manual (1972).

6. Palmer, K.N., Loss Prevention, 8, CEP Technical Manual (1974).

7. Naidus, E.S., Loss Prevention, 10, CEP Technical Manual (1976).

8. Explosion Venting, N.F.P.A. #68 (1978).

9. Chappell, W.G., Loss Prevention, 11, CEP Technical Manual (1977).

10. Palmer, K.N., "Dust Expolsion & Fires" Chapman & Hall, London (1973).

11. Naidus, E.S., Explosion Venting, N.F.P.A. meeting (May, 1977) unpublished.

12. Rust, E.A., Expolsion Venting, Chem. Eng'g. PP 102-110 (November 5, 1979).

13. Model Grain Elevator Dust Expolsion Studies - Privately published by N.G.F.A. (1979). Work done at Fenwal facilities with APC explosion vents.

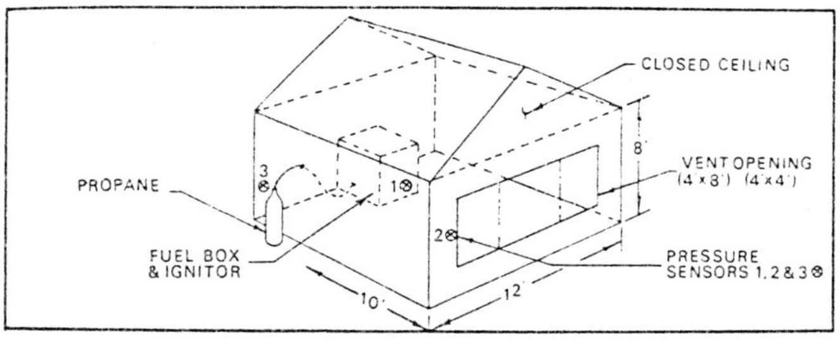

Figure 1. Experimental configurations.

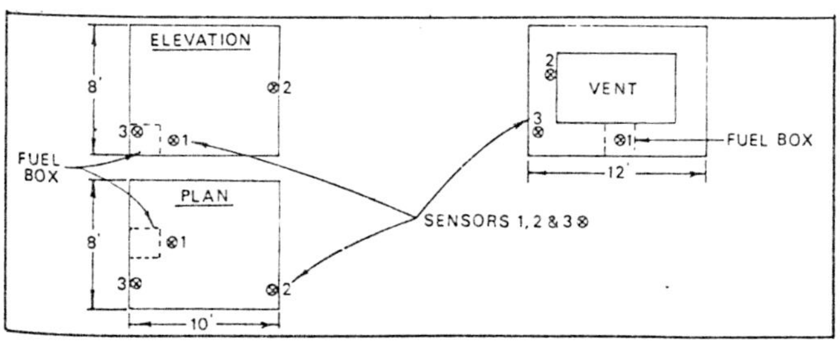

Figure 2. More experimental outlines.

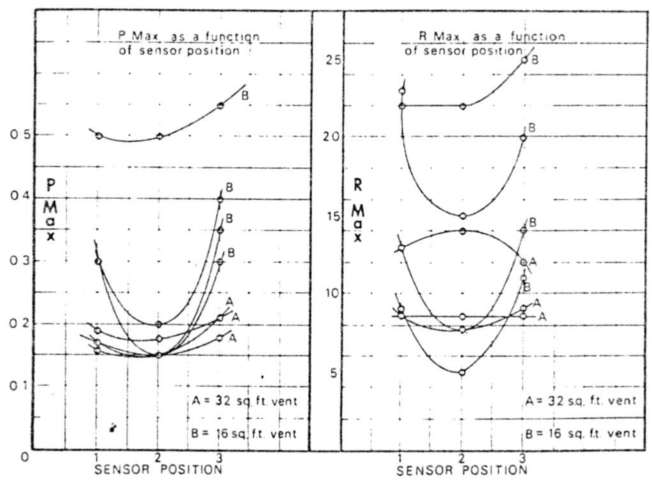

Figure 2a. Data in Table 2.

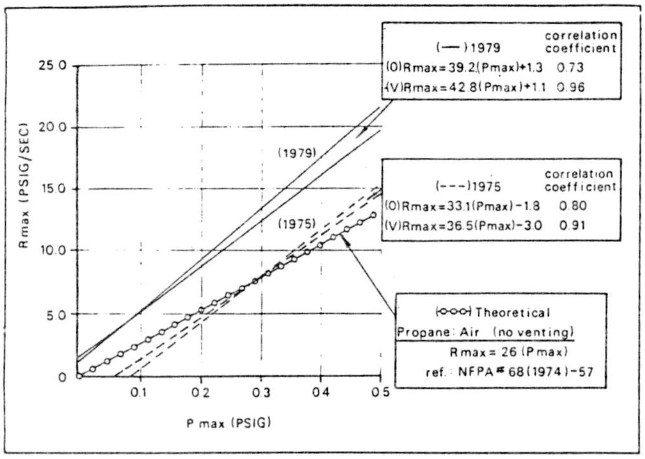

Figure 3. Regression analysis plots for 1975 and 1979 explosion venting studies.

NADIUS, E.S.

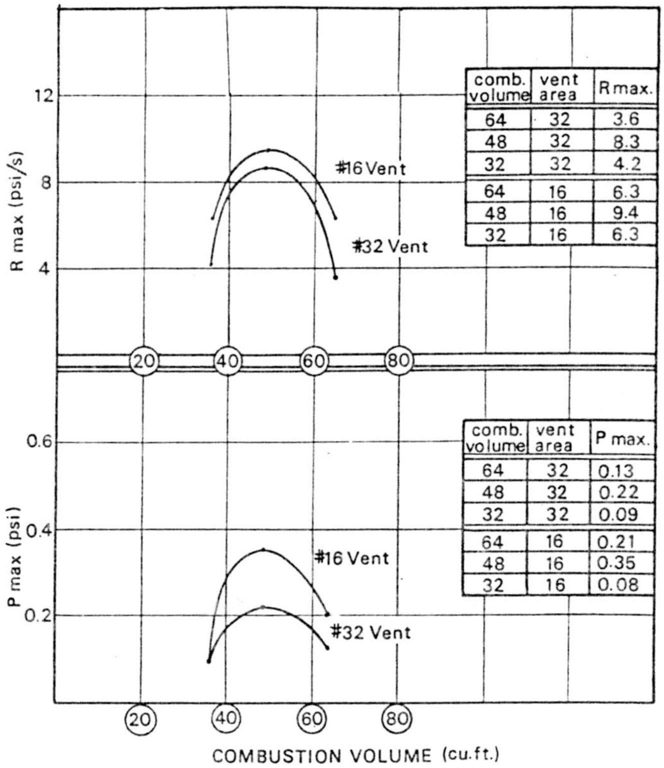

Figure 4. Partial volume explosions.

DISCUSSION

W.B. HOWARD, Monsanto: Did you analyze the mixture in your cell with some sort of analytic instrument to see what the concentration actually was?

NAIDUS: Well, since it was a sealed cell, I had them checked to see that there were no leaks. The propane was actually weighed out and then added to the fuel cell to form the gas mixture. The propane was weighed by difference so there was a known weight of propane introduced. We assume that there might be some minor variations, but the data are available.

HOWARD: When you have ignition, did you have any initial turbulence inside the box or was the gas mixture stationary?

NAIDUS: Well, it was mixed for one minute prior to the ignition by an electric fan. It is quite possible that some residual turbulence might have existed, but there was no indication. As a matter of fact, we did vary the mixing somewhat on the preliminary trials, mixing and letting it stand for longer periods. There was no evident difference that could be ascribed to that particular problem of turbulence. The mixture is such that we believe it's stationary at ignition.

HOWARD: At a previous Loss Prevention Symposium, there was a discussion of factors influencing turbulence development when combustion occurs inside a regular factory building. We have many obstructions of all sorts which constitute bluff bodies. These quite considerably increase turbulence and thereby quite considerably increase flame speed which in turn is related to your R max. In other words, if you had turbulence in your fuel cell, you would have had a higher R max. Our industry have to face the reality that when we have combustion of the gas mixture in a building, we will in fact develop a turbulent combustion because of all these bluff body objects.

NAIDUS: Well, I did consider that. But, we couldn't find it. If it were true, we should have gotten a more random amount of turbulence per se. You know interior bluff edges on the fuel cell. Obviously, the edges of the fuel cell themselves would function in that way. We made the simplifying assumption that once the explosion left the fuel cell, it wouldn't have anything else to work on, because it was essentially out and away. The problem then becomes one of what is happening after the explosion takes place, assuming that the explosion in the fuel cell is essentially homogeneous in the structure in which we enclosed it. That the front would be uniform enough in such a small volume and the question of the edge effect once the front left the enclosure would be negligible.

That's why we looked for "resonant effects." We thought we were just measuring pressure, not flame effects. What you are suggesting is possible; the pressure effects we observe are actually due to or enhanced by turbulence. Of course, there was indication of that because of the reproducibility of the data that we present, that if, in fact, we were getting turbulence, the effect would be greater than the experimental error actually observed.

HOWARD: As a matter of fact, I think you were getting relatively little turbulence, whereas in combustion in a building we would have a great deal of turbulence. We have found, for example, a considerable influence of turbulence in our own building venting tests as indicated by the values of dp/dt which we observed. This is known basically in combustion technology.

NADIUS: There is no doubt about that. The problem we were facing was to achieve a standard condition for explosion, not necessarily the worst condition. The standard condition should be such that the influence of venting ratios and fuel volume ratios to the total buidling volume could be examined. If we had introduced turbulence as a specific factor, I am sure that the magnitude of the explosion would probably have been greater. But, I don't know if it would have affected the relative information except that it might have been more difficult to measure if the explosion level were considerably higher.

What we hope is that in a real life situation, where an accident or a break-line gives rise to an unconfined vapor or dust cloud, some of the conditions that existed in this work will occur giving rise to a pressure wave, regardless of whether it occurs at a higher level because of a turbulent mixture or at a lower pressure from a quiescent mixture. The problem is how to protect the building against the pressure load, in a situation where we have not completely filled the enclosure. I would certainly agree with you that we should specifically look at the question of turbulence which we can do. In principle, we can report the same experimental explosion under conditions of severe turbulence and see whether the data essentially reproduced themselves but at a higher absolute level. This would be some confirmation to you.

HOWARD: I can assure you that just doing it with turbulence being greater will cause greater pressure resulting in the building during the explosion. You just can't help having turbulence, and we will confirm that in the paper we give next year. Thank you.

IAN SWIFT, Union Carbide: What was the geometry or rather the difference in geometry between the cells that you use? Were they always the same geometry although you just increased the volume?

NAIDUS: Well, this is one of the experimental problems which we are a little unhappy about. In the first instance in 1975, we used a single volume fuel body; we didn't vary it; it wasn't possible to vary it at that time. Our main variable was propane air ratio. We didn't even think in terms of varying the volume. That was a question that came out of that earlier work which we attempted to answer in this work. The fuel cell in this work that was constructed was actually a rigid box which had a movable solid back. If you recall the location of the fuel cell, it was actually at the back of the building at the floor against the rear wall. What we did was to move the back of the fuel cell forward in order to make a smaller volume from the full fuel cell (64 ft^3). So, we did change the geometry somewhat; we believed that wouldn't be too significant except for resonant effects. If we did have a resonant effect that was real, we may have been "tuning in" the fuel box to the building.

SWIFT: Yes, that's what I was trying to get out of. You just lengthened it or shortened it. Was the ignition source in the box? Is it in the center?

NAIDUS: It was in the center of the body in each case. We moved the ignition spark gap somewhat; it was actually mounted in the wall so we could always adjust it; so it was halfway between the front and the back walls of the cell. It was always in the center of the fuel cell. The problem was that we had a maximum in the data and that's what really required an explanation. Not the probable random effects of varying turbulence with each test explosion.

SWIFT: I thing Morse's comments on the turbulence is pertinent to this question. When you ignite your gases in the center of your box, you are not going to burn them completely before the sides burst and you are venting it so you are getting jets of unburned material passing away from the box and very high degrees of turbulence even so. So you are changing it each time, you are introducing turbulence on a degree; turbulence will change, I thing, with your volume and size of the box. Even though you are venting very quickly, you are still getting these fingers of unburned gas leaving the box ahead of the flame.

Here's another phenomenon which I happen to be working on and that's flame induced turbulence. You get it if you have a flame ball above a certain critical size which we can calculate for a given material. A spherical flame starts to get wrinkles in it. When the flame ball reaches a critical size, the turbulence and the wrinkles gets out of line and the increase in flame area accelerates the combustion process.

Chemical Transport Over the Highway: Equipment Design and Regulations

Anyone getting into highway shipping or carriage business should study the appropriate regulations and trade practices, and obtain the advice or services of an experienced professional in the field.

L. A. Botkin
Fruehauf Corp., Omaha, NB

Safe and efficient transportation of chemicals over the highway is the result of a lot of people doing their job right. Of key importance is the professional truck driver who knows his equipment, his route, and his product so that he gets the rig from the shipper to the customer, using good safe driving practices. First of all, he would in many cases have first loaded the trailer safely. He then must get the trailer unloaded safely and the equipment returned to the condition where it is ready to go on the highway again.

Each of the various interests involved in highway transportation sees the total picture from a slightly different angle. As a tank manufacturer, our primary interest is in the knowledge of tank design and manufacturing techniques which enable us to economically produce a trailer which will give long trouble-free service. In addition to this, we must maintain a library of regulations which covers both trucking equipment in general and those which apply to tanks for handling hazardous materials. Recent expansion of regulatory coverage requires us to get into areas of air pollution for example, which brings up the requirement of vapor collection systems. While such systems generally focus on handling gasoline, there are indications that the same type of regulations may be seen in the chemical field.

0149-3701/81/3942 $02.00

I believe that it is important to study in some detail the concept of hazardous material regulations in the United States as we see it. This is not to imply in any way that the regulations or the design and construction of the tank is more important or even as important as the safe conscientious driver. The fact is that we in this forum can work with the regulatory items and other safety features of this nature, whereas I doubt if any of us really have any control over the actual operation of the truck fleet.

The basic concept of hazardous material regulations in the United States appears to me to be that each party involved in the total transportation picture has certain responsibilities. These responsibilities involve diverse areas such as product hazard identification, labeling of tanks, reinspection and items of this nature, as well as a number of regulations involving the manufacture of the tanks themselves. As a matter of fact, the trailer tank is considered only one form of packaging for a hazardous material.

It is probably best to illustrate this by an example. Let's assume that we want to ship some hydrochloric acid. The hazardous material table, Paragraph 172.101 of the Code of Federal Regulations, Title 49, of the Department of Transportation, indicates that hydrochloric acid is considered a corrosive material with specific packaging requirements called out in Paragraph 173.263. Turning to Paragraph 173.263 we find that it is entitled

"HYDROCHLORIC (MURIATIC) ACID, etc." and lists such diverse packages as glass carboys in boxes, kegs, or plywood drums, but we find that we cannot ship this packaging by air. Boxed carboys may be used for single trip export only. Going on through boxed carboys, metal barrels and drums, wooden boxes protecting inside containers, rubber-lined metal barrels, lead-lined barrels, and rubber drums. Skipping on through the paragraph, we find that we can ship hydrochloric acid over the road in MC-310, MC-311, or MC-312 tank motor vehicles. Checking around the industry, we find that non-pressure tanks are seldom used, so this leaves us with MC-310, MC-311, or MC-312 pressure vessels as our logical choice. Only specification MC-312 is appropriate for new tank construction as it superseded MC-310 and MC-311 on December 1, 1967.

Our search for the right kind of trailer tank package does not end here. Anyone handling hydrochloric acid knows that it is highly corrosive and generally requires a rubber-lined container when we are talking of something the size of a trailer tank. Turning to specification MC-312, we find that various things such as the density of the product and the tank capacity govern minimum tank head and shell thicknesses. We also find that the tanks must be appropriately lined unless the tank has a corrosion allowance of ten years life, or the tanks become passivated in service.

This illustrates the relationship between the different parties involved in chemical transportation. To briefly sum it up, we might say that:

(1) The tank manufacturer is obligated to follow the MC regulations regarding design, manufacture, testing, and certification of the tank.

(2) The carrier is responsible for proper maintenance, reinspection, and placarding of the tank.

(3) The shipper is responsible for identifying his product and to make sure that he does not put a product into an identified tank which would cause corrosive damage and to be certain that hazardous products are loaded only into the authorized type of tank. In regard to MC-312 tanks and corrosive products, this care includes consideration of tank material, any lining used, and the concentration and temperature of the product.

At this point, things can get rather technical but I believe that they must in order to give adequate coverage to safe transportation. For example, it is my understanding that rubber-lined tanks are ordinarily used for the transport of hydrochloric acid as was used in this example. It would therefore appear that it would be the shipper's responsibility to make sure that the hydrochloric acid offered for carriage in the rubber-lined tank would be of such purity, concentration, and temperature that it would be compatible with the rubber lining. It would appear to me that this must be the shipper's responsibility since neither the carrier nor the tank manufacturer have any way of knowing exactly what is being shipped, let alone having any control over it.

A brief review of the DOT-MC (Motor Carrier) regulations discloses some of the highlights which will be helpful. Many of these are of specific interest to the shipper in selecting the right package or tank for his product. For those of you who have interest in more specific details, they can be found in Title 49 of the Code of Federal Regulations, Parts 100-199. The construction requirements for trailer tanks are found in Paragraphs 178.340 through 178.343 for liquids and Paragraph 178.337 covers specification MC-331 tanks for compressed gas. An excellent reprint of these regulations plus the obsolete tank regulations and other general information is published by the National Tank Truck Carriers, Inc., 1616 "P" Street, N.W., Washington, D.C. 20036, as their publication NTTC No. 17.

The liquid tank regulations are MC-306 for non-pressure service, MC-307 for low pressure service, and MC-312 for corrosives. In general practice, MC-306 tanks are generally used for gasoline and the various petrochemicals and solvents which we generally call "clean oils". MC-306 tanks are usually made of aluminum but there are quite a few which have been made of stainless steel.

Specification MC-307 covers what I would call the workhorse general purpose chemical trailer. A substantial portion of these are insulated stainless steel tanks. The working pressure is usually 25 or 30 psig.

Specification MC-312 is for corrosives, primarily acids. Although non-pressure MC-312 units are included in the concept of the specification, I do not recall that we have ever built one. Usually these are ASME (American Society of Mechanical Engineers) pressure vessels with a working pressure of 35 psig. The three common types are both carbon steel and stainless steel for sulfuric acid and rubber-lined tanks for hydrochloric acid.

The specifications are broken into two sections. Paragraph 178.340 covers general requirements applicable to all three of the specifications, while the subsequent sections cover detailed requirements for each type of tank. The general requirements cover materials of construction, structural integrity, minimum joint efficiency, shell reinforcement, accident damage protection, and certification. The item of certification is of particular importance to the carrier and the shipper since a metal certification plate containing much important information must be displayed on the tank and the carrier must retain a certificate of compliance in his file. Note particularly that the certification calls out the material used for the shell and heads as well as welds and that any lining which might be used is identified. Thus the shipper can look at the certification plate and make sure that this tank is suitable for the product that he intends to load into it.

Specification MC-306 requires certain minimum thicknesses for shell and head material based upon the cross-sectional area of the tank shape. The tank must be equipped with the low pressure normal vents and then have additional venting as required to avoid excessive pressure in event of fire exposure. These vents must be marked. Product discharge openings must be equipped with emergency valves which are automatically closed in case of fire exposure and have a means for remote tripping of the valve in an emergency.

Specification MC-307 was originally worked up to handle casinghead gasoline. As the petrochemical industry grew, it was found that this specification was also quite appropriate to general purpose chemical tanks. Today the quantity of stainless steel general purpose chemical tanks under this specification far exceeds other types. These units are typical clean bore and quite often equipped with insulation and heating provisions for handling rosins, waxes, and similar products. Being pressure vessels with the minimum design pressure of 25 psig, we find special formulas for calculating the shell and head thicknesses for tanks up to 50 psig. Above 50 psig they must comply with the ASME code. Because of company practices, a number of tanks in the 25 to 35 psig range are also built to the ASME code.

In addition to pressure calculations, there are minimum requirements for head and shell thicknesses based upon the cross-sectional area of the tank as was found in MC-306. Likewise, both spring-loaded service and fire exposure venting as well as emergency valve requirements are found. From a practical standpoint, it is important to frequently inspect the venting provisions on MC-307 trailers since these tanks are frequently cleaned and the vents and fusible caps are removed in the process. It is always possible to mix these items before replacement. Inspection to make sure that proper venting has been installed on the tank is, in my mind, a worthwhile precaution.

Specification MC-312 follows the usual pattern of minimum thickness tables for shell and heads. The new factor found here is different minimum thicknesses based upon the maximum product density. A large portion of the MC-312 trailers are equipped with rupture discs. This means that it is very important to equip any air supply system with proper pressure control fittings such as those per the Truck Trailer Manufacturers Association's recommended practice, TTMA RP #34-77. Until recently, a top outlet was required but now bottom outlets are acceptable and frequently used. Bottom outlets on MC-312 tanks must be equipped with a remote operator but are not required to be of the self-closing type.

The compressed gas tank, MC-331, is possibly of less interest to most in the group. The specification goes into some detail regarding pressure vessel design requirements and material properties, particularly regarding quenched and tempered steel. Other portions of the regulation detail piping requirements for liquid and vapor outlets. Special requirements are found for tanks handling sulfur dioxide and chlorine.

Our years of experience in the manufacture of trailer tanks for various purposes indicate that the DOT-MC regulations are sound and practical. As a matter of fact, the current regulations which have been in effect since December 1, 1967, were finalized after extensive meetings between representatives of the Department of Transportation and those of the tank manufacturers, shippers, and carriers.

A number of chemicals are shipped as dry products rather than liquids. The same general safety precautions are in order, even though the type of trailer tank is considerably different. No federal Motor Carrier hazardous material specifications are involved other than they be "sift proof". From a practical standpoint, driver training to efficiently operate the usual pneumatic unloading system is important. Aluminum is the most popular material for dry bulk chemical tanks. Some chemicals are sensitive to moisture or certain gases and therefore are handled in a controlled atmosphere.

The trip handling hazardous materials has not been a safe one until the tank has been unloaded at the destination. Based upon accident reports with which we are familiar, our experience indicates that unloading the tank is probably as hazardous as any other portion of the operation. Proper instruction of the driver and provisions and use of appropriate safety equipment are absolutely essential. Equipment maintenance appears to be a significant factor when reviewing problems during unloading, particularly the maintenance of gaskets and fasteners for the manhole or fill opening when pressure unloading. We feel that it is very important to maintain the manhole so that it seals easily and the fasteners are not corroded or broken. Drivers should never go up on the tank to try to tighten down a leaky manhole while the tank is under pressure.

Another important item during unloading is the handling of hose. While the practice appears to be spreading, we in the manufacturing industry do not feel that quick couplers should be used when handling corrosive materials because of the possibility of loosening the coupling while some pressure remains in the hose. We also feel that receiving facilities should have provisions for handling the air flow required to "blow down" or release the pressure in a tank after pneumatic unloading.

The truck driver is usually involved in loading and unloading the trailer. While doing these tasks, he is often handling hazardous materials. When corrosives or hot products such as molten sulfur are involved, the shippers usually have certain requirements for use of protective equipment or clothing and access to eye wash or shower facilities. The driver should be subject to the same safety requirements as any plant employee and the use of proper equipment enforced.

These sound like simple steps but we find that many of the problems that we hear about involve leaky manholes and improper handling of hoses.

This brief summary of highway chemical transportation regulations and practices in the United States interrelates at least three and probably four parties who must each perform their responsibilities in order to make the system work properly. In practice, the details involved in transportation are such that no brief summary can give adequate information. Anyone contemplating getting into highway shipping or carriage should take adequate time to study the appropriate regulations and trade practices, as well as obtaining the advice or services of an experienced professional in the field.

Figure 1. The insulated stainless steel MC-307 SS tank with a smooth interior which is readily cleanable is a very versatile workhorse of the chemical transportation industry.

Figure 2. MC-307 AL tank with 25 psig working pressure is popular for casinghead gasoline and light petrochemicals.

Figure 3. Aluminum pneumatic discharge hopper type tank popular for dry chemicals, fertilizers and plastic pellets.

Figure 4. Stainless steel MC-312 SS tanks are popular for acids except hydrochloric acid. The usual working pressure is 35 psig.

Figure 5. Gasoline and petrochemicals are typical commodities shipped in an MC-306 AL tank. This tank is nonpressurized.

Figure 6. Each compartment of an MC-306 AL tank is equipped with a remotely controllable internally mounted emergency valve.

Figure 7. The rubber lined MC-312 MS tank is most popular for transporting hydrochloric acid.

This Certifies that Fruehauf Tank-Trailer serial number ____OMZ 794701____,
manufactured by Liquid and Bulk Tank Division, Fruehauf Corporation on ________
__5/30__ 19_78_, and having a nominal capacity of ____9200____ U.S. Gallons, has been designed, constructed and tested on ____5/30____ 19_78_, in accordance with Department of Transportation Specification No. MC-__306 AL__, ________, ________,
________, and is structurally designed for a max. product load of ____57000____ lbs. or a max. product density of ____7.5____ lbs/U.S. gal.

SPECIFICATION EQUIPMENT NECESSARY

SPECIFICATION EQUIPMENT		MC QUANTITY REQ'D	MC QUANTITY SHORT*	SIZE	MC QUANTITY REQ'D	MC QUANTITY SHORT*	SIZE	MC QUANTITY REQ'D	MC QUANTITY SHORT*	SIZE	MC QUANTITY REQ'D	MC QUANTITY SHORT*	SIZE	MC QUANTITY REQ'D	MC QUANTITY SHORT*	SIZE
VENTS	Press Actuated															
	Fusible															
	Frangible															
PRODUCT DISCHARGE	Top															
	Bottom															
	Press Unloading Fitting															
COVERS	Manholes															
	Fill Opening															

* Equipment not provided by factory:

LIQUID AND BULK TANK DIVISION
Fruehauf Corporation
LBT 73-62

Figure 8. A typical certificate of compliance.

BOTKIN, L.A.

DISCUSSION

STAN SKINNER: With storage tanks we are very conscious these days of the necessity for good inspection programs. I wonder if you can say anything about how vessel inspection is applied to transportation.

LAWRENCE BOTKIN, Fruehauf: In the regulations, a visual inspection is required every two years. Being real honest with you, from a practical standpoint, a two-year inspection period is rather ridiculous because so much can happen in two years; especially in chemical transportation tanks, it is relly meaningless. I believe there are moves that have been started, recommendations are being formulated to reduce this period of visual inspection.

From a practical standpoint, the good operators which you might call the pro's in the business, inspect their tanks on a trip basis. This is particularly true of the general purpose chemical tanks where they will be switchloaded and things of this sort. They are pretty thoroughly inspected on every trip. A dedicated tank, like a hydrochloric acid tank or something of that sort, will probably see much less frequent inspection. But, it would vary so much among operators that a general statement can not be made. Some people are very good at it and some give some problems.

ANONYMOUS: We try to keep our trailers in good condition with respect to covers and inner valves. I would like to underscore what Mr. Skinner has brought up about inspection in trying to maintain these vehicles in safe and proper working order.

We've inherited the gasket on inner valves from the fuel oil and gasoline transport people, so we in chemicals have to check our inner valves once a month in an effort to keep them from leaking. This is pretty tough. I would like to point out the fact that there isn't an inner valve that really stands up. The other thing has to do with the cover. The gasketing in the covers has the same problem. I suspect, looking at the vapor recovery setup that you've shown on your slides, that we will again inherit the same gasoline-type arrangements which will be attacked by various solvents. I wonder when these types will be looked at by other manufacturers. I realize you are trying to make a good trailer, but we must have representation from the inner valve, dome cover companies.

One last important point in public safety relates to design of vessels, and that is off-loading them and handling them in a situtation where they've been involved in an accident. Provisions perhaps must be made to remove material through prelocated openings that are designed specifically for that purpose. This might also apply to tank cars.

BOTKIN: To comment on the first part of your question this business of the seals that they use in valve construction, as well as gaskets, is a real corker. The valve and fitting industry, as well as manufacturers, have looked into this and about the only material that seems to have a broad spectrum of acceptability is teflon. But, unfortunately, teflon will attract every bit of grit within 100 yards and it isn't very resilient. You do have very definite limitations and it's tough to keep a chemical valve so it will seal properly; there's no two ways about that. It seems like every time somebody invented a good gasketing material, somebody invented a chemical that ate it up. So, I might ask you when you are going to quit making chemicals that eat up the valves.

We're both in transportation, but I can tell you that we've been talking with inner valves manufacturers and it really is a problem.

Yes, it's a tough deal and the industry is continuously working on this and the problem is that there are so many different chemicals, so many different situations that you run into, that it's a tough one to get a handle on, but I think the equipment today is a lot better than it was 10 years ago. Ten years from now, hopefully, we will take another step forward. I wish I had an answer for it; if I did, we wouldn't have the problem because you'd have different equipment.

Removal of product from a tank in the event of an overturn has been discussed quite a bit. There was recently a meeting at a DOT contractor in Pheonix. All facets of the industry were there, and this was one of the items discussed at length. It was also discussed at a recent meeting of the API Committe on Transportation; one concept is to provide access ports in the side of the tank so that if a tank was rolled over you would always have a hole on top.

When a fellow asked: "Don't you think these holes would be a good idea?" I said; "Sure, just put them on the top side of the tank; if you're going to tell me which side is going to be up, that's where we'll put the fittings." But, if you just run around and put fittings on the tank, you're going to probably have a greater possibility of a spill because of those fittings which more than offsets the benefit you would expect to gain from it.

If you have the pipe fittings, you can draw down a tank from the discharge piping, and then you take the valve off and take the rest of it out with a suction stub or something like that. The problem you have with anything except the propane tank, your heavy pressure vessels, is that the skins on tanks are so thin that if you put access fittings in there you are creating sort of a hard spot. Structurally this can cause a lot of problems in an accident situtation because of the tremendous distortion which the tank sees.

R.H. JENSEN, Jensen Consultants, Inc.: With respect to access after the tank is turned over in an accident, I would just ask that the bottom flange not be put directly above the rear bogie so that you can at last pry the flange off without practically cutting the bogie off the back first. That is one thing. The other thing is with respect to chemical transport. Has anything been done with respect to the possibility of transport under, say, two to five pounds pressure so that we would not have to have the possibility of sloshing or splashing through a vent in the event of deceleration/acceleration?

BOTKIN: To answer your first question, if you get everybody to order double conical tanks, you won't have a valve back there in the chassis area, back by the rear wheels. To answer your second question, this problem of tank pressure of the gasoline-type tank, the MC-306, has been discussed at length since the vapor collection business started up. Jokingly in th industry, we call it an MC-306 and a half tank, half way between a 306 and a 307.

But right now, there is no regulation or specification covering an intermediate pressure tank; but we're seeing an increased interest in it, particularly in the oil field country down in Texas. Whether industries are going to request DOT to establish a new regulation for this type of tank or not, frankly, I don't know. But, there has been a lot of serious discussion about it.

It may be with gasoline transport that you can afford to have a little splash and it doesn't bother people too much. But, when certain chemicals are being transported, if they have a very low-odor threshold, a little splash will sometimes alert an awful lot of people that a chemical transport is going through. In that instance, all a shipper has to do is to specify an MC-307 tank, and then he will be corked up to 25 psi.

TREVOR KLETZ, ICI: I have two questions. First, to return to the question of whether you have detailed regulations or performance regulations, I'd like your comments on the compromise which we're trying to achieve in the United Kingdom. We are moving towards a situation where we have a very brief regulation setting out the objective you must achieve and this is accompanied by a Code of Practice or Guidance Note which suggests one way in which it can be achieved. Most companies, particularly the smaller ones, will take the advice of the Code of Practice or Guidance Note but they don't have to. If you think that there is a better way or if you think that the Code of Practice is inappropriate to your needs, you don't have to follow it, provided you can show you are doing something which is just as safe. You are not, therefore, confined by regulations which are inappropriate or out-of-date. Regulations take time to change when technology changes. This is why we like the system I have described.

The second question is: Have you any explanation for the very different experience of the United Kingdom and America so far as accidents to road vehicles carrying hazardous chemicals are concerned? I have kept records for 10 years in the United Kingdom, for road accidents involving chemicals and there have been very few. In fact, there has been an average of less than two people killed each year, one of whom is a driver. Now, we know from the reports of the National Transportation Safety Board that there are many more incidents here. This is a bigger country and distances are greater but that doesn't seem to explain the difference. I wonder if it is the result of different standards of construction or something like that.

BOTKIN: I have become by necessity familiar with ADR/RID, the Blue Book, and several of these other things that govern in Europe. It is very difficult for an outsider to really come up with a specific design which he can pretty well predetermine, will be accepted, or acceptable to the responsible authorities in England and particularly with regard to tank containers. We've had a lot of fun in this regard, even though we do work with our English affiliate on that. So, we find that flexibility is a two-edged sword.

Another distinguishing feature you have in the U.S. is what you call product liability. Every time you get called on the carpet with a law suit, you've got to be able to prove that your piece of equipment meets a specific regulation. When regulations call our some general objective like "haul everything safe forever," the fact that you had an accident proves that you didn't. So, it puts you in a very embarrassing spot and it would make a lot of lawyers richer.

With regard to the second point, I thing that a big difference between the U.K. and the American experience is probably a different traffic pattern, and the driver education. When you see some of the hours that some of the drivers drive (and I hope I am not stepping on anybody's toes) it is very seldom that you go into a particular accident case and find out that a driver had a full eight hours sleep before he started, the truck was in perfect condition, all the brakes were hooked up, and things of this sort.

You get into details and usually you will find that that was an accident just waiting for some place to happen. I think actually that you can control things like that a lot tighter in England than you can do in the United States. It's possibly because of volume. Maybe, you don't have any more bad apples percentage-wise, but in the U.S. you have got a lot bigger barrel to put them in. Thank you very much.

NTSB Hazardous Materials Spill Maps: a New Safety Information Resource

As more sources become increasingly involved in reporting and using hazardous materials accident spill maps, it eventually will be necessary to establish a clearinghouse for map publication and distribution to users.

L. Benner, Jr.
National Transportation Safety Board, Washington, D.C.
R.A. Rote
4312 Blackbeard Rd. , Virginia Beach, VA.

A National Transportation Safety Board (NTSB) special investigation, undertaken in response to observed hazardous materials emergency response problems, revealed a lack of practical, currently available information about hazardous materials behavior in transportation accidents. Researchers, mathematical modellers, emergency response personnel, and others voiced a need for data to support predictions about how hazardous materials could be expected to behave in accidents. The NTSB Hazardous Material Accident Spill Map program was developed to fill this need. NTSB spill maps now report such behavior in a tentatively standardized format. The maps feature a time-sequenced display of dispersion patterns and ranges, weather at the time(s) reported, injury/fatality exposure locations, and a synopsis of the accident scenario. Although these maps are currently used by the Safety Board to report behaviors of hazardous materials released in transportation accidents, applications can be seen for a much broader range of releases.

THE NEED ADDRESSED BY MAPS

The Safety Board's development of a reporting standard for hazardous materials releases supports an accelerating shift in the basic philosphy for controlling hazardous materials transportation safety risks. In the past, the basic safety objective was to contain hazardous products safely under "conditions normally incident to transportation." However, accidents and failures of containment systems frequently resulted in emergencies that had to be handled by local emergency response personnel. Resulting casualties among response personnel and other personnel enlarged the Safety Board's concern to include not only accident prevention objectives, but also casualty reduction during accidents. One way to reduce these casualties was to improve the decisionmaking process, so emergency response personnel could logically think their way through an emergency to achieve the safest possible outcome. This process requires predictive estimates of the likely behavior of hazardous materials present.

0149-3701 81 3949 $02.00 © 1981 AIChE

At a public hearing held by the NTSB in April 1978, emergency response personnel reported problems encountered in developing pre-emergency plans, and making tactical decisions during emergencies involving hazardous materials. Many voiced the belief that existing technical manuals, emergency action guides, and other information sources do not provide adequate information needed to identify how a hazardous material might behave in a specific emergency. The Safety Board reported·

> "Accidents resulting in the tragic deaths of at least 50 firefighters since 1968 clearly demonstrate the need for improved methods of handling hazardous materials emergencies. Emergency services cannot be expected to adequately plan for and make life

or death decisions at hazardous materials emergencies when information which will help them identify how a given material will probably act in a specific emergency situation is not available."

An analysis of the hearing testimony, previous accidents, and subsequent interviews with emergency response personnel identified the need to determine how a hazardous material will behave before tactical decisions towards the safest outcome can be made. That is necessary to assess exposures risk and safe operational distances. Five aspects of the hazardous materials' potential behavior must be understood before effective pre-emergency plans can be formulated, or tactical on-scene decisions needs can be resolved:

1. Where is the hazardous material or container likely to go if released during the emergency?

2. Why is the hazardous material likely to go there?

3. When is the hazardous material likely to go there?

4. How will the hazardous material get there?

5. What harm will occur when the hazardous material gets there?

Already available information sources were surveyed to determine the quality and quantity of information provided with respect to these points. Accident reports, the most common source of information on past hazardous materials accidents, contained data which helped determine probable accident cause, yet lacked information regarding hazardous materials behavior and casualty relationships. Research information such as mathematical modelling provides one basis for forecasts of hazardous material behavior; however data for validation of predictions is limited. Emergency action guides, which usually provide lists of various known properties and characteristics, can be linked to hazardous materials behavior only in general terms. As a result, instructions are brief, general, and vague. No presently available information source provided adequate information required for the emergency response decision process.

During the investigation it was noted that some mapping was already employed in accident reports and some of the more comprehensive manuals to display dispersion boundaries, casualty location, or damages. However, no general standard to record the events was established. A variety of methods, scales, map bases, and information content were being used.

NTSB APPROACH TO MAPS

The NTSB approach towards displaying hazardous material behavior information was suggested by data obtained in an investigation of a tank truck accident in 1976 involving release of anhydrous ammonia. Infrared photographs, U.S. Geological Survey (U.S.G.S.) maps, and events-sequence analysis were used to record the accident events. Analysis of data recorded by the Safety Board, the Houston Fire Department, and the Houston Air Pollution Control Board established relationships between product dispersion, environmental conditions, and injuries which could be mapped. To record these relationships, timing of events, timing of the dispersion pattern, a timing of exposures to the hazardous materials had to be shown. A graphic display on a standardized map base was selected to present this information. The relationships could be displayed on time-sequenced maps, constructed to illustrate hazardous materials dispersion patterns for the progressive phases during the emergency, and showing weather data and injury/fatality information.

To fulfill users' needs, the NTSB established some basic criteria for map base choice, information content, and layout. The selection of the most desirable map base was governed by several considerations. A map base with a uniform scale, available nationwide, at low cost to all users, was considered essential. A consistent map scale would enable comparison of hazardous material effects, regardless of individual accident locations. The map base should include transportation corridors and topographic features. Display of exposures in the potential path of released hazardous materials would be useful. Descriptive information about the accident should be included in a consistent form, as should weather conditions and fatality location information. A uniform means of representing hazardous materials dispersion effects had to be formulated.

A short synopsis summarizing the

sequence of events, and a summary showing the locations and degree of injury resulting from hazardous material exposure was also needed, to support the behavior prediction effort.

The U.S. Geological Survey 7-1/2 minute series, 1:24,000 scale, topographic map series was chosen as the standard mapping base. Scales 1:12,000 or 1:6,000 scale were also provided, to permit reporting of small release patterns. These maps are available from U.S.G.S. for most parts of the country. Mapping of other areas is to be completed soon; in such areas, a 15 minute series, 1:62,500 scale map is available and may be enlarged to the 1:24,000 scale or greater. U.S.G.S. topographic maps include relief features, waterways, towns, and buildings. Roads, highways, railroads, pipelines, airports, and docks are shown. Thus, an accident location on the transportation corridor can be displayed as accurately as data permit. The inclusion of buildings on the U.S.G.S. maps also helps indicate potential exposures in the path of a hazardous material. The U.S.G.S. maps are readily available for $1.25 each. They are presently used for a variety of related applications, including local civil defense and planning, so that many users already may be familiar with the map scale and symbols.

Using the U.S.G.S. topographic map as a base, the behavior of the hazardous material is displayed at various times during the emergency. Time sequencing is important for emergency response personnel, because needed response actions change with time. Selecting the times at which maps will be prepared is related to the changes in the hazardous material behavior pattern. Maps try to show the growth of the hazardous material dispersion pattern, changes in area affected due to environmental conditions such as wind direction, changes in number of people affected, or other changes in the behavior which affect emergency response actions.

Both surface and airborne effects are displayed. These displays are prepared from observed data, supplmented by estimates suggested by witness' comments. Estimated surface effects appear as shaded areas. Solid lines enclosing surface effects represnet observed patterns. An estimated boundary of airborne effects to represented by a dotted line; and observed boundary is represented by a dash line. A key on each map identifies these representations.

Weather conditions at the times mapped are displayed in a block that indicates wind direction and speed, sky condition, present weather, visibility, and air temperature. ("Present weather" is the form of precipitation present.) Data in this block may change with time on each map in the sequence.

An x,y, coordinate grid system is used to pinpoint the location of fatal exposures. Scales are numbered along the horizontal (x-axis) and vertical (y-axis) edges of the map. Using x,y coordinates, approximate locations of hazardous materials exposures leading to death are tabulated on the map.

Descriptive data about the accident is provided in a box containing: type of accident, date and time of accident, time of event displayed, time after initial hazardous material release, product released, container type, quantity of product released, map base, and map scale. This information is shown on each map in the sequence.

USES OF NTSB SPILL MAPS

These maps present observation of hazardous material behavior in actual accidents. From such information, many applications are foreseen. Model verification, release comparisons, and accident simulations are some of the expected uses. The mapping exercise may be expanded from recording releases due to transportation accidents to mapping other releases, such as at storage or manufacturing facilities. Maps eventually may be referenced on-scene during hazardous materials emergencies, providing a foundation for critical decisions.

For each application, precaution must be taken to accommodate the variables of weather, product, amount released, population density, and topography in different situations. Locations, environmental conditions, terrain features, release mechanics, etc. are unique for each release. The dispersion of a released product in a mountainous location, influenced by specific weather conditions, may be compared only in general terms to a similar release in an area with little topographic relief. Accident simulations in areas with like characteristics may also be accomplished when the variables are considered.

Accident comparison, based on spill size and injurious effects as related to commodity and transportation mode, is helping to distinguish the more dangerous or lethal shipments in accidents involving hazardous materials. Using the standard scale, spill maps will facilitate such comparisons. They provide estimates of fatal exposure areas and other exposure ranges which should help emergency response personnel to appraise potential lethality to guide their response decisions.

Information from a past accident may be used to simulate that accident on a local community map. Simulation of mapped accidents in a local community can provide insights about the nature and magnitude of local problems which might arise in information gathering, emergency assessment, obtaining access to the site, logistics, evacuation, treatment of victims, resource limitations, communication and command needs, and other decision dilemmas or data needs. Participants in simulations can obtain experience in resolving actual problems without risking exposure to actual on-scene dangers. Planners in local communities may use simulations to identify "target hazard" areas where emergency service tactics, command actions, coordination, and possible courses of action could be practiced.

The variables of topography, weather, released product quantity, etc. can be used to test models which make predictions such as downwind dispersion patterns. Such predictions may be tested against actual occurrences using information provided by the hazardous materials accident spill maps. Through such verification, the confidence level in mathematical models could increase. Subsequently, emergency action guides which rely heavily on models, could be upgraded.

Maps depicting hazardous material releases may also be used as a basis in establishing or amending regulations. For example, the Department of Housing and Urban Development is interested in maps in relation to siting under its housing programs. Significant differences between commodities behavior when spilled will be evident on maps. For various materials, the timing of the dispersion pattern formed and the harm that results has been noted to differ significantly; slower dispersion rates seem to produce lower losses in comparable situations. Such comparisons may lead to a rationale for changes in regulatory requirements for high loss-producing commodities.

In the future, hazardous material accident spill maps are expected to aid on-scene response activities. Stored maps and associated data from past accidents at a one-call information center could be recalled to help local emergency service personnel assess the dangers during an actual emergency. Emergency response coordinators would be able to call the center and obtain information about accidents with similar circumstances. Local variables could be interpreted and probable hazardous materials behavior then could be predicted. Recall and display capabilities within emergency response time frames seem to be technically feasible if libraries of the maps were to be developed in computer-compatible form.

The basic map elements needed to fulfill the range of applications are (1) time sequencing, (2) standard scale, and (3) an adequate number of maps. The first two elements are available on the NTSB hazardous materials accident spill maps. The third, however, can only be accomplished with the help of others. Standardized hazardous materials map preparation is presently performed solely by the NTSB, to our knowledge. Involvement of outside organizations would multiply the effectiveness and utility of the maps.

WHAT'S NEXT

Other organizations have expressed interest in this mapping project. Between 1500 and 2000 parties on the Safety Board's present mailing list receive copies of each map. In addition to interested users of the maps, a number of individuals have requested additional information for preparing the maps. Written procedures for preparing maps have been drafted. Presently the Board is soliciting comments as to how these maps might be improved to meet particular needs. In addition, attempts are being made to involve other organizations in mapping hazardous material releases.

Any significant release of a hazardous product in transportation, workplace, industry, etc. could be recorded on U.S.G.S. maps to provide standardized spill maps. The current goal to increase the hazardous materials data base is attainable with the cooperation of all who share an interest in understanding their accident behavior. Maps could be provided by those interested workplace releases of hazardous materials. They could be prepared by those responsible for recording environmental insults. Maps could be constructed by those who respond to incidents where hazardous materials are released, an interest which already has been expressed. Each new map of a hazardous material release will add to the degree of certainty in predicting the behavior in future releases.

As more sources become increasingly involved in reporting and using hazardous materials accident spill maps, it eventually will be necessary to establish a clearing-house for map publication and distribution to users. The Safety Board is not equipped to handle such a voluminous project, which will likely include maps for non-transportation uses. Public or private agencies may have to share this task, as useage becomes more widespread.

The idea of hazardous materials accidents spill maps is being accepted with enthusiasm by several users. As other users become familiar with this new data source and possible applications, we are seeing more and more excitement about the project, and are getting more and more comments and suggestions about them. This is the most important step for successful use of the maps. Because a successful program depends so greatly on outside participation, comments from all those who might see a possible application of these spill maps for their own purposes are genuinely welcomed by the Safety Board. Any comments and suggestions to make these maps more useful for specific applications should be forwarded to the Safety Board, Washington, D.C. 20594, Attention TE-40. The current standardized way of presenting the data is only tentative until these comments have been reviewed and worked into a final map system. The final standardized map system is expected to be proposed by the Safety Board by the end of 1980. We hope your inputs will be available to help shape the future of this program.

BENNER, JR.. L.

ROTE, R.

DISCUSSION

BOB COFFEE, Eastman Kodak: When the National Transportation Safety Board investigates an accident and prepares a report, the availability of that report is mentioned in the Federal Register. Are you proposing to announce the availability of these spill maps in the same way?

ROTE: That is not the procedure at this time. If you think that it should be, you could put that in for a suggestion.

COFFEE: Well, the Federal Register is currently the only place (other than direct contact with the NTSB) where we find out whether you have investigated a particular accident and prepared a report. Thus, if a spill map has been prepared, the Federal Register would be a logical place to announce it.

ROTE: I'm not really sure how notices of the maps are disseminated but we do have a mailing list; that's the initial distribution. At this point, I don't think the Federal Register includes any notice as to when the maps are coming out.

COFFEE: As an alternative, you're preparing a mailing list of those people who would like to receive them?

S.S. GROSSEL, Hoffman-LaRoche: Would you know which mathematical models were used for the calculation of how quickly the vapor clouds spreads?

ROTE: The maps are not based on dispersion models; they are based on actual observation or estimates that are suggested by witness observations. That's the difference between the map and the model because they are what was observed after an accident.

GROSSEL: In other words, if we want to prognosticate ahead of time with a chemical which is not yet mapped by you, we are out of luck?

ROTE: That's why we need more input for all these releases.

GROSSEL: Thank you.

HOWARD FAWCETT, GMF: What is the longest

distance in either kilometers or miles whatever scale you use that you have been able to map so far from an actual real world incident? I am talking about the Pensacola, the Youngstown and the pipeline incidents so far. This is prompted by reading the other day that crypton 85 had actually been detected in Albany, New York from Three-Mile Island which is some 275 miles away. How far have you actually mapped, based on real world data the dispersion of either a flammable or a toxic material?

ROTE: We haven't made maps for every release, you're aware of that first of all. The largest map that we have made covered about 26 miles from the source, and I believe it was ethyl acrylate. There was no plume that was observed, so that map was based on witnesses observations of an odor.

FAWCETT: Twenty-six miles based on odor from the source is what you are saying?

ROTE: Yes, that's right.

FAWCETT: That is the maximum that you have mapped so far. Now my request, tell your boss Ludi Benner to bring you out to the conference in Columbus, Ohio. Did he tell you about this on the 13th and 14th of July?

ROTE: He hasn't said anything to me.

FAWCETT: He, like I am, is a member of the advisory committee for a National Science Foundation's study on acute chemical emergencies and their response mechanisms to it. You ask him about it when you get home.

W.B. HOWARD, Monsanto: When we have sudden releases, let's say, from a tank car like the ammonia accident in Pensacola, I wonder whether it is better to evacuate people, or to tell them to stay in their homes, turn off the air conditioning, close the windows, and maybe even breathe through a wet rag if necessary, something like that. Do you have enough experience in how exposures have occurred to give us any advice on whether it is better to get people outside of their homes, take them through the cloud, and "evacuate" them, or to keep them in their homes? The same thing for office buildings as well. There was a significant accident in Texas City where there was a major ammonia release again. This was picked up in an office building through the air-conditioning system's fresh air intake. This was very serious; the people inside the building were basically captives; the air conditioning was not turned off

ROTE: I know what you are asking; you are asking if we have enough information. At this time, we've just begun. But, there has been some work done on the question that your are asking. Are you talking about an accident in Houston? A safety board report tries to consider these decisions to be made—whether to evacuate or to keep people in.

TREVOR KLETZ, ICI: I'd just like to take this opportunity to say publicly how outstandingly good the U.S. National Transportation Safety Board reports are by world standards. There is no other country that provides such comprehensive information and they are the only source of reliable data on what has happened. Just to give an example. there is considerable interest at the moment on the heat radiation from fireballs. Fireballs occur when a vessel containing flammable liquid or gas burst as the result of exposure to fire or just as a result of impact. There are a variety of theoretical models which give widely differing answers. The only source of actual data, so distinct from theory, are your reports and these are the only ones that can be used to validate the equations. So the plea I am making is to keep up the good work and if anything give more detail. Can you show the locations of injured people and the extent of their injuries?

Safety and Loss Prevention in the Tank Truck Transportation of Chemicals

The wider use of chemicals has necessitated better defined distribution patterns often including deliveries to small users.

C. J. Harvison
National Tank Truck Carriers, Inc., Washington, D.C.

Initially, let me note that as one of the distinct minority of non-technically educated and non-technically oriented attendees at this session, I fully intend to take advantage of my culturally-deprived status and speak to you in layman's terms.

Actually, this should be good practice for all of us, for when you deal with tank truck companies you will find that their management lacks sophistication in your discipline, also.

Nor, do they expect you to conceptualize hours of service rules, axle loads or traffic lanes.

Now, I realize that all I've done so far is to outline two types of individuals and industries that have no reason to even talk to each other; but, obviously, such is not the case.

Fundamentally, the nation's tank truck industry is broken down into three economic units. First there is private carriage where one company buys its own trucks to haul only its own products. Secondly, there is the element which I represent -- the for-hire segment -- which holds itself out to serve all shippers at published rates under the economic regulatory authority of the Interstate Commerce Commission and/or state agencies. Lastly, there is the jobber (or commission agent) type of operator who primarily engages in a buy/sell operation with bulk transportation provided.

With that overly-simplified outline of the tank truck industry, let me note three very good reasons why, indeed, we should be talking to each other.

1) This nation's life style (including the provision and maintenance of some essentials) is increasingly dependent upon the chemical industry);

2) The wider use of chemicals has necessitated better defined distribution patterns often including deliveries to small users;

3) The demand of the public that if we are going to use their railroads, highways and waterways to assist in this distribution -- we had better damn well do it safely.

The net result of these three factors (with emphasis on the latter) is that you, the shipper, and, we, the carrier, had best do a considerable amount to planning and good, old-fashioned homework -- together -- before the first load of product is dispatched.

Shipper/carrier recognition of current Federal transportation regulations -- and the responsibilities they invoke upon both parties -- is the first step in this encounter. Such regulatory oversight is being duplicated and complicated by individual States; and, in many cases, even by counties and towns.

Now, I recognize that regulations are nothing new to you. I know that, today, you are measuring and calculating hazards that you didn't even know existed five years ago.

Transportation safety regulation is however different, because it is written and promulgated with the premise that the person in control of your product -- a truckdriver -- will know next to nothing about that product. Thus we see that transportation regulation emphasizes packaging and paperwork. Granted, the regulations presume that there will be a certain amount of pre-dispatch training as to the obvious and overt characteristics of the product and initial steps (only) about driver responsibilities and actions should a product-release incident occur.

At this point I should note that there is quite a debate, within our industry concerning driver training and what detail it should accomplish.

We are becoming increasingly aware that while many of your product names (literally) sound alike, a simple difference in suffixes for example ITE vs. ATE) may mean vastly different hazard characteristics and response techniques.

In this context, then, we are quite concerned about the possibility of over-training our drivers. Let's say that a driver has been to training sessions at DuPont, Allied, Monsanto, etc. and is given selective information about 20 or 30 products that he might be called upon to haul.

That's fine, but remember, when an accident does occur, that driver is very likely to be in a highly emotional, somewhat distraught state of mind, and -- training take the hindmost -- he might not be the most reliable source of information at the accident site.

Therefore, our industry is asking that you both continue and accelerate the development of written product information aimed at emergency personnel and designed to be carried in the vehicle. Personally, I have yet to see the training aid or regulatory theory which will serve as an acceptable substitute for hard copy.

Please keep in mind, that up to this point, all I've discussed (with respect to the term "regulation") is that of the Department of Transportation.

I don't for one minute want to dismiss our collective obligations to EPA and OSHA.

As transporters, we bear full obligation for reporting and consequent civil forfeitures and liabilities for in-transit spills of your products. Remember, that one of the legal bases for the concept of "common carriage" is that the carrier accepts "care, custody and control."

Therefore, if there is a spill -- we are responsible. However, if we are informed about possible environmental consequences we can take some action. For instance, a newer vehicle with improved roll-over protection might be utilized for a particular dispatch; routing might be changed to avoid proximity to large bodies of water; and many of our carriers work with local fire and emergency personnel who should be advised of the hazard potential.

Since social regulation so heavily impacts both of our industries, both carriers and shippers must recognize that such regulation carries with it a "guilty until proven innocent" connotation (when incidents occur), thus we must adopt an idiom that I'll call "systemwide" compliance and cooperation. In short, it's not enough to adhere just to the letter of the law... both letter and spirit are compelling.

For instance a shipper may test a product, determine that it is flammable or corrosive, duly prepare shipping documents, etc. and be in "compliance".

If, however, those same tests (or in-plant experience) reveal that the product produces pitting in certain grades of stainless steels or produces noxious fumes when mixed with caustic compounds -- our equipment men and cleaning rack personnel would certainly appreciate such knowledge.

The same goes for temperature-sensitive products. Insulated tanks are available, but it does have limitations. I might add that those limitations are not necessarily increased by loading products at temperatures higher than those called for in your specification sheets.

Allow me to interject, here, a plea on behalf of our industry, regarding both loading and unloading.

I recognize that few of you, in this audience, become involved in the operational aspects of the transportation scenario, but, you may serve on plant safety committees or have contemporaries so involved.

In this context we urge the shipping community to institute more active programs in marking intake and discharge piping, positively and plainly; and reviewing those markings, often. When blending or plant mixtures change, please be sure the pipe markings are changed.

In the past seven years our industry has been involved in two disastrous incidents, involving multiple fatalities in each, where the real cause was simply putting a product into pipes which were either mismarked or not marked.

I think it is significant to note, at this point, that for the past two years, my Association has done a rather in-depth study of the DOT hazardous materials incident report forms (DOT Form 5800.1) that have been filed following product-release accidents and incidents.

In each of those two years, the survey results have been practically identical in that over 80 percent of the industry's losses occur off the highways, specifically at loading or unloading sites. Of the off-highway accidents, over 70 percent are at the unloading point.

We followed-up (particularly on the unloading accidents) with the people actually involved and were able to evolve two, broad-based conclusions as to "cause" (applicable to the predominate number of incidents). First and foremost is a lack of communications (or a failure in communications) between the driver and plant personnel. Common among such failures would be an example wherein the driver (who initiates the unloading), for some reason believed that the intake valve is open, when, in fact, it is closed. The net result here is generally a ruptured hose or overpressured fitting.

The second general "failure" involves a continuation of the human species' fantastic ability to ignore common sense (on the one hand) and training (on the other). I won't give you any specifics on this -- I'm sure you can supply your own examples.

While I'm dealing with the operational aspects, please allow me to add a few other situations that might be discussed at the plant safety meeting.

First, I hereby issue a fervent prayer that you carefully research and test if you plan to blend product in the trailer or load from more than one product line in the same trailer. Unlike you, we are not necessarily enthralled by seeing the magic of chemistry in action.

Next, some admonitions about the human factor. Preliminary to transportation, you and the motor carrier should have agreed as to what protective clothing is required and who will supply it. I will be the first to confess that many tank truck drivers exhibit a "macho" complex which often manifests itself by a disdain for protective clothing. This is particularly true in the summer months. Full, protective gear is hot, goggles can cause the eyes to water and heavy gloves are ungainly.

Nonetheless, we ask you to instruct your operating personnel not to authorize loading until they are satisfied that he or she is properly attired.

Remember, also, that during the loading and unloading of any hazardous product the driver is mandated, by regulation, to stay with the vehicle. Invitations by your personnel to the driver to come inside a plant building for a spot of tea and a walk around the hors d'oeuvre table -- while socially gracious -- are contrary to law and can be most counterproductive.

Also, we ask that the carrier be advised of any known characteristics of the product that may produce problems during loading and unloading. Products that congeal or the settlement of particulates can cause unshirted hell, unless precautions are taken.

Time your transportation. Inclement weather, traffic jams and detours are part of the American way of life. If you can so determine double check with the carrier on the working hours of the receiving facility. If that receiver won't accept deliveries after 5 p.m. or on Saturdays and Sundays -- work with the carrier to plan the original dispatch accordingly.

Please remember that our drivers don't work for the Postal Service. If you have packets of material for the receiver -- such as analyses or special instructions -- mail or phone that information. Don't entrust it to the driver for delivery. Keep in mind that the driver who picked up the load, may not be the one who delivers it.

The same goes with samples. Some carriers will allow small samples to be carried in the truck cab -- some will not. If they are accepted, they must be properly packaged and labeled according to DOT regulations. This does not mean a corked vial in a paper envelope with a rubber band around it.

Obviously, I'm not here to render a litany of the industry's horror stories. I've tried to be positive, today, and I'll attempt to continue by noting that the newly-published EPA/DOT regulations on the transportation of hazardous wastes might well prove to be a prime example of how government and industry can cooperate in a regime of meaningful social regulation.

As you know, the regulations call for a "cradle to grave" accounting of the generation, transportation and disposal of hazardous wastes and in many circumstances, those of us in industry would conjure up thoughts of mountains of government regulatory do's and dont's together with the accompanying paperwork and zillions of reports. Those of you who interact with OSHA probably know such a scenario.

However, I believe that the cooperation between industry and government in formulating the RCRA structure might well serve as a model for the future. After much hard work we have a "livable" system for cataloging generators and transporters. Contemporary DOT shipping documents, markings and placards were blended into the regulations, as were existing resources for emergency notification and emergency response. So, while it might be heretical for a representative of the private sector to openly applaud government, I, nonetheless, do so.

Ladies and gentlemen, I harbor a grave fear that is probably common to non-technical speakers who appear before technically sophisticated audiences such as yourself. I fear that the information is so basic, so simple, that it might be "primeval" in the context of this session.

Let me assure you, however, that each one of the examples, admonishments, suggestions and cautions I've given you today has been borne from an actual incident involving one of my members and a major shipper, within the past calendar year.

Therefore, if my presentation prompts nothing more than to prompt a review of basic, elementary procedures -- and if I have convinced you that today's society views regulations as total responsibility rather than sheets of Federal Register pages, then I've served my purpose.

HARVISON, C.J.

DISCUSSION

RUSSELL HILL, Oxirane-International: Comment, not question. This is certainly one paper in which the impact has been enhanced by the speaker's very excellent delivery I would beg of him to make a television cassette tape of this talk and make it available to us, so we could deliver it to our operating facilities. I think that would do more good than distributing free prints or anything else.

HARVISON: Thank you very much, I think you just told me that I talk better than I think, in which case you are probably right. Thank you, your comments are most generous.

W.B. HOWARD, Monsanto: This is really not about this talk but about Mr. Botkin's talk. There was a comment about the cold flow property of teflon when used in gaskets. I'm with Monsanto, not du Pont. Du Pont does have another material, now available for gasketing, which fully overcomes the cold flow properties of teflon and this is a true elastomer which they sell under the name Kalrez. It is normally used in the form of O-rings; O-rings constitute exceedingly good gaskets. Thank you.

HARVISON: I might note on the same subject. In the transportation of Gashol, we are having very significant problems in gasketing and overall valve protection. I don't know whether that particular product would be the be all to end all in the transportation of Gashol, but we have not found a solution yet. Teflon has not been quite the answer in that area either.

TREVOR KLETZ, ICI: The speaker has very eloquently said a number of things that the manufacturer ought to do to help the transportation company. We find that one of the most common accidents we have in transportation is not on the road but in the factory. It is the driver driving away before loading is complete or before the hose is disconnected. In *Loss Prevention* (Vol. 10, p. 151), there is a photograph of a very simple and effective device that prevents this happening. In front of the hose connection on the tanker is a small metal plate. In order to connect the hose, you have to move this plate to one side; that locks the brakes on the tanker and you cannot drive it away until the hose is disconnected and this plate is moved back into position. We have now fitted most of our own tankers with this device; it is very cheap; it uses standard components and costs only 100-200 per compartment; obviously, you need separate one for each compartment.

HARVISON: I do remember the product being marketed over here. There was some concern not only by the industry but by the regulators themselves as to the possible scenario where a unit had to get out of a particular area, let's say, because of the danger of fire impingement in a very big hurry. There would, of course, under those circumstances, be the second idiom where he would have to go through this job of removing the safety device which under some circumstances might be counterproductive. We have also thought of tying a string from the end of the hose to a certain portion of the individual's anatomy, we have often felt that that would probably be the best.

The accident scenario that you have described is all too common. Again, as I noted when we went back through the DOT reports, we felt that we picked up a good dozen of these which the industry terms "drive offs." It was basically a communications problem. First of all, it almost always occurred in some degree of inclement weather. Because the weather was inclement, the plant personnel as well as the driver, in order to give each other a break, were really sharing the unloading duties. So, there was always the common understanding of O.K. when she is dry and that airline is out, the plant man disconnect and put the hose in and I'll do this, that and the other thing. These little informal communication arrangements have just been the foundering point of the industry in many of these accidents. That is just one example of people reaching assumptions, thinking that they have communicated with each other when in fact they have not.

STAN GROSSEL, Hoffmann-LaRoche: I have just spent a few months looking at problems like this and there is quite a bit of feedback from the LPG industry, where this happened many times. There is a piece of apparatus called a S-Trong Wall bulkhead which is recommended by people in the industry which anchors your lines so that even if the driver drives off without disconnecting the hose, the lines will not be yanked away from your vessel. Now, I have also been informed by our operating supervisor that for tank truck deliveries our procedure specifies that the truck driver is not allowed to hook up or unhook the lines. It is the responsibility of our own operating supervisor and that is one way of avoiding problems with the truck driver.

HARVISON: I am really pleased to see that the shippers take that attitude and I will add one other thing to it (not only in compressed gas). Let's face facts. Depending on what the Supreme Court does with the benzene decision, we on the for hire side are really going to have to sit down with you and the private carrier/shipper side. You hire our customers and our competitors at the same time and with respect to the entire aspect of carcinogens or the transportation of identified carcinogens, we are going to have to start talking to you folks about, "Do we bring our driver into your work area?" or "What is the restricted area dealing with carcinogen transportation?" If we bring him in to perform loading, unloading, hookup, etc., what are the cost parameters. Going to what some of the compressed gas shippers have done and say O.K. drop the truck right at the front gate and we will take responsibility for putting it in, loading it and delivering it back out of that restricted area. It is something that we as carriers and you folks as shippers are just going to have to sit down and talk about.

OWEN KUBIAS, SCM/GLIDDEN: One of the concerns that I have on tankers coming into the plants is the disturbing number that come in with the safety devices inoperable; the fusable links and the remote shutoffs. Does your organization have a program of educating the carriers on the need for and the maintenance of these devices?

HARVISON: The answer is yes.

Very candidly, this is probably one of the prime targets of the current DOT enforcement campaign where they bring carrier managers in and then they say you had a fusable link replaced by a bolt on a 306 and that bolt is going to cost you $40,000 in civil forfeitures. We are trying to educate the industry along that line. We are also getting into an area where particularly working with two or three of the major chemical companies, we are finding that some of our carriers have been slipping up on putting the properly-sized and the proper capacity relief valve in a particular trailer. We are going out (hopefully) with preprinted information designed for the terminal use where the carrier will be able to put in his equipment number, put in the spec number, and put in exactly the type of capacity and the proper venting and valving that is required. I might suggest that we like a fail-safe fusable. The reason, very frankly, those nuts and the bolts and up on there in place of the fusables is that we just can't get them to last.

ANNONYMOUS: I would like to come back to immobilizing a vehicle from moving. I see no account of things you use. I've seen such things as taking the keys off the driver and locking them up, but any commercial trucker who hasn't got spare sets of keys is probably not in the business. I've seen big signs put in front of vehicles and they drive over them. The thing I like, the thing we use widely in Europe, is the lockable chuck. It is a pretty heavy thing which our people in Luxemburg developed and you put it onto the actual wheel and you lock it and the operator, supervisor puts the key in his pocket and only when he is satisfied and things are finished can you unlock it. Now, they stole the idea from the Parisian police and they in turn stole it from Denver in the States because in Paris they call it the Sabo de Denver.

HARVISON: I am well familiar with it.

ANNONYMOUS: I strongly recommend it. There are too many trucks that drive away, snap the hose and spread fire so I am strongly in favor of the lockable chuck; I think it is the only 100%-effective method.

HARVISON: For those of you not familiar with the lockable chuck or the "Denver Boot," come to Washington, D.C., get about six traffic tickets, and don't pay them, because the police will put them on your automobile and you can get an exactly first-hand view of what they are.

Again, and not to treat the question lightly, the question here is, I don't think, the necessary way to or not to mobilize the truck. I think there is a regulatory/philosophical argument that has to be counted first. In other words, do we want to mobilize that truck and possibly risk a greater (configuration) under some circumstances, or do we want to look it up and say, hey we're done with this hose-break bit? Once that debate is over, I would suggest that the actual mechanics of mobilizing the vehicle would be a simple matter. Thank you again ladies and gentlemen. I enjoyed being here.

POLLUTION CONTROL PRACTICES:

Incinerator Problems & How to Prevent Them

What makes proper incinerator operation imperative is that if it doesn't operate the plant doesn't operate.

D. F. Ready and R. F. Schwab, Allied Chemical Corp., Morristown, N.J. 07960

In an environment-conscious age, appropriate waste disposal is part of proper operation of a chemical plant. One of the methods of handling wastes, particular liquid or gaseous streams (including those that are nothing more than a bad smell) is incineration. It is now a fact of life that if the incinerator does not operate, the plant also does not operate.

Case 1: Phthalic anhydride plant incinerator

An explosion occurred in the heat exchange section of our phthalic anhydride (P.A.) incinerator under normal operating conditions with no unusual conditions being experienced by the operators. Figures 1 and 2 show the process flow.

Basically the phthalic anhydride tail gas from the condensers at approximately 76°C (165°F) passes down through a vertical heat exchanger. Those tail gases were heated to about 315°C (600°F) by the combustion gases from the incinerator. These tail gases then were passed through a cross-flow horizontal heat exchanger, where they were heated to about 590–595°C (1,100°F) before being piped into the combustion chamber where the fumes were incinerated by a natural gas flame. No additional outside air was needed, the tail gas contained enough oxygen to accomplish the incineration.

0149-3701/81/4237 $02.00

The explosion was centered in the duct work between the vertical heat exchanger and the horizontal cross-flow heat exchanger, Figures 1 and 2. The bottom part of the vertical heat exchanger and the piping-ductwork were insulated on the inside. When the explosion tore open some of the ductwork, some of this internal insulation was seen to be burning (somewhat persistently). The question is: What went wrong here? As usual, there isn't a simple answer.

This incinerator had just been installed about a year ago replacing an older unit. One of the considerations was to maintain surface temperatures of the equipment and ductwork to a sufficiently low level so that personnel would not be burned, if they accidently touched one of the incinerator surfaces. In order to do that, insulation was installed on the inside. The original engineering specifications called for a nonporous refractory type. Between specification and installation, a rock wool type insulation which was quite absorbent, was installed. We have reason to believe that we had carryover of phthalic anhydride into the area where this insulation was installed so that eventually it soaked the insulation and we had a classic phthalic anhydride insulation fire.

It appears that the supplier of our heat transfer oil gave us some replacement oil which was incompatible with our older material. This resulted in an unusual amount of sludge which impaired heat transfer leading to higher temperatures in the discharge duct from the condensers

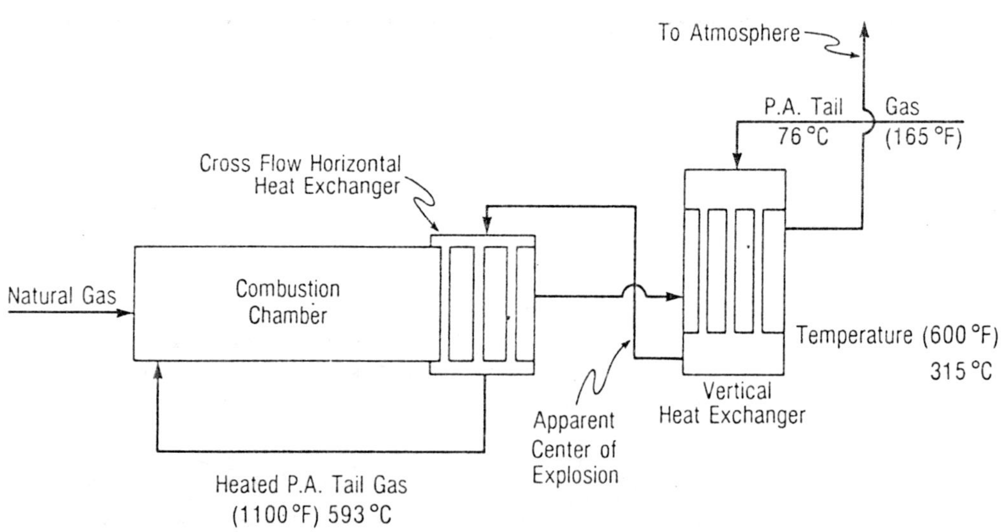

Figure 1. P. A. incinerator flow sheet.

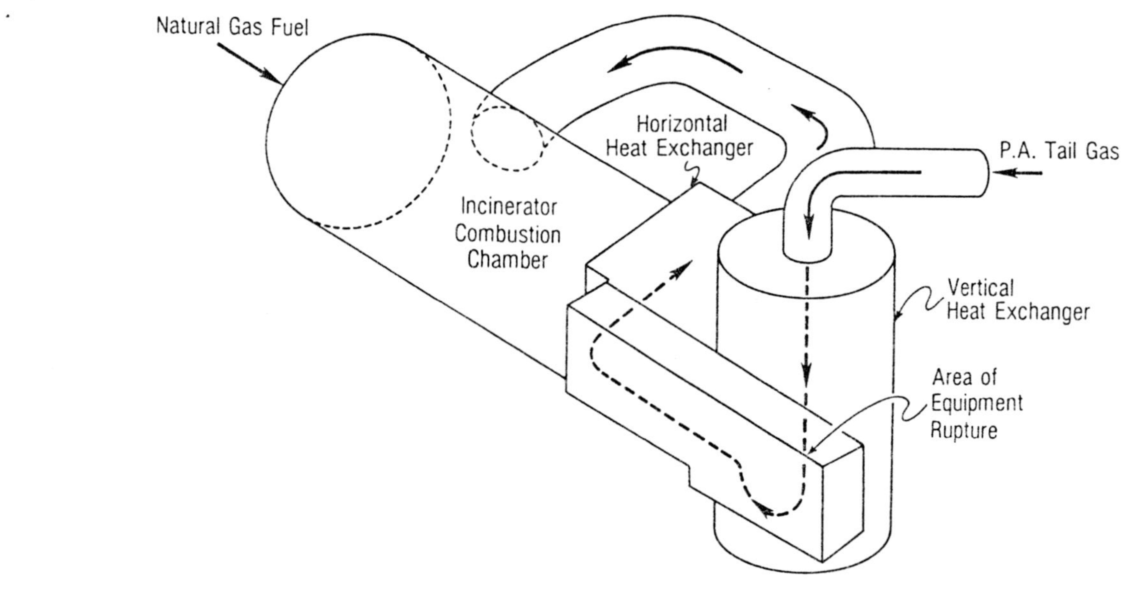

Figure 2. P. A. incinerator.

than we usually like to see. Condensation of P.A. on the line between the condensers and the incinerator was excessive, despite heat tracing. This caused occasional solid pieces to get into the heat exchanger section. This manifested itself by occasional "heat spikes" in the incinerator. Apparently a sufficient amount got into the heat exchanger section to result in a mild explosion.

As a result of this event (which was not a major explosion), several changes were promptly implemented:

1. The absorbent internal insulation was removed and replaced by a nonabsorbent type at the bottom of the heat exchanger.
2. The ductwork was replaced by stainless steel with outside insulation.
3. Explosion relief was provided for the ductwork and heat exchanger equipment.
4. The heat exchange system of the condensers was improved to allow lower condensing temperature.
5. Improvements in steam tracing, jacketing, and insulation were made in the transfer line to prevent excessive P.A. condensation in the line between the condensers and heat exchangers.

Additional long-range improvements to reduce the P.A. load on the incinerator are being contemplated.

Case 2: Phthalic anhydride incinerator

This particular case occurred several years ago. Again, it

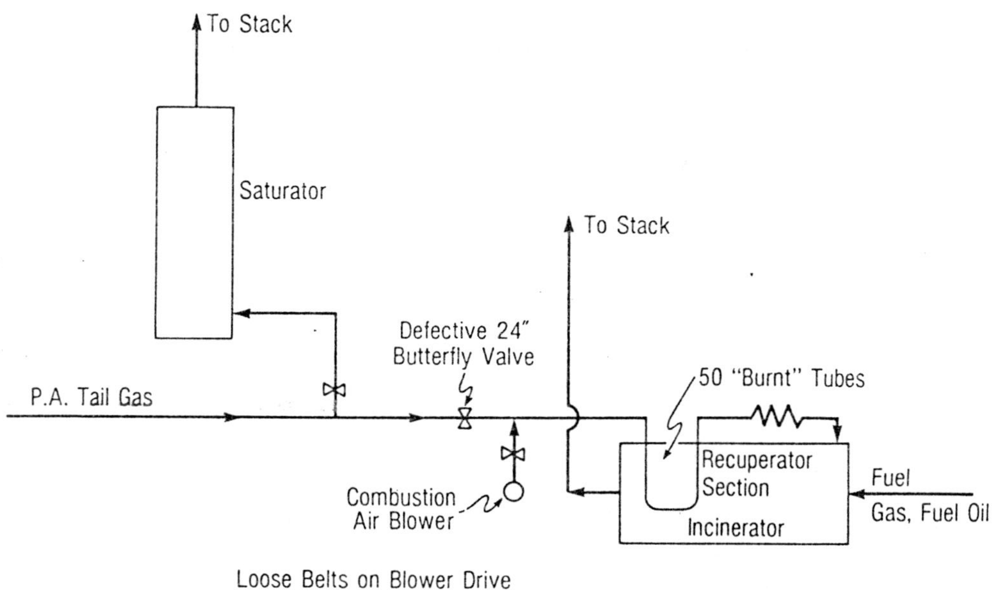

Figure 3. P. A. incinerator failure.

involved a phthalic anhydride plant, but with a different arrangement. The process flow is as shown in Figure 3.

In this arrangement the P.A. tail gas is sent to a "recuperator" or "heat exchanger," which heats the stream before feeding into the incinerator section where it is incinerated. In parallel with an incinerator is the saturator, which washes out the tail gas fumes before discharging to the atmosphere. In this case, for 15 hours before the incident, the tail gas was being fed into the saturator while the Maintenance Dept. was repairing the gas pilot system for the incinerator. After the repairs had been completed, the incinerator was restarted on oil and preheated to 650°C in approximately 8–10 minutes using the startup air fan to circulate air for preheating. After this preheat cycle, process gases were switched from the saturator into the incinerator. Very shortly after the switchover, an explosion occurred, blowing out the expansion joint in the line between the recuperator section of the incinerator and the firing end of the incinerator.

Investigation after the loss revealed several deficiencies:

1. There was a leaking 610-mm (24-in.) butterfly valve leading to the incinerator.

2. The belts on the startup fan were loose, thus the startup fan did not function properly.

3. There were approximately 50 burned-out tubes on the discharge side of the recuperator.

4. There was a cake of solid P.A. approximately 38 mm (1.5 in.) thick covering the rupture disk.

It was found that the 610-mm (24-in.) butterfly valve leaked and deposited a considerable amount of P.A. "hay" in the inlet vestibule of the recuperator. Because of the loose belts on the startup fan, the air flow was not enough to move the hay. When the butterfly valve was opened, the P.A. hay was moved into the incinerator where it was ignited. The tail gas mixture, along with the deposited P.A. "hay" (crystals) was in the combustible range. It should be noted here that the phthalic anhydride crystals are a dust explosion hazard. Thus, it would not be a requirement to have the P.A. hay vaporize in order to propagate the deflagration.

Several steps were taken to prevent this from recurring:

1. Installation of two 610-mm (24-in.) butterfly block valves with a means of pressurizing the space between with inert gas, with the bleed discharge between the valves discharging into the saturator system.

2. A maintenance program was established to periodically check the drive belts for the startup fan.

3. An air flow indicating switch was installed to ascertain if appropriate amounts of startup air are indeed flowing.

4. Additional rupture disks were installed in the outlet side of the recuperator, providing the appropriate explosion relief.

Multipurpose incinerators

Allied Chemical's multipurpose incinerator system serves a plant that produces organic intermediates. It is used to

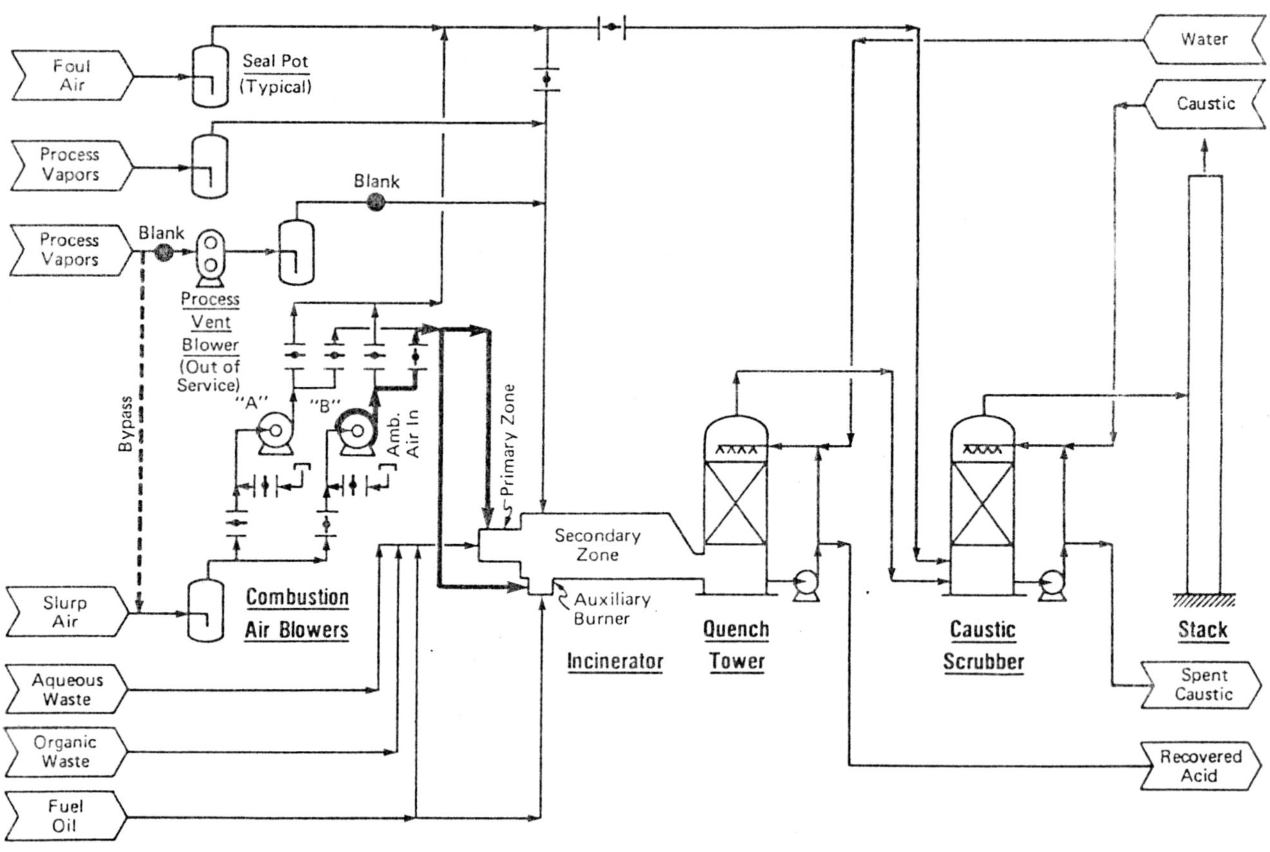

Notes: 1) Heavy lines indicate area of explosion
2) Dashed line indicates modification of original system
3) Positions of valves at time of explosion indicated by symbol ⊸

Figure 4. Explosion in combustion air piping.

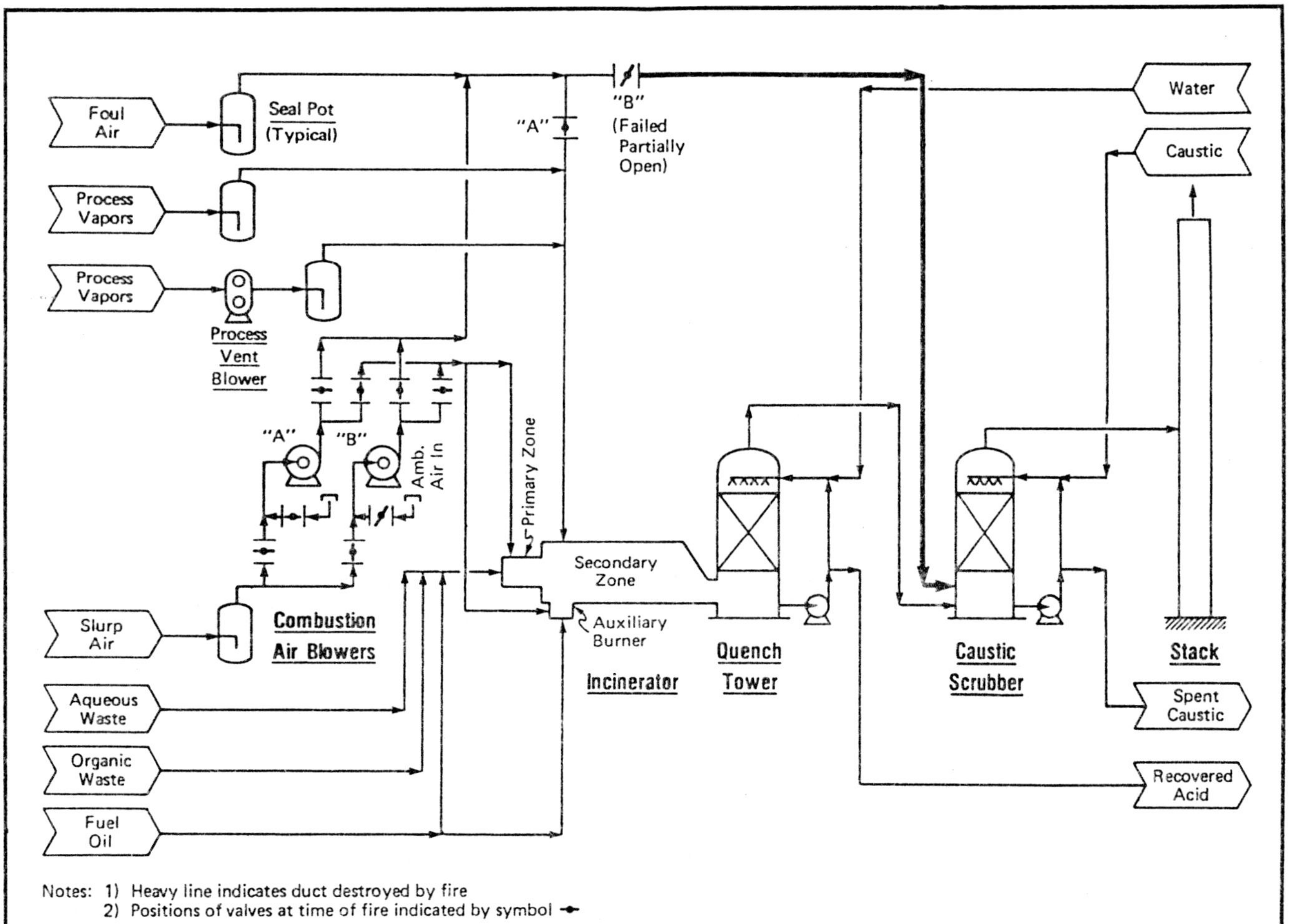

Figure 5. Fire in bypass line.

destroy several types of liquid and vapor wastes generated by the plant. This unit is essential to the operation of the rest of the plant, since it provides the only means of disposal for a number of toxic and malodorous materials which cannot be discharged.

Figure 4 shows the flow scheme of the incinerator system. The various liquid waste streams from the processes are accumulated in two tanks: one for organic and the other for aqueous-based liquids. From these tanks, the liquid wastes are pumped at controlled rates to an injector nozzle in the primary combustion chamber of the incinerator. Much of the fuel value required for operation of the unit is supplied by the organic waste liquid with most of the rest being furnished by fuel oil which is co-fed with the waste.

Combustion air for the primary fuels is supplied by a pair of centrifugal blowers (normally with only one on line). These blowers also provide suction for a plantwide "slurp" system which collects contaminated air from packaging stations, sumps, vessel "breathing," etc. This air stream normally has no significant fuel value.

Another major contaminated air stream from the plant enters the secondary combustion chamber of the incinerator. This stream normally has insignificant fuel value and is incinerated primarily for odor destruction.

Finally, there are two small vent streams from the process which carry significant concentrations of flammable materials. These vapors are fed to the secondary combustion chamber. Although they are relatively rich in combustibles, these streams normally contribute little fuel value due to their low flow rates.

Each of the vapor feed streams passes through a separate liquid-filled seal pot intended to preclude the possibility of flame flashback from the incinerator to the process.

The wastes handled by this system contain substantial quantities of chemically bound chlorine, nitrogen, and sulfur. Thus, the combustion gases require scrubbing prior to atmospheric discharge. This is accomplished in two stages: first, a combined water quench and scrub, followed by a second caustic scrub.

In addition to the main burner, the incinerator is fitted with a small auxiliary fuel oil burner which is used only for startup and standby operations.

Case 3: Fan Explosion

After the plant had been in operation for several months, an explosion occurred in one of the combustion air blowers and the associated discharge piping. The discharge end casing of the blower was broken by the force of the explosion and the combustion air line to the auxiliary burner was destroyed, Figure 4. The violence of the explosion in the air duct was so great that sections of the line were propelled with sufficient force to tear away several feet of handrail nearby. The explosion occurred after restarting the incinerator which had been down briefly due to a flameout.

Investigation of this explosion indicated that it had resulted from the passage of a "slug" of combustible vapor/air mixture through the slurp system, fan, and combustion air line. The source of ignition may have been the incinerator itself or the blower. In either case, the explosion definitely resulted from the presence of unexpectedly high concentration of flammables in the combus-

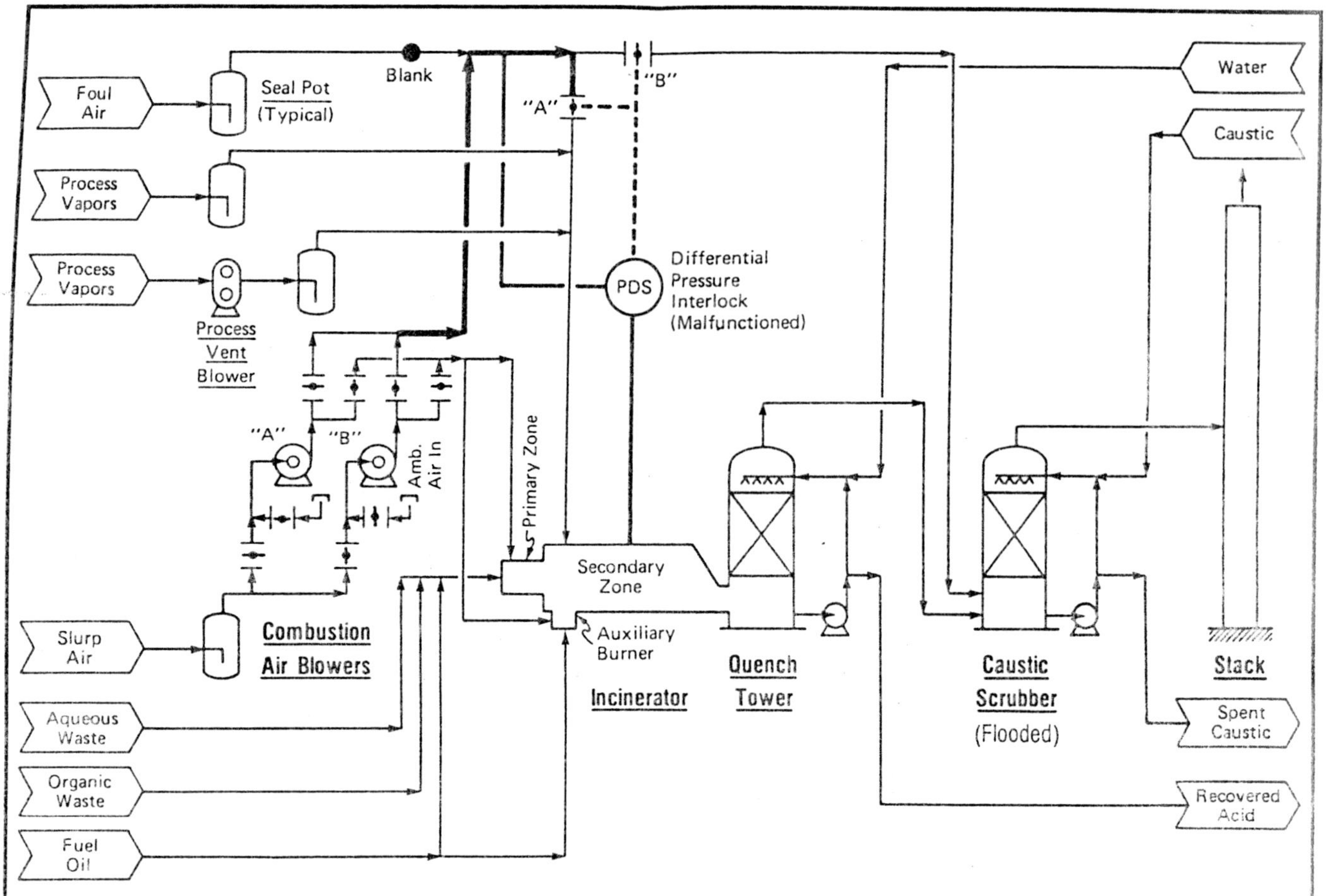

Figure 6. Fire in fume inlet line.

tion air. Further investigation of the events preceding the incident indicated two critical contributory factors:

1. At the time, procedure for restarting the incinerator following a flameout permitted valving off the slurp system. This allowed the operating combustion air blower to draw ambient air for purging. Upon completion of the purge, the burner would be relighted using fuel oil, and the slurp system reactivated by reopening its valve to the blower suction. This sequence would typically result in the slurp system being "dead" for approximately 10 minutes during a restart. During this dead time, it would be easy for explosive concentrations of vapors to collect in various legs of the piping network.

2. Since plant startup, severe difficulties had been experienced with the small blowers used to feed one of the the low volume flammable-rich vent gas streams from the process to the incinerator. Eventually, these blowers were removed from service and the vent line tied into the slurp system. This action was deemed safe due to the small flow of the vent stream in comparison to the vastly larger sweep of air through the slurp system. Accumulation of flammables in the stagnant slurp system during burner restart was virtually assured by this modification of the originally installed vent system.

To minimize recurrence of a similar incident in a number of measures were taken. Among these were:

1. The burner restart procedure was modified to eliminate any period of stagnation in the slurp system. In the present scheme, the second blower is started and used to supply purge and combustion air during the restart cycle. Meanwhile, the blower which had been supplying combustion air and drawing through the slurp system continues to maintain slurp flow, but now discharges to the caustic scrubber rather than the combustion chamber. By establishing stable burner conditions, the blower drawing the slurp air is valved back into the incinerator and the backup blower is shut down. This procedure is operationally more complex than that originally used and requires that both combustion air blowers be operable at all times, but it does keep the slurp system from stagnating and thus, minimizes the potential for flammables to build up to hazardous levels.

2. The flammable-rich vent stream which had been diverted to the slurp system was returned to its original routing in order to minimize the base load of combustibles in the slurp system. Additionally, the entire slurp system was reviewed and various steps taken to further minimize the discharge of combustibles into the slurp stream.

3. A systematic monitoring program was instituted to periodically check for hazardous concentrations of flammables in the slurp air stream.

Case 4: Fire in fume bypass line

Approximately nine months after the explosion in the combustion air system, the multipurpose incinerator suffered a pair of incidents in which sections of the fume inlet duct were destroyed by fire. The first is described here.

Following the explosion discussed earlier, it was decided that both combustion air blowers should be run at all times (not just during burner startup) to further minimize the risk of another such incident. In this operating mode, blower A was used only to draw in ambient air for primary combustion and blower B to draw from the slurp system. The slurp air from blower B was discharged into the secondary combustion chamber of the incinerator. During incinerator outages, the slurp air was diverted from the incinerator to the caustic scrubber by simultaneously (via pneumatic actuators) closing valve "A" and opening valve "B," Figure 5.

This incident occurred after restarting the flow of slurp air to the incinerator following a period of several hours during which this flow had been diverted due to instrumentation difficulties. A few minutes after switching the valves back to their normal positions ("A" opened, "B" closed), the plastic bypass line between valve "B" and the caustic scrubber caught fire and disintegrated. The metallic line between valve "B" and the incinerator was observed to glow "red hot" but was not destroyed.

It was concluded that the direct cause of this fire was a failure of valve "B" to close properly in response to the switch-over signal. With both valves open, a low resistance path was created for the hot combustion products to flow from the incinerator directly to the caustic scrubber through the bypass line. There may or may not have been an accumulation of combustible material in the line to feed the fire, but this would have been of relatively minor importance since the hot combustion products would have been more than sufficient to destroy the plastic line in any case.

Modifications to avoid recurrence were implemented as follows:

1. The plastic pipe section was replaced with one of a suitable metal for improved high temperature resistance.
2. A limit switch and alarm were installed on valve "B" to give warning of any future failure to close fully. (Valve "A" had already been equipped with a similar device.)
3. A high-temperature alarm was installed in the line between valve "B" and the caustic scrubber.
4. An air purge was installed in the line segment between valve "A" and the incinerator. This purge was to be activated whenever valve "A" was closed to keep this segment free of combustibles which might otherwise collect in such a stagnant line.

Case 5: Fire in fume inlet line

Not long after the fire in the fume bypass piping, a second fire burned out the plastic line between fan "B" (slurp air) and valve "A" in the fume inlet line to the incinerator. The first sign of trouble in the system was an abnormal increase in the pressure drop across the caustic scrubber which produced a high backpressure in the incinerator. The pressure increased erratically over a period of approximately one hour, at the end of which the line ruptured and burst into flames, Figure 6.

It was concluded that the high backpressure in the incinerator had caused the fire either by producing a backflow of hot combustion gases toward the blower or by causing a reduction in slurp air flow and velocity to the point of near stagnation. With slurp flow near stagnation, a buildup of combustibles could have occurred and set the stage for a flashback into the line from the incinerator.

The high pressure in the incinerator was almost certainly caused by flooding of the caustic scrubber. This column had been repacked some time previously following difficulties with the original packing support grid. At the time of repacking, a mesh pad was installed between the support members and the packing bed to eliminate the sifting of packing into the bottom of the tower. This modification served its intended purpose, but created a restriction at the bottom of the packed bed which made the column prone to flooding. Furthermore, this pad tended to act as a filter, trapping particulate matter and gradually becoming even more restrictive. During this period of time, episodes of poor air/fuel ratio control were common and frequently gave rise to high particulate levels in the combustion gases. This further aggravated the pluggage problem in the bed support mesh.

The incineration system was originally installed with an interlock designed to prevent just the type of backflow or flashback responsible for this incident. The interlock was designed around a differential pressure switch between the blower discharge and the incinerator. In the event of low forward-flow through the line, the differential pressure would fall, tripping the switch and automatically diverting the air stream from the incinerator to the caustic scrubber.

Prior to the incident under discussion, a portion of the process had been temporarily shut down and the foul air flow from this source to the incinerator eliminated. With only the slurp air flowing in the fume inlet line, the pressure drop was reduced, leading to troublesome "nuisance" tripouts of the interlock system. Substantial "fiddling" with the pressure switch was then required to eliminate these trips. Since the switch was destroyed in the fire, it was not possible to ascertain its operational state before the incident, but it seems likely that it was set too low or even disabled completely and thus unable to perform its intended function of preventing a flashback.

Finally, it was discovered that blower "B" had been running with reversed rotation for some time prior to the incident. Running backwards, the blower would have developed significantly less pressure than when rotating in the correct direction. This reduced output pressure undoubtedly made it easier for backflow or flashback to occur.

Preventive measures taken to avoid repetition of a backflow or flameback of this kind included:

1. Holes were cut in the caustic scrubber-packing support mesh to minimize the tendency of this column to flood and produce high backpressures in the incinerator.
2. Reversed rotation of combustion air blower "B" was corrected.
3. The differential pressure interlock was re-installed and checked for proper operation.
4. A high pressure shutdown switch was installed on the incinerator.

In conclusion

The incidents recounted in this article are typical of what can be expected in similar installations. It demonstrates the need for careful examination of all aspects of the design and operation of the units and the consequences of process excursions, both short term and long term.

An essential point to remember, where we incinerate a maloderous or toxic potentially combustible process stream, is the necessity of keeping this stream out of the flammable range. It is necessary to exercise a vivid imagination on process stream variables and equipment malfunctions to predict problems. The consequences of not doing so can lead to some very expensive incidents which may also involve extensive plant shutdowns and production loss. #

D. F. Ready, who earned his B.Ch.E. at Cooper Union and his M.S.Ch.E. at New Jersey Institute of Technology, is a process engineer in the Chemicals Co. of Allied Chemical Corp.

R. F. Schwab, who earned his B.Ch.E. from CCNY and his M.S.Ch.E. from the Polytechnic Institute of Brooklyn., is manager, process safety and loss prevention for Allied Chemical Corp. He is a Professional Engineer in New Jersey and Connecticut.

DISCUSSION

JOHN ELINGTON, Mobil Chemical: Your kind of problems are not surprising; we've had the same kind of problems. I was surprised, though, that you mentioned the water seal as a method of preventing flame propogation. Similarly, you have flame arresters. Both can foul sometimes, but I am surprised that no mention was made of fast response extinguishment systems such as powders. Pressure and temperature rate of change can be used to trigger them. Fenwal makes some of these systems. Also, the inclusion of interlocking with massive amounts of steam and inert gas such as nitrogen is good. The use of extensive grounding (such as duct work with high carbon content, making it electrically conductive is appropriate. I would be interested in your comments.

SCHWAB: The system is well grounded. We are dealing here with materials that contain chlorine, sulfur and things of this nature. The corrosion problems that we see in a system like this are almost unbelievable. It is the Materials Engineers' paradise. The products of combustion attack most materials of construction. That in itself precluded the use of these suppresion systems. This matter was very carefully reviewed before decisions were made to go the way we went.

TREVOR KLETZ, ICI: I would like to ask a general question about almost the very last sentence of your paper: "It is necessary to exercise a vivid imagination on process stream variables and equipment malfunctions to predict problems." I would suggest that the vivid imagination by itself is not sufficient and that you need to supplement this with a systematic survey of the way things can go wrong: for example, of the type described by H.G. Lawley in *Chemical Engineering Progress* (April, 1974, p. 45). Many companies which carry out these systematic studies of malfunctions on their mainstream plants do not always bother with them on what they describe as "only an incinerator," "only a waste disposal unit" or "only an off-plot." Similarly, you sometimes find operators who operate their mainstream plants splendidly but tend to neglect the little units or the equipment that deals with the wastes. We have a phrase that rot starts at the edges, not literally at the edges, but at the peripheral activities. In the incidents that you described, were either of these features at work; did the designers or the operators fail to give this equipment the detailed attention they would have given to a mainstream unit?

SCHWAB: Yes, let me say this. We recognized from the very beginning that we were going to have many problems with this unit. We spent many, many engineering hours going through the system and we thought we had most of the problems solved. We spent a considerable amount of effort in operator training and paid pretty good attention to the problems inherent in operating this incinerator. Despite all of these things, we still had the problems that we described in the paper. Some of them we did foresee but we did not foresee all of them. We had much debate about the problems of flashback in that system but we frankly did not foresee the buildup of flammables to the extent we have seen in the operation of this system. We did all the calculations we possibly could do, but we do get into the flammable range quite frequently, despite all the calculations. I think I know where it comes from but we're really not 100% sure yet. Does that answer the question?

KLETZ: Yes. Thank you.

SCHWAB: Sometimes the chemistry of a brand new process like this can fool you.

HAL FORREST, FMC Corp.: It has been recommended to us (and we comply) to put a half-inch perforated plate just upstream of where the fuel-air mixture enters the combustion chamber. This plate is designed so that the boundary velocity gradient through the holes in the plate is greater than the boundary velocity gradient for flashback. Thus, even if you get a flammable fuel-air mixture entering the furnace through the combustion air duct, it cannot propagate back upstream past the plate. We also put a thermocouple just downstream of the plate; so if a flame stabilizes on the plate, we divert the combustion air and shut down the furnace. Would you like to comment?

SCHWAB: The problem of flashback has been a very troublesome one; we haven't got that solved yet. We have considered such things as flame arresters and we rejected that because flame arresters would corrode and act as sieves to collect crud and other corrosion products. The problem associated with the flame flashback by maintaining velocity in that line is a difficult one. Does that answer your question? We haven't solved all the problems yet.

SIDNEY BROOKS, Dupont: Your paper referred to the installation of additional rupture discs on the Falcon (Thalcon) .Hydry Incinerators. What was wrong with the data and the design basis for the initial installation and what additional data did you get to provide the basis for the second installation?

SCHWAB: Alright. In the first place you're talking about incident number one. Well, this incident we put in rupture discs and we used as a basis of design NFPA 68, *Guide for Explosion Venting*, for the duct work based on Bartknecht's work. On Explosion Venting, we had sufficient venting area to do a decent job.

Procter & Gamble's Environmental Incident Program Loss Prevention Aspects

This program was developed from plant experiences and risks to meet EPA regulations. Provisions for spills, air emissions, and hazardous waste disposal are discussed. Key parts of the program are: facility design and upgrading; written incident prevention plans; incident response procedures; evaluation of incidents; and informal audits.

H. Schwartzman
The Procter & Gamble Co., Cincinnati, OH

Procter & Gamble's environmental incident program has developed in scope and complexity in response to the ever-increasing profusion of U.S. EPA regulations and to actual or potential incidents at Company facilities. It was started to prevent spills of oils and hazardous substances to waterways and publicly owned treatment works (POTW's). It has since been expanded to include prevention of similar spills to ground water, to control of hazardous air emissions resulting from spills and to assure proper disposal of hazardous wastes.

In this paper, I will describe the functional parts of the program and illustrate the loss prevention aspects with examples of parts of the program which address potential incidents with combined environmental and process/personnel safety impacts.

A word on my credentials. The environmental group I lead in P&G's Corporate Engineering has among its responsibilities, the design of spill protection facilities, the development of environmental incident prevention plans and practices and coordination of environmental incident response. For a number of years, I also participated in task groups of the Chemical Manufacturers Association (CMA) which inputted and responded to the EPA in the development of regulations on hazardous substance spills and Best Management Practices (BMP's) for control of discharges of toxic pollutants and hazardous substances. Earlier in my P&G career, I led an engineering group which expanded our domestic process safety program to cover our international plants.

Although our program is primarily proactive, it is, of course, reactive with respect to actual incidents. Key parts of the program are:

1. Facility design and upgrading.

    a. Identification of potential risks.

    b. Determination of the risk probability and seriousness.

    c. Development of and obtaining Corporate approval for cost effective plans or designs for risk prevention utilizing facilities, equipment and human activities.

0149-3701/81/4776 $02.00 © 1981 AIChE

d. Engineering and implementation of detailed plans and designs.

2. Written incident prevention plans incorporating SPCC (Spill Prevention Control and Countermeasure) plans and Best Management Practices (BMP) programs.

3. Formal incident response procedures with clearly delineated responsibilities for plant and corporate staff personnel.

4. Identification of and follow-up on gaps or oversights in the program by evaluation of incidents and by periodic, informal audits.

Our management has traditionally been quite proactive with respect to environmental protection. Thus, we usually have a high level of protection installed well ahead of regulations. For example, when we prepared Oil SPCC (Spill Prevention Control and Countermeasure) plans in response to EPA regulations in 1974 we did not have to expend any significant additional capital to implement the plans as we had already installed adequate protection at almost all locations. The majority of our plants already have adequate facilities and procedures to assure compliance with the forthcoming EPA hazardous substances SPCC plan and BMP program regulations. We have found that our environmental proactivity makes good business sense. By not waiting until after regulations are in place, we avoided higher costs of compliance due to inflation and have not had to unduly divert management attention from business matters to compliance with increasingly complex regulations.

Some parts of the program were adapted from a long standing, proven Company approach to process safety based on formal studies, plant audits and incident reports. For background, I refer you to an article in Chemical Engineering (10/28/78) by R. E. Cocks and J. E. Rogerson of Procter & Gamble, titled, "Organizing a Process Safety Program" and a paper I presented at the 1976 Conference on Control of Hazardous Material Spills, titled, "Spill Protection Engineering for Industry--An Integrated Approach".

FACILITY DESIGN AND UPGRADING

Our engineering responsibilities include environmental aspects of design of new facilities, evaluation of the introduction of new process materials into existing facilities and the retrofitting of existing facilities based on recent findings, incidents or regulations. Risk identification and prevention varies with these various situations but requires engineering judgment and creativity in all cases. For new installations, a vast amount of data on the site, the materials, processes, services and incident history for similar units must be assembled to assess the environmental risks. With existing installations, at about five year intervals, a team of corporate and plant engineering personnel reviews plant processes and procedures to identify environmental hazards and propose solutions.

Identified risks are evaluated via a matrix of probability and seriousness. A conclusion that both those parameters are rated as "high" usually leads to a built-in solution while a rating of "low" on both parameters most often leads to risk management through human training and response to incidents. The steps of developing solutions, obtaining approvals and detail engineering design are relatively straightforward. We try hard for cost effective integrated solutions to control wastewater discharges and prevent spills and other incidents.

INCIDENT RESPONSE PROCEDURES

As with any other loss prevention program, reporting, analysis and follow-up of incidents and failures is of great import. It enables us to learn and improve from experience. A description of our incident reporting policy may be of interest.

We define an environmental incident as an out-of-limits, abnormal or potentially harmful emission or discharge to the environment. Emissions include gases, vapors, particulates, odors or noise. Discharges include spills, slugs (short-term surges) or runoff of liquids or solids to a waterway, a POTW (publicly owned (wastewater) treatment works) to ground water or to the ground.

All environmental incidents at P&G facilities must be reported to our corporate environmental engineering section immediately by telephone and later via a one-page form (copy attached). Our section then provides immediate guidance to the facilities on mitigation and/or clean-up and disposal and on the needed notification of regulatory authorities. All reported incidents are evaluated to determine if and what changes in procedures, equipment and facilities can prevent future incidents at that plant. We then repeat the evaluation for other similar processes or plants to avoid a repetition of that type incident there. The appropriate plant and staff groups are informed and a cost-risks-benefits analysis is carried out. Finally, we seek corporate approval of any required funding and procedural changes to implement our recommendations.

CONTROL OF DISCHARGES OF HAZARDOUS AND TOXIC SUBSTANCES

The key to our approach is an arrangement with product development, plant and other staff groups to give us early, adequate information on new materials, formulations, projects, production tests, etc. Procter & Gamble conducts extensive human and environmental safety testing to insure that all of its products and materials are safe as used by the consumer. The environmental testing provides information on the threshold levels at which treatability and toxicity effects are seen. This testing, plus other physical, chemical and biological data, facilitates an evaluation of the affects of discharges of materials and manufacturing use (and spill) levels.

With a few exceptions, all P&G plants in the U.S. discharge their wastewaters to POTW's. Thus control of discharge of slugs (short-term surges) from process upsets or wash-outs and from run-off of possibly contaminated stormwater, either of which could interfere with POTW operations, must be considered in addition to the prevention of classic spills. Where feasible, we design facilities which combine spill protection with control of wastewater discharges. The typical discharge control utilizes containment/equalization for slugs and run-off.

This slide has a typical flow sheet for an integrated system for spill protection and control of wastewater. Storage tank spills and initial washouts flow to a concentrated waste sump. Dilute washouts and stormwater contaminated by minor drips and leaks flow to a separate sump. The layout and elevations are arranged to capture spills from tank truck and tank car unloading in the same system.

Three different levels of spill protection are used. With primary protection, containment is provided for at least the volume of the largest tank in an area plus a statistical allowance for accumulated stormwater. Secondary protection consists of containment (via spill basins, use of natural terrain or emergency gates in run-off ditches) and/or automatic controls such as valve and pump shutoffs. The third level utilizes human controls and countermeasures including alarms, manual valves and shutoffs, sandbags, earth moving equipment, portable pumps, boats and booms and mutual aid agreements. For new plants, all three levels of protection are selectively applied.

Conventional spill protection via diking of storage tanks might be rendered ineffective by a tank explosion which could expell the tank contents beyond the confines of the dike. Our process safety group has initiated a weak top tank program. The program's objective is to insure that all new process tanks which could contain flammable vapors are constructed with weak tops which will relieve in an explosion. This minimizes the risk of a tank being lifted off its foundation or having its side seams split due to an internal explosion. Where feasible, existing tanks which could contain flammable vapors are being retrofitted with weak tops.

SPILL PREVENTION UPGRADING--EXISTING PLANTS

Periodically, a team of environmental and process engineers visits each liquid process plant to identify spill risks and develop prevention alternatives. The probability and seriousness of each risk is evaluated and costs for the several prevention alternatives are estimated. A prioritized list of recommendations for funding is produced via a cost-risk-benefit analysis. Human response procedures are identified for the lower priority items and training is initiated. The plants incorporate descriptions of the prevention facilities and human countermeasure procedures in Oil SPCC plans, where applicable, or in similar plant documented plans.

PROTECTION OF GROUND AND SURFACE DRINKING WATER SUPPLIES

Because of the nature of our products, we utilize few, if any, listed toxic chemicals at our plants. Even so, we have instituted policies and procedures to protect drinking water supplies from accidental discharges to both ground and surface waters of those toxics we do use. We no longer install underground pipelines for hazardous or toxic chemicals because of the difficulties of inspection and risks to groundwater. Additional inspection procedures for buried storage tanks were developed and implemented. Plants which could have spills of toxics to groundwater will obtain data on the volume and direction of groundwater flow and the location of nearby wells which utilize the groundwater. A response program was developed to notify drinking water plants of discharges to surface waters which could endanger drinking water supplies.

AIR EMISSIONS FROM SPILLS OF LIQUIDS

Some mineral acids in contact with moisture in the air, produce copious clouds of acid mist. Normal spill protection via diking for liquid containment does not address the generation of mist from such acids. Because of the potential risk, we evaluated the situation at each of our plants which handle such acids. Subsequently, we developed an approach for fume control utilizing containment with minimum surface area and suppression with a suitable mineral oil. In case of one acid where there is no cost effective liquid for fume suppression, we replaced an outdoor storage tank with a new tank located inside a building. The building module includes liquid spill containment and a fume scrubber.

These examples demonstrate that it is virtually impossible to foresee all possible risks. However, a proactive program combined with good incident response can provide a good balance of costs, risks and benefits.

SCHWARTZMAN, H.

Environmental Incident Information

Person Calling: ______ Date ______ Time ______

Location/Address ______ Tel No. ______

Location of incident ______

Is this a reportable incident? ______ Has it been reported? ______

Agency & person to whom reported ______ Tel No. ______

Other agencies & persons involved ______ Tel No. ______

What happened? (Include times, names, companies, materials, quantities, environmental aspects/locations/streams affected, etc.) ______

What actions are under way? (and by whom?) ______

Who (other than outside agency) should be/has been notified? (name, function, phone number) ______

What specifically is now needed? ______

Distribution:	Initials		Name and Location
USJ Assoc Director	______	Industrial Hygiene	______
ECS Sect Head	______	External Affairs	______
ECS Group Leader	______	Legal	______
ECS Respons Eng.	______	Prod. Development	______
Prod Tech (Name & Location)	______	Originating Plant	______

Person receiving information ______ Location ______

Telephone number

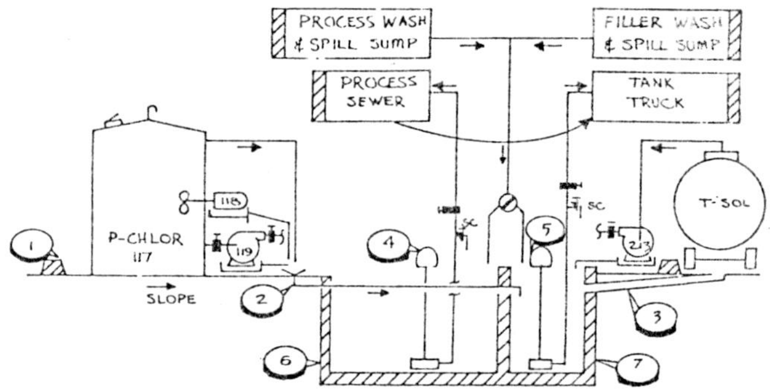

ENVIRONMENTAL CONTROL INDEX

PROTECTION	OPERATION/EQPT.	HAZARD
1. DIKE	TANKS, UNLOADING	SPILLS WASHING
2. RAISED DRAIN	P-CHLOR O'FLO, DRIPS	CONC. SPILLS
3. TRENCH DRAIN	T-SOL TANK' CAR	UNLOAD, SPILLS
4, 5. MAN. START PUMP	PROCESS SEWER	SLUG DISCHARGE
6. DILUTE SWAMP	TANK MAT	SPILLS, WASHING
7. CONC. SUMP	PROCESS SEWER	CHEMICAL DISCHARGE

Figure 1. Wastewater & Spill Control Flowsheet.

DISCUSSION

BOB COFFEE, Eastman Kodak: Yours has been a very interesting presentation. However, we don't always know what chemicals are hazardous to the environment. Up in upper New York State at the present time, someone accidentally diverted 10,000 gallons of molasses into a stream and, of course, molasses is soluble. Most people wouldn't think that molasses presents any hazard but apparently it requires more oxygen than normal and therefore removed all the oxygen from the river and all the fish died. EPA says that it's going to take about three years before that river is regenerated. Thus, we have to look at not only the known hazardous chemicals but also have to examine practically all the chemicals we are handling to find what their effect may be on the environment.

H. SCHWARTZMAN: I assume that was a statement and not a question. My only comment is, sure. We look at all chemicals, and we are very aware of the BOD loads. I'd be surprised if the fish are really gone for three years because if an oxygen deficiency was all there was, the fish will come back pretty fast. In fact, in the incident on the slides I showed, we paid a fine for the loss of the aquatic life. When we asked the state if they were going to use that money to restock the stream, they said there wasn't any need to restock; it would naturally restock itself. If you have real toxics and have actually poisoned the benthic organisms on the bottom, the stream may be dead for quite a while.

We had a recent incident in which we were accused of killing some fish in a golf course near one of our plants. On investigating, we learned the cause was simply spring turnover due to rising temperature. Very often in the springtime, bottom sediments of lakes and ponds turn over and rise to the top. This anaerobic material uses all the oxygen, and the fish die. By the end of that summer, the fish had been breeding without restocking. I would really be surprised if it takes three years for the fish to come back in the stream with molasses spill.

TREVOR KLETZ, ICI: You referred to the fact that on one or two occasions your tanks have failed at the bottom and in reconstructing them you decided to go for a tank with a weak seam roof. Presumably you followed the design procedures of API 650. Now calculations show that if you follow the design procedure of API 650, or the corresponding British standard, the tank may still fail at the bottom unless you bolt it down. Some companies do bolt their tanks down. On the other hand, in practice whatever theory says, tanks built to API 650 usually fail at the top even though they are not bolted down. We don't bolt down a tank, if calculation showed that the bottom joint would go before the emergency vent, usually a hinged lid. I wonder in fact if you had any views on this and what in fact your practice is.

H. SCHWARTZMAN: I'd like to defer to our experts because luckily they happen to be here; Jack Rogerson, I think, can answer that. We have a division of responsibility; I take care of the environmental side and Jack and Dick Cox handle process safety.

J.E. ROGERSON, Procter & Gamble: The incident that Howard illustrated started what we call our Weak Top Tank Program. This was aimed at having two types of tanks. One, a 50-lb pressure vessel which will retain an explosion, and the second type, a tank with a weak top which will fail in the case of an explosion. The bottom of the tank will remain intact and retain the contents. Tank failure is defined as a rupture of the bottom or lower shell of the tank. In many weak top tanks, overpressurizing will cause the bottom to bulge outward. The shell will rise several inches. It is still possible to use the tank by packing the space between the bottom and the foundation with sand and concrete.

API 650 indicates that the top angle ring be 2-1/2" by 1/4". A large tank of about 50 ft in diameter built to API standards will always fail at the top. As the tank diameter decreases, a given explosion pressure produced a much lower force on the top of the tank and the angle ring does not buckle as it would on a larger tank. Most of the pictures of tank failures indicate that the top angle ring has buckled. What we did with the existing tanks was to weaken the top angle ring by cutting notches in the horizontal portion of the ring. In new tanks, we made the angle rings much lighter. For a tank under 25 ft in diameter, we use an angle ring measuring 1-1/2" x 1-1/2" x 3/16" thick for a top compression ring. One can still attach the roof with a 3/16" fillet weld on this angle and it results in a relieving pressure that is lower than the one if a heavier compression ring was used. When tanks are between 25 and 50 ft in diameter, we use a 2" x 2" x1/4" angle ring. For tanks over 50 ft in diameter, a 2-l/2" x 1/4" angle is satisfactory. I think API 650 was recently modified to permit the use of a lighter angle ring. In prior years, this problem was not considered by the API.

We found that it was almost impossible to control what you might find in the vapor phase of tankage. In the particular case that Howard cited, we found decomposition products, practically all of which were flammable. As an illustration of this, we found ethane in the vapor phase of one tank on sampling. We used no ethane whatsoever in our processes, but discovered that the ethane was used as a blanketing material for ethylene oxide at a vendor's plant. The vendor made a product from ethylene oxide which was shipped to us. Ethane was found in our storage tanks. This can be very disconcerting, if you are not prepared to handle a flammable vapor.

H. SCHWARTZMAN: Let me add one thing apart from safety. We've retrofitted spill protection into existing areas that previously didn't have dikes. In that case, we consider the spilled liquid floating the tanks off their foundations. In this situation, we've had to go in and bolt down some tanks, and also very often raise the pumps and piping up above what the maximum containment level of a spill would be.

Case Study in Reactor Design for Hazards Prevention

This paper discusses how active and passive safety systems were designed and integrated into the plant, and how the features were simulated and tested to insure that personnel, property and the environment are adequately protected.

J.L. Silverstein, B.H. Wood, and S.A. Leshaw
American Cyanamid Co., Wayne, NJ

The American Cyanamid Company has recently developed and commercialized a new continuous adiabatic nitration process for the manufacture of nitrobenzene (MNB). A simplified flow sketch for the nitration section of the process is given in Figure 1. This process is inherently safe when compared to existing processes because the heat of nitration is not removed by external cooling. Instead, the heat of reaction is removed as sensible heat by raising the temperature of the reaction mixture to a controlled level dictated by the amount of sulfuric acid/water mixture circulating through the system. The acid is required for the nitration to take place and serves the important function of acting as a heat sink. Various control strategies are utilized to insure that the heat sink is always provided when reactants flow into the system. If the sulfuric acid pumps fail, the nitric acid and benzene pumps are automatically shut off. However, sulfuric acid is still present in the reactor to control the temperature. In contrast, if the external cooling system fails in a traditional design, there is great potential for a runaway reaction to occur resulting in vessel rupture.

In routine hazard evaluation of the process, it was found that violent exotherms occur when MNB and 85 wt. % H_2SO_4 are mixed and heated above 200 °C; mild exotherms were found for mixtures of MNB and 69 wt. % H_2SO_4 at 150-170 °C. In the Cyanamid process, quantities of MNB will be in contact with 65 wt. % H_2SO_4 at 135-148 °C. This will occur during normal operation in reactors 3 and 4, which are continuous stirred tank reactors, and in the continuously operated separator, which is virtually unagitated. Mathematical simulation of the nitrators indicated that with certain feed ratios, well away from normal operating conditions, the temperature could reach 180 °C. The discovery of this exotherm presented evidence of a potentially significant hazard which warranted further testing and evaluation. It is the purpose of this paper to summarize:

(1) The analysis of the data which was done to understand the nature of the exotherm and to deal with it successfully to protect the plant.

(2) The approach decided upon for dealing with the exotherm.

(3) The experimental program undertaken.

0149-3701/81/4255 $02.00

Outline of Experimental Program

Cyanamid initiated two programs to study the exotherm. The first involved a series of thermal stability analyses on small samples (~20 mg) using a Perkin Elmer Differential Scanning Calorimeter (DSC).

To conduct the second program, Cyanamid commissioned an independent testing organization, Hazards Research Corp. In this program, larger scale (~250 - 300 ml) samples were evaluated for their thermal stability and their ability to be vented through various sized orifices under various upset conditions. The geometry of these vent tests were then scaled to plant dimensions to determine the vent sizes required for successful pressure relief. Details of both experimental programs are discussed in later sections of this paper.

The results of the studies were used to design the control system interlocks and safety relief devices for the plant.

DSC ANALYSIS

Description of Apparatus

DSC experiments are carried out in a small sealed vessel using about 10 - 20 mg. of reactants. Temperature, but not pressure, is measured. A second empty vessel is used as a blank to compensate for heat losses. The temperature of both vessels are initially programmed to increase at a rate of 5°C/minute. A feedback control system maintains the temperature difference between the vessels at or near zero. If an exothermic reaction occurs in the filled vessel, the heat supplied to this vessel is cut back to maintain a steady 5°C/minute increase. It may also be necessary to add heat to the blank to maintain a zero temperature difference. The difference in heating rate to the two vessels is the rate of heat produced by the reaction. This rate of heat production is plotted vs. temperature as shown in fig. 2.

The measured heat of reaction ranged from -98 to -281 cal/gm. This means that if the decomposition reaction were to occur adiabatically, potential temperature rises of approximately 200-580°C are possible. The accompanying pressure rise would surely rupture the containing vessel. It is noted that in all cases except two, the exotherm initiated at temperatures well above 200°C. In the other cases, the initiation temperatures were 197°C and 177°C respectively. Since the maximum temperature anticipated during normal operation is 135-148°C, no hazard as a result of the $MNB\text{-}H_2SO_4$ exotherm is anticipated unless some upset condition arises which causes a significant increase in reactor exit temperature.

Worst Case Analysis of DSC Data

To obtain an estimate of the speed of the decomposition reaction if it should occur, a technique was utilized for analyzing DSC data based upon the Townsend method developed by D. I. Townsend(1). The approach is based on the assumption that for the major part of the decomposition, the reaction behaves as though it is zero order; it is independent of reactant concentrations and depends exponentially on temperature. It is usually only when the reaction is within minutes of achieving its maximum rate of heat production that the zero order assumption breaks down. However, by this time, accurate prediction of the reaction rate are unnecessary because a runaway situation exists. It is only when the reaction is still controllable that good kinetic predictions are necessary; in this range, the zero order approximation is good. The DSC run having the lowest initiation temperature was analyzed. The initiation temperature is 177°C and the heat of reaction is -158 cal/gm.

This represents the most conservative set of data obtained. A relationship between initial temperature and adiabatic time to maximum rate under adiabatic conditions is given in fig. 3. This relationship states, for example, that if the reaction mixture is initially at 170°C and is maintained adiabatically, it will reach its maximum temperature in about two hours. The pressure will rise during this time as well, due to increased vapor pressure and formation of noncondensables. As the initial temperature rises, the time to maximum rate decreases exponentially.

The DSC was used for two purposes. First, it was used to screen samples with various compositions to determine the most hazardous mixtures. These were then further evaluated in small scale venting tests. The second purpose was to obtain a decomposition rate expression that could be used in mathematical simulations of the process under upset conditions.

HAZARDS RESEARCH TESTING

Summary of Large Scale Tests

To provide guidance in the sizing of rupture discs for the $MNB\text{-}H_2SO_4$ separator and the crude MNB hold tank, Hazards Research Corp. (HRC) was commissioned to run venting tests on the

$MNB-H_2SO_4$ decomposition reaction. HRC conducted venting simulation tests for the $MNB-H_2SO_4$ separator and the MNB hold tank. Several sized vents were used for the separator vent tests, equivalent to 4", 6", 8", 14" and 18" on the plant vessel. It was found that a 4" vent with disc burst pressure 15 psia above normal operating pressure is adequate. In addition, to comply with existing company guidelines, the separator is also equipped with an 18" vent closed off by a disc set to burst at 30 psia above normal operating pressure.

In addition to vent testing of the separator, tests were also conducted on the MNB hold tank. It was postulated that if the level control failed in the separator, acid could be spilled over into the MNB hold tank, and the exotherm could now take place here. This situation was simulated by testing a two-phase mixture consisting of 90 vol. % organic and 10 vol % spent acid. These tests were run with a vessel loading of 60% of full volume and with a vent pressure 70 psi above the operating pressure These tests were conservative on both counts since the vessel loading will be kept below 25%, and the vent pressure will be 30 psi above operating. It was found that a 10" vent with disc burst pressure 30 psia above normal operating pressure is adequate.

Thermal Scans

The first experiments run by HRC were their standard thermal scan experiments. For these tests, the initial reaction mixture was placed in a small thick walled stainless steel reactor (300 cc volume), which was then closed off and heated by means of an oil bath. The temperature and pressure were recorded during the heat-up.

For a non-reacting system, the temperature would rise linearly throughout the run. For an exothermic system, a temperature peak occurs, the magnitude of which reflects the severity of the reaction. HRC would normally size a rupture disc from the pressure rise curve taking the maximum rate of pressure rise and converting it to a rate of vapor generation for which the vent must be sized. Since the 2000 psi disc was ruptured, it is clear from the pressure curve that the $MNB-H_2SO_4$ exotherm can be severe if no attempt is made to control it or to vent the system. The thermal scan for the MNB hold tank (90% organic, 10% acid) indicated that the rate of pressure rise is less severe than for the separator.

Isothermal Stability Tests

In these tests, the oil bath temperature is held constant while the sample temperature and vessel pressure are monitored.

For a series of fixed bath temperatures, HRC determines the time required for decomposition of the sample. This is then plotted in a manner similar to that described earlier for the adiabatic time to maximum rate calculated from the DSC data. It must be emphasized that the HRC system is not adiabatic, since heat will be lost to the oil bath as the reaction proceeds. However, as shown in fig. 3, the plot of the HRC data shows a faster reaction at elevated temperatures than that predicted by analysis of DSC data. It is the HRC data which were used in this temperature range to obtain kinetic rate constants for the simulation to be described in the next section. For lower temperatures, the DSC results were used.

Vent Tests

Vent tests are conducted exactly like the thermal scan runs except that, at a designated pressure, a valve is opened which exposes an open vent orifice to the reaction mixture. The size of the orifice is determined by scaling down from the chosen vent size for the plant vessel. The scaling criterion is that the ratio of the orifice cross-sectional area to the vessel volume is constant.

Several vent sizes opened at various pressures were tested for the separator by HRC. In each case a sucessful vent resulted. The results of the tests used to size the vents for the plant are of the form given in fig. 4.

VAPOR VENTING SIMULATION OF SEPARATOR

Vapor Venting Simulations

Simulations of the HRC vapor venting experiment for the separator with a 4" equivalent diameter orifice were made using the vapor venting model described in the Appendix. The applicability of the vapor venting model (as opposed to a two phase venting model) was tested using preliminary criteria developed by the AIChE Design Institute for Emergency Relief Systems (DIERS). As of this writing, the criteria are still being tested for their general applicability. The rates of decomposition used are those derived from the "worst case" DSC run and from HRC isothermal runs. The equation for flow through the vent pipe is that for isothermal flow of a compressible fluid. In these simulations the plant scale separator was modeled.

As shown in fig. 5, the prediction of the pressure response is very good, while that of the early stages of the temperature response is low. This is due to the fact that in the early stages of an HRC run, the vent is opened and rapidly closed to purge air from the system. However, it is believed that sizeable quantities of benzene and water vapors are also purged, having the effect of lowering the system vapor pressure. It is noted that the shape of the predicted temperature curve quite closely follows the shape of the experimental curve.

A slight pressure peak occurs in both the experimental data and the simulation before the pressure drops to atmospheric. This occurs because at this point, all of the organic phase has boiled off, and the vapor-liquid equilibrium shifts from one for a three-phase system to one between the acid and vapor phases. At the beginning of the simulation, the organic phase consists of MNB and a small amount of benzene. With the opening of the vent, the benzene rapidly boils off within five minutes. Now with continued heating, the rate of boil-off drops and the temperature rises. As long as the MNB phase is present, it will exert its own vapor pressure. With increasing temperature, its vapor pressure rises. The effect of heating on the acid phase is to drive off water at a faster rate than the vapor pressure rises as a result of increasing temperature The result is that the acid phase becomes more concentrated and its vapor pressure drops. The net result on the whole system is that the total vapor pressure increases slightly. When the MNB is finally driven off, the pressure of the system is due only to the vapor pressure of the acid phase, so it drops. This is a successful vent.

Two other cases were run to determine how parametric changes would affect venting through a 4" vent line. In the first case, the heat of reaction used in the base case is doubled to determine how a larger adiabetic temperature rise affects the venting process. In fig. 6, it is shown that a successful vent will also occur.

In the second case, the reaction rate was increased by ten times the value used in the base case. The results of the simulation are shown in fig. 7. It is noted that both the temperature and pressure rise rapidly until all of the MNB is driven off. After this, the pressure drops quickly to 0 psig and the temperature drops off as well before rising again due to continued external heating.

Two other venting simulations were made. In both cases, it is assumed that the system pressure reaches the disc burst pressure due to some control system or operator failure and not because of an external fire. As shown in figs. 8 and 9, the system vents successfully, both with the design volume of organic present in the separator and, also, with twice this volume being present. The rate of decomposition at the temperature of venting is too slow to allow the exotherm to be rapidly self-accelerating. It is only with an assist from an external fire that this occurs.

DRAINING OF VESSELS

The action taken prior to venting of the separator is a drain of the nitrators, separator and crude MNB hold tank. The MNB layer in the separator will drain back through the reactors. The purpose of draining is to evacuate the vessels before significantly higher temperatures are reached. The fluids are collected by being dumped into a drain tank containing cold concentrated sulfuric acid. A derivation of the mathematical expressions used to calculate drain times is given in the Appendix. These expressions were used to determine if the drain lines were adequately sized to rapidly drain the vessels.

It is noted that the drain time of the separator MNB layer through the reactors is very much faster than the drain time of the separator itself. The time to drain the MNB layer plus the entire reactor contents is only 1.45 minutes. This is sufficiently fast to prevent the decomposition reaction from occurring in the nitrators during draining. Indeed, the drain times for each unit of the reaction section of the plant are sufficiently fast to avoid a decomposition reaction.

Finally, instantaneous flow rates on the order of 3000 gpm are possible upon draining. The proper design of pipe supports to allow the piping to withstand the accompanying forces in on important consideration.

The drain system was tested during start-up. It was found that the predicted drain times agreed well with the observed value for atmospheric drains. The pipe supports were found to be adequate.

QUENCHING OF REACTION IN DRAIN TANK

When the vessels are drained into the drain tank, the hot fluid will contact a cold heel of concentrated sulfuric acid. The resultant temperature will be considerably below that of the draining fluid, and will be sufficiently low to preclude the exotherm from occurring in the drain tank. A graph of the relationship between hot fluid draining temperature, total volume of fluid drained, volume of heel and final cooled temperature is given in fig. 10.

A drain tank half-filled with cold acid is being provided. Consequently, during a manual drain when the separator temperature reaches 20C above normal, the fluid will be cooled to about 90C. As a result of an automatic drain at 30C above normal, the resultant temperature will be about 95C.

ATMOSPHERIC DISPERSION OF RUPTURE DISC DISCHARGES

Although experimental results and computer simulations demonstrate that the system can be vented safely, the instantaneous flow rates accompanying these vents are large. Until the pressure falls to atmospheric, large quantities of MNB and benzene are vented to the atmosphere. In an attempt to assure that the vapor discharged from a ruptured disc on the MNB phase separator would present no explosion hazard, atmospheric dispersion calculations were made on the vented material.

The calculations are based on equations presented in the "Workbook of Atmospheric Dispersion Estimates"(2). From discharge rates and vapor compositions determined by computer simulation, and from assumed atmospheric and weather conditions, the concentration of the vented material at any point surrounding the vent can be calculated.

Three typical discharge rates and vapor compositions were examined. These correspond to the "beginning, middle and end" of a vent. It was determined that the "end" case represents a worst case for any given wind speed. A typical envelope for gas concentrations above the lower explosive limit is shown in Figure 11.

From the attached figures and the layout of surrounding buildings, it was determined that no sources of ignition exist within the flammable region shown.

SUMMARY OF SAFETY STRATEGY

The only foreseeable process upset which would cause a high temperature excursion in the nitrators is a deviation in the feed ratios, specifically, a high ratio of reactants (nitric acid and benzene) to sulfuric acid. Several basic control features and interlocks were installed to virtually eliminate this possibility. The sulfuric acid flow control station is a fixed orifice and trim control valve in parallel, with most of the flow through the orifice. As a result, flow of the process heat sink is not stopped upon complete failure of the flow controller. In addition, low sulfuric acid flow as detected by redundant flow sensors causes automatic closing of the control valve and block valve in both the nitric acid and benzene feed lines. All critical interlocks are on an uninterruptable power supply and all critical air actuated valves have an independent air reservoir.

The following additional control features were provided as a result of our study of the exotherm problem to further protect the plant in the unlikely event of high nitrator temperature, due possibly to external fire or multiple simultaneous control failures:

1. First Stage Temperature Deviation - High temperature alarm sounds, warning operator that some corrective action is required.
2. Second Stage Temperature Deviation - Standby Shutdown occurs during which the nitric acid and benzene feed pumps are automatically turned off. The sulfuric acid heat sink will continue to circulate and the system will be cooled with heat being removed by the evaporator.
3. Third Stage Temperature Deviation - Alarm sounds. Operator is advised to actuate a system drain.
4. Fourth Stage Temperature Deviation - Automatic system drain occurs.
5. Fifth Stage Temperature Deviation - Rupture Discs relieve pressure build-up in system.

CONCLUSIONS

The control system as conceived for protecting the MNB plant from the potential effects of the occurrence of the MNB-H_2SO_4 exotherm is adequate, and should prevent any hazard without the need for venting through the rupture discs provided. Because of the automatic controls designed for the plant, the probability of the rupture discs bursting due to increased pressure caused by the MNB-H_2SO_4 exotherm is extremely low. However, experimental tests and computer simulations show that if the discs are forced to burst by the addition of external heat, the integrity of the plant will be maintained.

Literature Cited

References

1. Townsend, D. I., AIChE Meeting, Houston, Texas. March 20 - 24, 1977, Paper 44e.

2. Turner, D. B., "Workbook of Atmospheric Dispersion Estimates." U.S. Department of HEW. Publication No. PB-191-482.

APPENDIX

VAPOR VENTING SIMULATION MODEL

The Venting of the MNB separator is treated as a semi-batch reactor with material leaving the vessel through the emergency vent line, but with no fluid entering A schematic drawing is given in Figure 12.

The important parameters for this system are:

m_I = Total instantaneous moles of organic phase.
m_{II} = Total instantaneous moles of aqueous phase.
m = $m_I + m_{II}$
X_B = Mole fraction of benzene in organic phase
X_M = Mole fraction of MNB in organic phase.
Z_A = Wt. fraction of H_2SO_4 in aqueous phase
Z_W = Mole fraction of H_2O in aqueous phase.
Y_M = Mole fraction of MNB in vapor phase.
Y_B = Mole fraction of benzene in vapor phase.
Y_W = Mole fraction of water in vapor phase
Q_g = Heat input to system either by exothermic reaction or external heat supply.
T = Temperature
P = Pressure
V = Molar flow rate of vapor leaving separator
H_v = Enthalpy per mole of vapor leaving
$P^o_B(T)$= Vapor pressure of benzene
$P^o_M(T)$= Vapor pressure of MNB
$P^o Acid\ (T,Z_A)$ = Vapor pressure of H_2SO_4 solution
C_B, C_A, C_M = Specific heat

The mass balances are:

Total Mass: $$\frac{dm}{dt} = -V$$

Organic Phase: $$\frac{dm_I}{dt} = -(Y_M + Y_B)\,V$$

Aqueous Phase: $$\frac{dm_{II}}{dt} = -Y_W V$$

Benzene balance: $$\frac{d(X_B m_I)}{dt} = -Y_B V$$

Water balance: $$\frac{d(Z_W m_{II})}{dt} = -Y_W V$$

Heat balance is:

$$\frac{d\sum_i(m_i X_i C_i)}{dt} + H_v V = Q_g$$

where:
$$Q_G = (-\Delta H)\alpha\, m e^{-E/RT}(1-x)^n$$

Additional relations which are used are:

Vapor liquid equilibria:

$$Y_B = \frac{X_B P^o_B(T)}{P}$$

$$Y_M = \frac{X_M P^o_M(T)}{P}$$

$$Y_W = \frac{P^o_{acid}(T, Z_A)}{P}$$

Closure: $Y_B + Y_M + Y_W = 1$

$X_B + X_M = 1$

The flow rate of vented vapor is calculated by using the equation ror the isothermal flow of a compressible fluid. The energy balance is:

$$\int_{u_1}^{u_2}\frac{u\,du}{g_c\alpha_c} + \int_{z_1}^{z_2}\frac{g\,dz}{g_c} + \int_{P_1}^{P_2} V\,dP + \int_{L_1}^{L_2}\frac{f\,\bar{u}^2}{2g_c D}\,dL = 0$$

Where: L = equivalent vent length
D = vent diameter
f = friction factor

$\underline{u}$ = average velocity

$\alpha_c = 1$

It is recognized that:

$$G = \rho \underline{u}$$
$$\nu = \frac{1}{\rho}$$
$$\underline{u} = \nu G$$
$$d\underline{u} = G d\nu$$

Substituting this into the energy balance gives:

$$\frac{G^2 \nu d\nu}{g_c \alpha_c} + \frac{g}{g_c} dz + \frac{4fG^2\nu^2}{2 g_c D} dL + VdP = 0$$

If we assume that the Reynolds Number is high. then f = constant. Let us now divide both sides of this equation by ν^2:

$$\frac{G^2 d\nu}{g_c \alpha_c \nu} + \frac{g\, dz}{g_c \nu^2} + \frac{dP}{\nu} + \frac{2fG^2}{g_c D} dL = 0$$

Integrating this equation from the vent entrance to its outlet gives:

$$G = \left[\frac{M}{RT}(P_1^2 - P_2^2) \Big/ \left(\frac{2f\Sigma L}{g_c D} + \frac{\ln\left(\frac{P_1}{P_2}\right)}{g_c \alpha_c} \right) \right]^{\frac{1}{2}}$$

Where it was assumed that:

$$\int_{Z_1}^{Z_2} \frac{g}{g_c} \frac{dz}{\nu^2}$$

is negligible compared to other terms.

DRAINING OF VESSELS

General expressions for the liquid drainage time of a vessel are derived in this section. A schematic of a typical vessel is given in figure 13.

The energy balance for liquid flow through the system is:

$$(1) \quad W + \frac{g}{g_c}(Z_2 - Z_1) + \frac{V_2^2 - V_1^2}{2g_c} + \frac{P_2 - P_1}{\rho} + \Sigma \frac{V_2^2 f \Sigma L}{2 g_c R_H} = 0$$

Where: W = shaft work

Z = vertical height above a reference level

$\underline{v}$ = average velocity

P = pressure

ρ = density

L = vertical height of drain pipe

R_H = R/2, hydraulic radius

f = friction factor

ΣL = equivalent length of straight pipe including all fittings

The assumptions which are usually made are:

1. Both points 1 and 2 are open to the atmosphere. so $P_1 = P_2$.

2. v_1 is negligible with respect to v_2.

3. There is no shaft work.

The following assumption facilitates later calculations if the friction factor. f, is a function of Reynolds Number. However, if the flow is highly turbulent and f = constant. then the assumption is unnecessary.

4. The contribution of kinetic energy at point 2 to the overall energy balance is small compared to that of potential energy and frictional losses. Thus, the total effect of kinetic energy is negligible.

If assumption 4 is made. then the energy balance becomes

$$(2) \quad \frac{g}{g_c}(Z_2 - Z_1) + \frac{V^2 f \Sigma L}{2 g_c R_H} = 0$$

The difference in height:

$Z_2 - Z_1 = (H+L)$

so:

(3) $$V^2 = \frac{rg(H+L)}{f \Sigma L}$$

For turbulent flow where $4000 > Re > 100{,}000$ and a hydraulically smooth pipe:

(4) $$f = \frac{0.0791}{Re^{0.25}}$$

and:

(5) $$V = \left[\frac{2 r^5 g^4 (H+L)^4 \rho}{(0.0791)^4 (\Sigma L)^4 \mu}\right]^{\frac{1}{7}}$$

The continuity equation for flow from the tank to the pipe is:

(6) $$\frac{dH}{dt} = -V \frac{r^2}{R^2}$$

When equation (5) is substituted into equation (6), it can be integrated to give:

(7) $$t_{efflux} = \frac{7}{3}\left(\frac{R}{r}\right)^2 \left[(H_1+L)^{\frac{3}{7}} - (H_2+L)^{\frac{3}{7}}\right]\left[\frac{(.0791)^4 (\Sigma L)^4 \mu}{2 r^5 g^4 \rho}\right]^{\frac{1}{7}}$$

If assumption (4) is not made, and f = constant, then

(8) $$V = \left[\frac{rg(H+L)}{f \Sigma L}\right]^{\frac{1}{2}}$$

and

(9) $$t_{efflux} = 2\left(\frac{R}{r}\right)^2 \left[(H_1+L)^{\frac{1}{2}} - (H_2+L)^{\frac{1}{2}}\right]\left(\frac{f \Sigma L}{rg}\right)^{\frac{1}{2}}$$

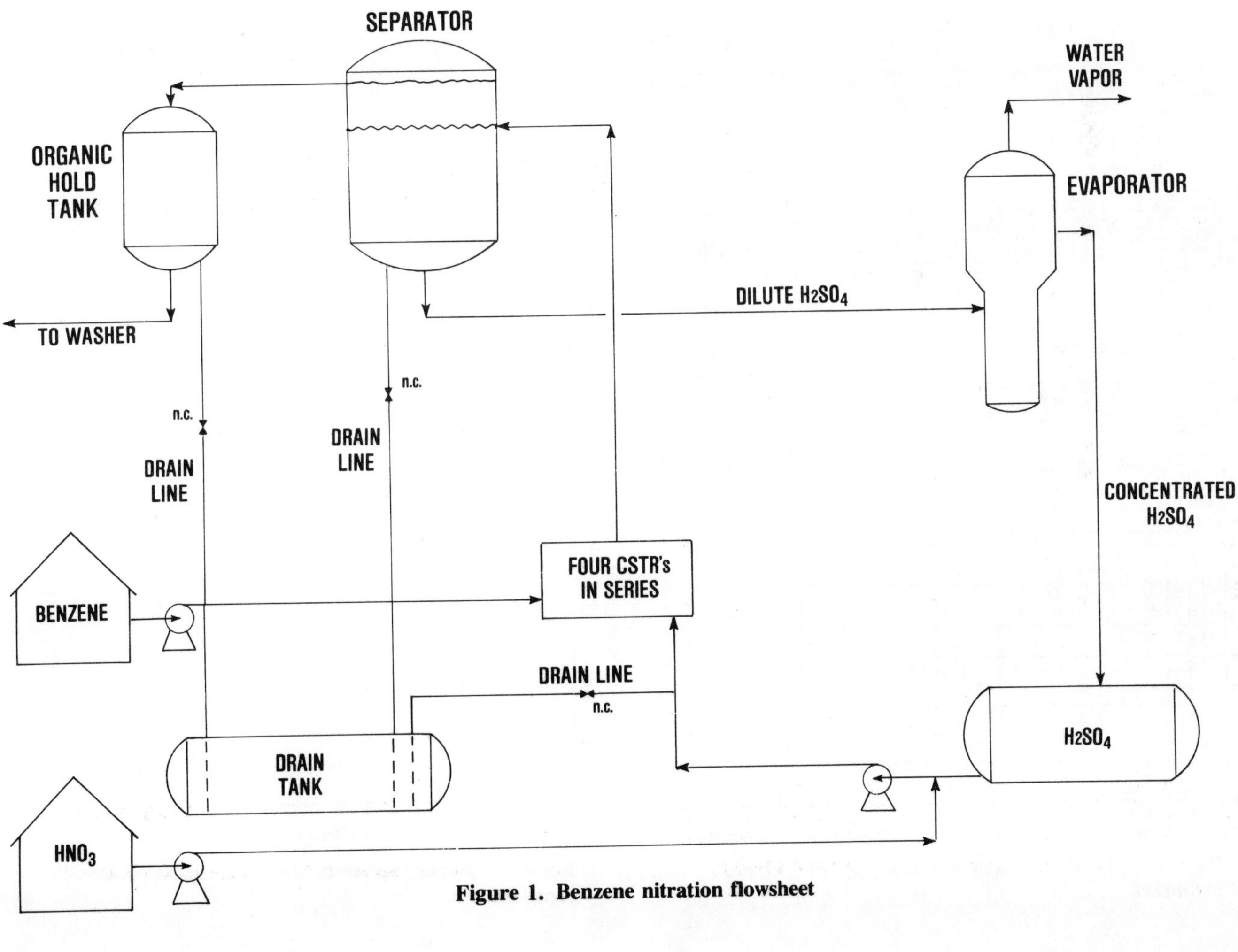

Figure 1. Benzene nitration flowsheet

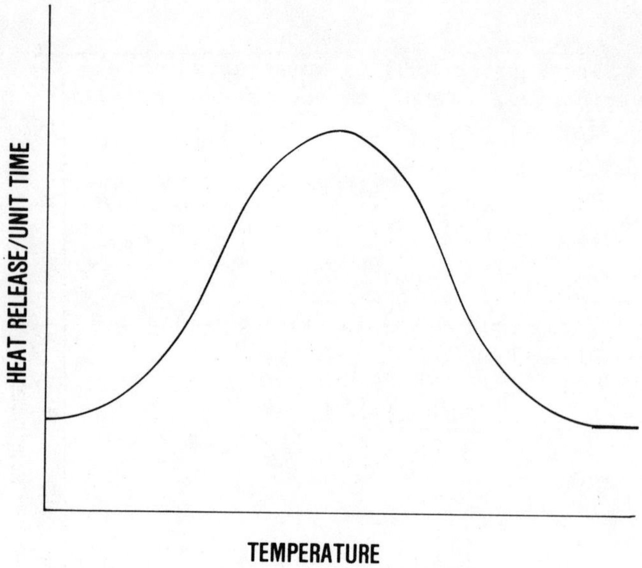

Figure 2. Typical DSC curve.

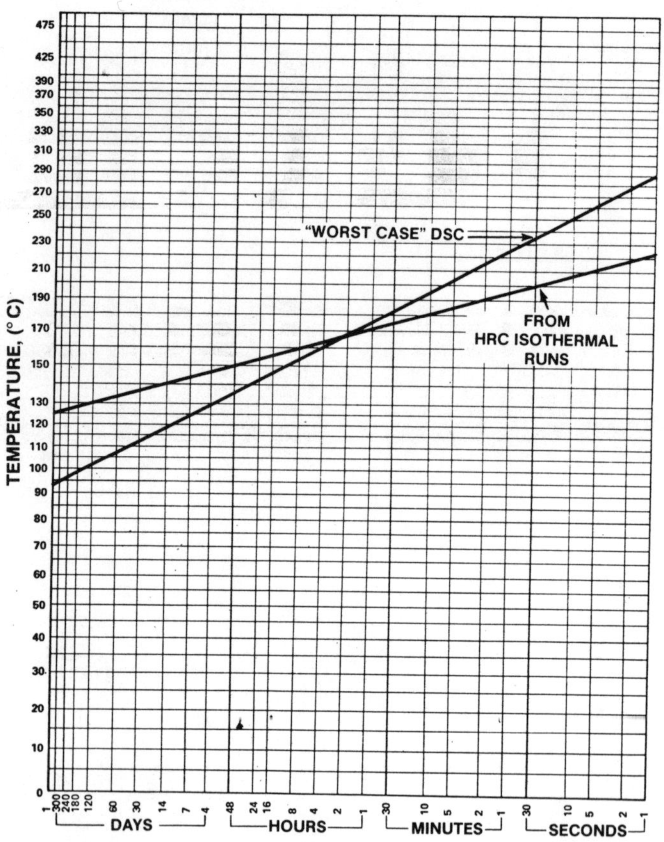

Figure 3. Time to maximum rate for MNB-H_2SO_4 exotherm.

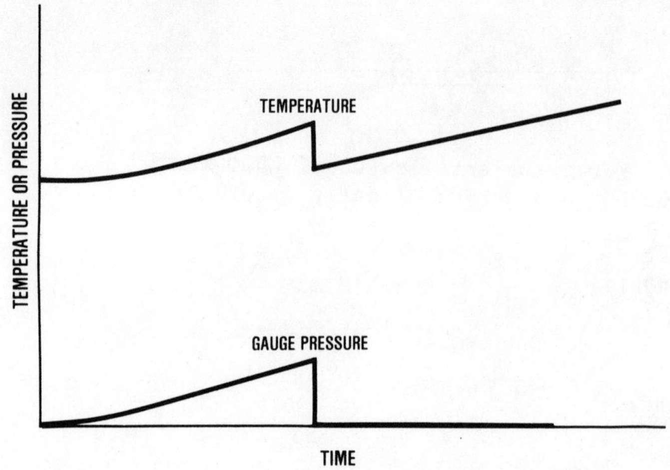

Figure 4. Typical results of successful HRC venting test.

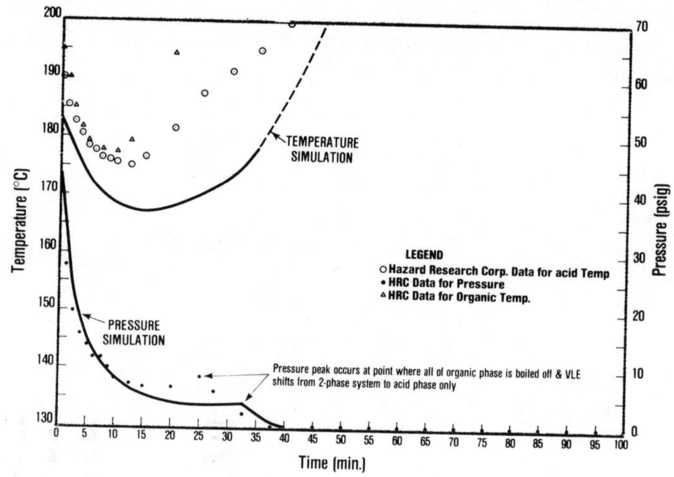

Figure 5. Vapor venting of MNB separator through 4" vent line (Hazards Rezsearch Corporation experiment).

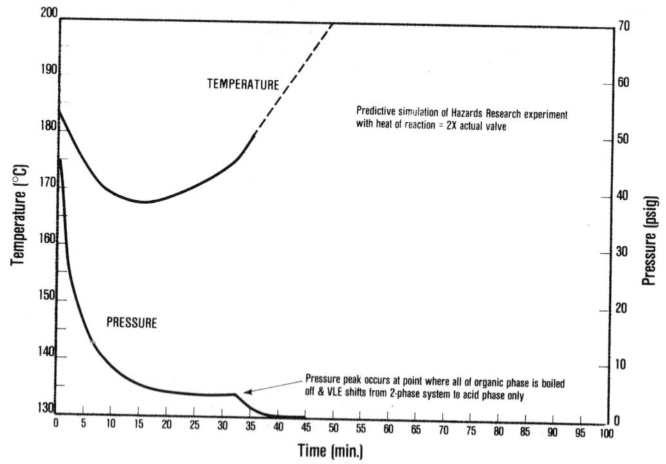

Figure 6. Vapor venting of MNB separator through 4" vent line.

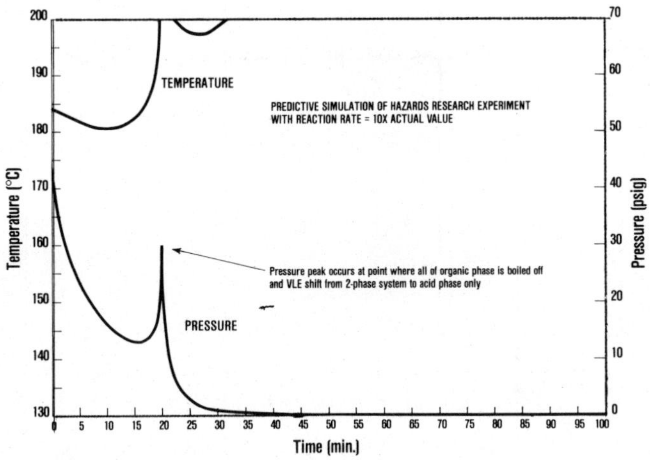

Figure 7. Vapor venting of MNB separator through 4" vent line.

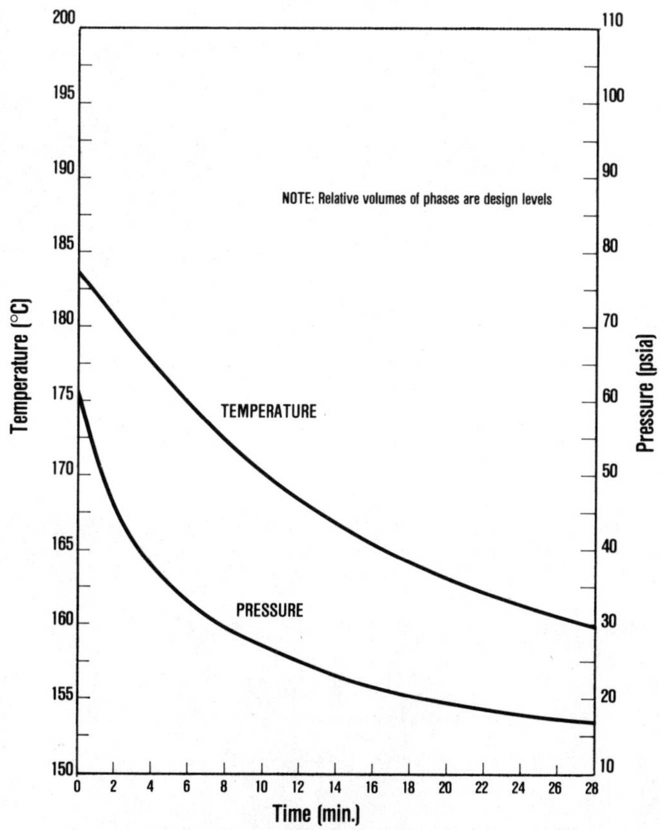

Figure 8. Vapor venting simulation of MNB separator through 4" vent line.

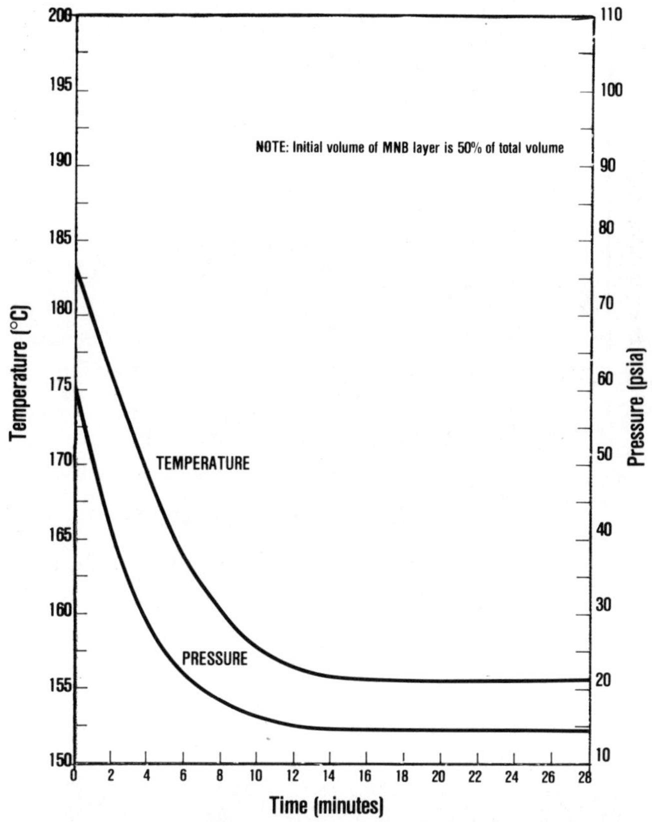

Figure 9. Vapor venting simulation MNB separator through 4" vent line.

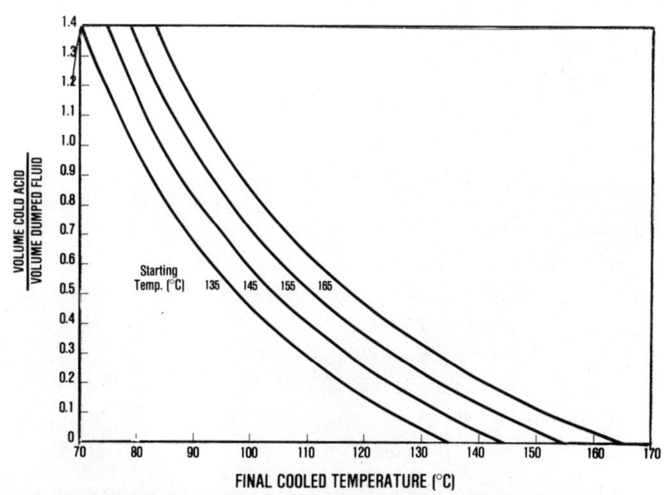

Figure 10. Volume of cooling acid in dump tank required for quenching process fluids.

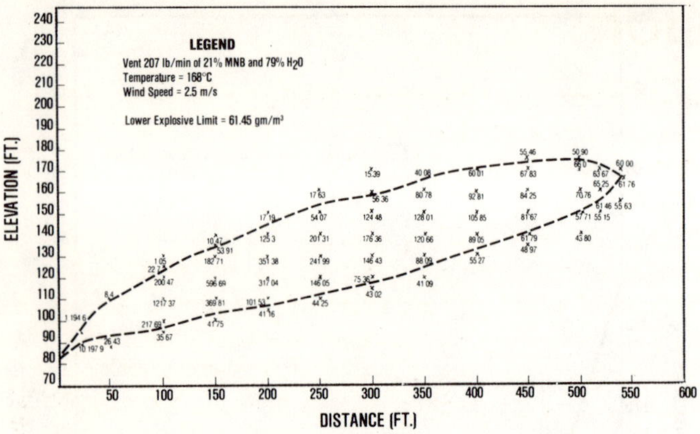

Figure 11. MNB separator vent atmospheric dispersion.

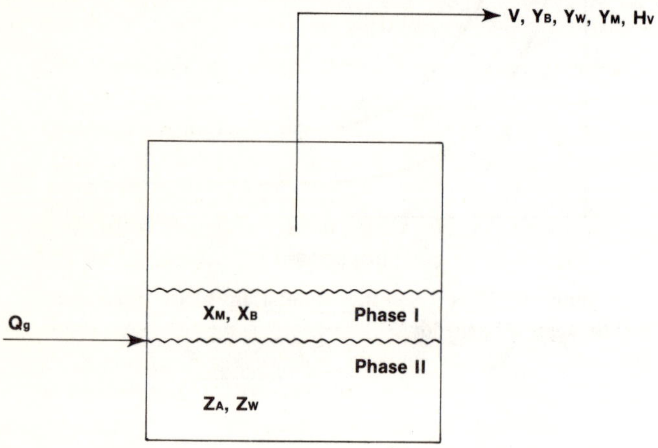

Figure 12. Schematic of phase separator.

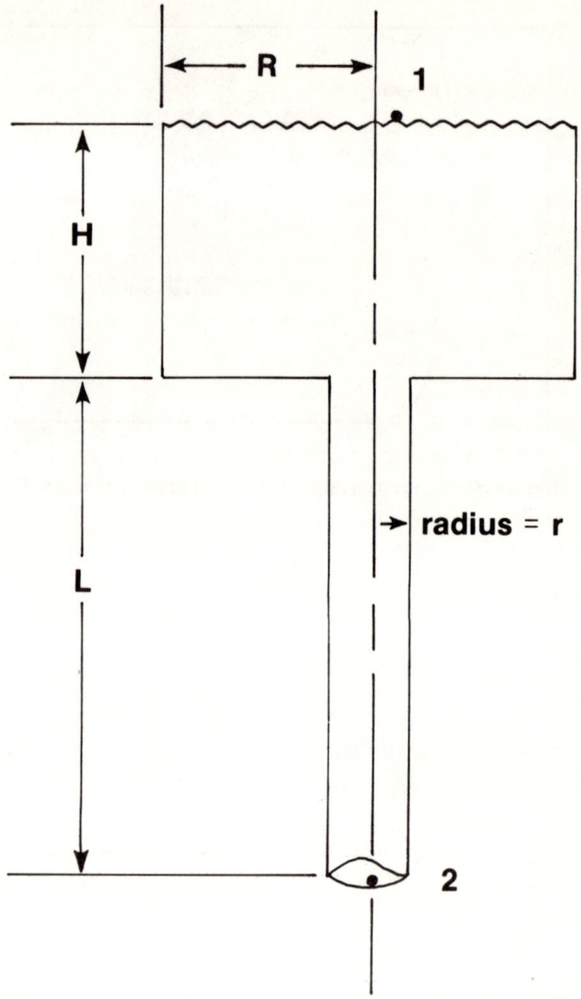

Figure 13. Schematic of draining vessel.

SILVERSTEIN, J.L.

WOOD, B.H.

LESHAW, S.A.

DISCUSSION

BOB BARKER, First Chemical: Was the hazards evaluation done solely by you or by the supplier or plant, or in conjunction with them?

SILVERSTEIN: The plant was developed by Cyanamid and we have a contractural relationship with Chemetics of Canada to actually contruct it for us. They got involved in some development work of their own. However, they were not involved in the hazads evaluation or with the design of safety procedures. Cyanamid consulted with them on a regular basis to keep them informed of our progress.

BARKER: Were these hazards pointed out by Chemetics or did you all work this out together? I understand they did their own hazards evaluation on the process.

SILVERSTEIN: ICI, the parent company of Chemetics, had done some routine hazards work. We got the original report from them that high concentrations of sulfuric acid in contact with MNB could cause a problem. We then took it from there and initiated our own program.

ANNONYMOUS: In the vapor explosions, you didn't show benzine having any effect.

SILVERSTEIN: The attached figure is just an example of what we did. We simulated various clouds that would occur at instantaneous points in time during a vent. At any point in time the concentration of the gases leaving the vent line is unique. We developed vapor clouds for the whole range of instantaneous gas concentrations leaving the vent. Since benzene is present in small amounts initially, its concentration decreases rapidly with time. MNB also eventually will be depleted and all that will remain in the separator will be sulfuric acid. Once the MNB is completely vented, the reaction ceases.

SILVERSTEIN: Yes. As indicated in the paper, we applied the tentative criteria develped by the Design Institute for Emergency Relief Systems of which Cyanamid is a member. These indicated that the system should vent only as a vapor. Consequently, our venting simulations are for isothermal gas flow.

RICHARD SCHWAB, Allied Chemical Co.: In the design of your vents, did that take into account the two-phase liquid and gas flow through the vent and the pressure drop over certain length to crank that information into the design equation and come up with that particular design?

One would expect in the composition of that nature to get a good deal of liquid coming out of that vent when that is happening.

SILVERSTEIN: That's correct.

SCHWAB: It would also lead to the conclusion that the vent size would have to be increased drastically.

HAL FORREST, FMC Corp.: I don't think this scales up. Just because small scale experiments show that you get vapor venting, it doesn't mean that if you have a vessel that's 10 ft high you will still get vapor venting.

CHARLIE SOLOMON, Essochem Europe Inc.: Longer ago than I like to remember, I worked at an explosive plant. There was an incident which involved drain lines to a drain tank similar to that described in your paper. When the temperature in the TNT reactor reached an unsafe limit, the operators retired to an emergency station outside the earth berm surrounding the plant and pulled the emergency lever. This caused a hole to be punched in a lead disc in the bottom of the reactor which allowed the contents of the reactor to drain via a closed pipe to a drowning tank.

On one occasion, when the operators pulled the lever, the whole house disappeared. The reason was that when the reactants entered the drain pipe, the reaction continued but the pipe prevented the reaction gases being freely released. The solution to the problem was simple: the closed pipe between the reactor and the drowning tank was replaced by an open channel so that there was no containment of the reactants. I wonder whether you had similar potential problems in your design?

SILVERSTEIN: Yes. We borrowed the conept of a system drain from TNT technology.

TREVOR KLETZ, ICI: You mentioned that the total quantity of material in the reactor and indeed in the whole of your process is very small, smaller than in alternative processes. This, of course, is the best way to minimize any potential incident by simply not having enough material there to give a big bang. In 1979, descriptions appeared (for example, in *The Chemical Engineer*, Feb., 1979, p.79) of the Swedish Bofurs Nobel-Chematur continuous process for benzene nitration in which the total iventory in the system was so small that even if it all went up you couldn't get a big incident. I wonder if you are familiar with tis system and how the inventory in your plant compares with that of the Swedish process, for a similar output.

SILVERSTEIN: Unfortunately, I am not aware of that work.

W.H. DOYLE, Consultant: I have a comment. I consulted a long time ago with a firm which was making a hazardous chemical called paracetic acid and I thought I did an excellent job. With their engineers we sat down and designed a good safe system and we were all through patting ourselves on the back when the chief engineer said: "I want this process instrumented to hell and gone so when it blows, I know what happened." Thank you.

Preventing Flammable Atmosphere In Electrical Equipment Rooms

Flammable gases or vapors can travel beyond the established hazardous boundaries. Unless preventive measures are taken, open arc-producing devices in standard electrical equipment rooms could ignite these flammable materials.

J. T. Woods, Du Pont Co., Orange, Tex. 77630

Large areas of most petrochemical plants are electrically classified due to the processing of flammable materials. Some plants use appropriately classified, expensive equipment in these areas to eliminate poteintial electrical ignition sources. National and industrial standards, however, permit the use of less expensive arc-producing equipment, provided approved measures are taken to isolate the equipment from the flammable atmosphere.

Paragraph 500-1 in the National Electrical Code (*1*) includes the following statement:

"Through the exercise of ingenuity in the layout of electrical installations for hazardous locations, it is frequently possible to locate much of the equipment in less hazardous or in non-hazardous locations and thus reduce the amount of special equipment required. In some cases, hazards may be reduced or hazardous locations limited or eliminated by adequate positive pressure ventilation from a source of clean air in conjunction with effective safeguards against ventilation failure."

This statement relates primarily to electrical equipment rooms located inside the hazardous area. However, a study of gas dispersion characteristics reveals the need to apply similar principles to standard electrical equipment rooms, located outside but near high-potential hazardous areas. In either case, the challenge is to obtain a source of clean air for room pressurization or ventilation when the surrounding area may be engulfed in gas or vapor.

Du Pont's Sabine Rivers Works in Orange, Texas, manufactures nylon intermediates, ethylene and polyethylene products. Large quantities of flammable materials are transported, stored, and processed throughout the plant. The most important priority in the prevention of fires and explosions is ensuring adequate design, installation, and maintenance of the process facilities that contain the flammable materials. Following the serious 1974 incident in Flixborough, England, involving cyclohexane on a plant that also produced nylon intermediates, we began a review of their electrical standards dealing with all hazardous areas. Specifically, the review was to determine if existing standards adequately covered area classification, potential electrical ignition sources within hazardous areas, and potential electrical ignition sources outside but near high-potential hazardous areas. The review led to the develop-

0149-3701/81/4262 $02.00 © 1981 AIChE

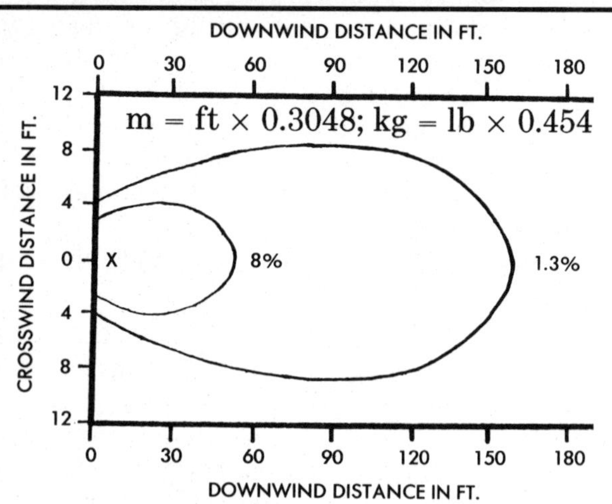

Figure 1.
100 lb. · mol/h release rate: 2 mph wind.

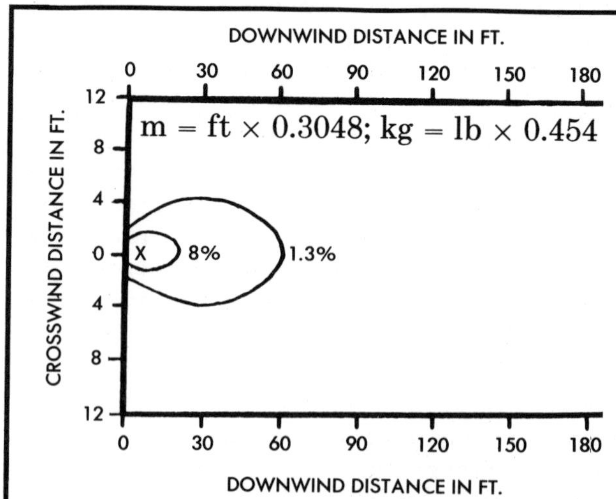

Figure 2.
100 lb · mol/h release rate: 10 mph wind.

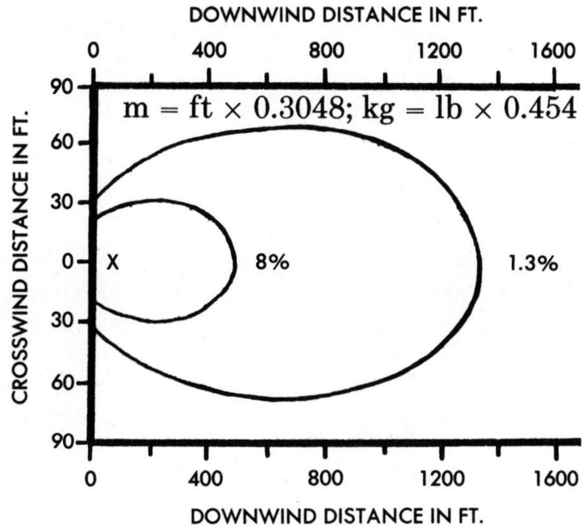

Figure 3.
5,000 lb · mol/h release rate:
2 mph wind.

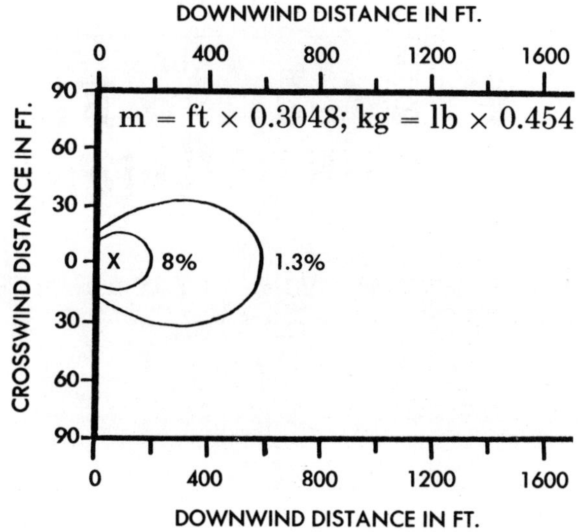

Figure 4.
5,000 lb · mol/h release rate:
10 mph wind.

ment of a new plant standard dealing with all these subjects. Key parts of the new standard are covered here.

Gas dispersion

A study of large gas releases may be made using the Gaussian dispersion equation with Pasquill–Gifford Type-E stability (2). The Gaussian equation for a release at ground level is:

$$C_{(X,Y,Z)} = \frac{Q_E(10^6)}{\pi\sigma_Y\sigma_Z U} \exp\left\{-1/2\left[\left(\frac{Y}{\sigma_Y}\right)^2 + \left(\frac{Z}{\sigma_Z}\right)^2\right]\right\}$$

$C_{(X,Y,Z)}$ = concentration in ppm at point X, Y, Z: X, distance downwind; Y, distance crosswind; and Z, vertical distance.

Q_E = release rate, m^3/s (scfs).

U = wind speed, m/s (ft/s).

σ_Y, σ_Z = standard deviations of concentrations in the Y and Z directions.

$$\sigma_Y = aX^m$$

$$\sigma_Z = bX^n$$

For Pasquill–Gifford Type-E stability, the values for a, b, m, and n are as follows:

X, m(ft)	a	b	m	n
0–300 (0–980)	0.10	0.09	0.92	0.84
300–4,000 (981–13,120)	0.10	0.62	0.91	0.59
4,000 (13,121) and greater	0.10	3.22	0.92	0.42

Type-E, consequently, was chosen for heavy gas dispersion because the σ_Y and σ_Z values simulate the flat, pancake dispersion pattern expected for heavier-than-air gases. Therefore, Type-E stability may be used for all actual stabilities for gases with MW ≥ 35, but only for stable

Figure 7.
Electrical equipment room: inside hazardous area.
m = ft × 0.3048

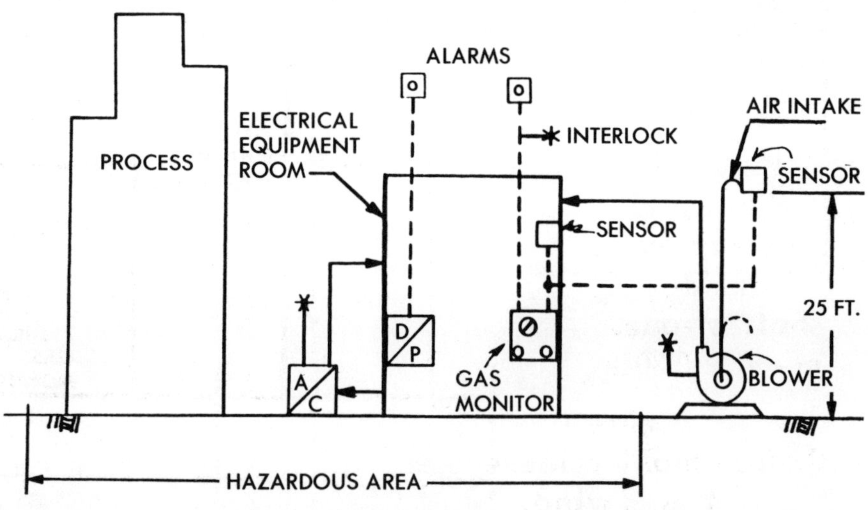

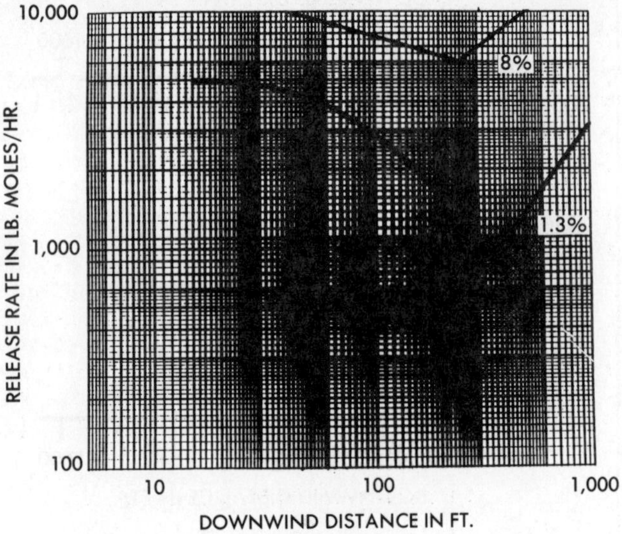

Figure 5.
Elevation 20 ft:
2 mph wind.
m = ft × 0.3048; kg = lb × 0.454

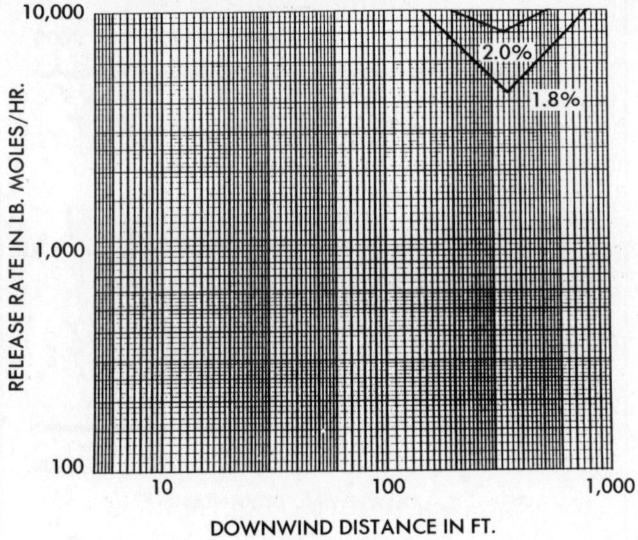

Figure 6.
Elevation 20 ft:
10 mph wind.
m = ft × 0.3048; kg = lb × 0.454

conditions for gases with MW 27–34.

Figures 1 to 6 show data developed with the Gaussian equation. All of the charts reflect ground-level releases of cyclohexane, one of the heavier and more common flammable materials. Ground-level can be considered anywhere within 3.7 m (12 ft) of the ground. The charts apply to flat, level ground. Higher concentrations might be expected around tall buildings or in ditches due to channeling. Only continuous, steady-state releases were considered during the study. While the concentrations represent steady-state, ten-minute values, it is believed they also represent peak values of approximately one minute for Type-E stability.

Figures 1 and 2 show ground-level concentrations at wind speeds of 3.2 and 16.1 km/h (2 and 10 mph) respectively for a release rate of 45 kg (100 lb) · mol/h. The 3.2-km/h (2-mph) speed was chosen to show potential concentrations on relatively calm days. The 16.1-km/h (10-mph) speed was selected since that is the approximate average wind speed at Sabine River Works. A release rate of 45 kg (100 lb) · mol/h represents a relatively small leak. However, it should approximate the majority of leaks experienced on a petrochemical plant.

Figure 1 shows a flammable concentration extending downwind from 15 to 49 m (50 to 160 ft) of the source. A crosswind distance of 2.4 m (8 ft) is shown. Figure 2 shows flammable concentrations extending downwind only 6.1–21.3 m (20–70 ft) from the source as the wind increases to 16.1 km/h (10 mph). A crosswind distance of 1.2 m (4 ft) is recorded.

Figures 3 and 4 show ground-level concentrations for a release rate of 2,270 kg (5,000 lb) · mol/h. A release of this magnitude may represent a line or vessel rupture. For a wind speed of 3.2 km/h (2 mph), Figure 3 shows flammable limits between 152–427m (500–1400 ft) downwind and 21 m (70 ft) crosswind. At a 16.1-km/h (10-mph) wind speed, Figure 4 reveals reduced dimensions of 61–183 (200–600 ft)

Figure 8.
Electrical equipment room: inside buffer zone.
m = ft × 0.3048

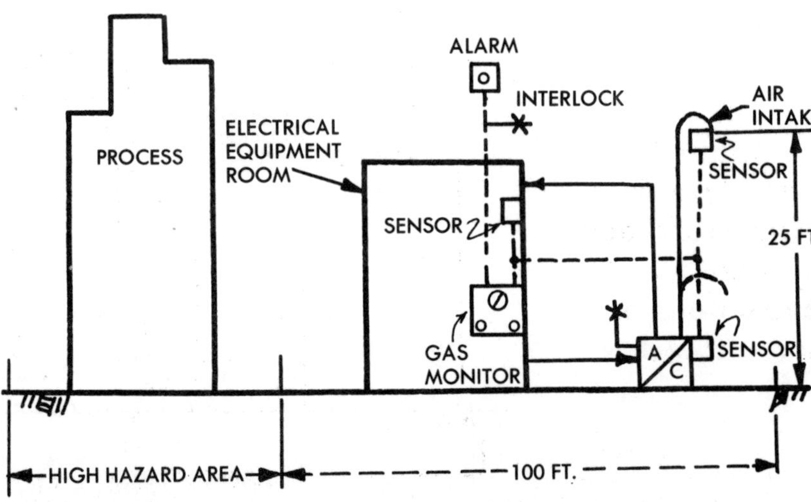

downwind and 9.1 m (30 ft) crosswind.

Flammable concentrations of heavy gases and vapors may cover great distances at ground level and at low wind speeds. This occurs because dispersion is slower with less turbulence. Lighter-than-air gases, such as hydrogen, disperse in much shorter distances.

Figure 5 and 6 show concentrations downwind at an elevation of 6.1 m (20 ft) above grade. Note that flammable limits are never reached at this elevation at the lower release rates. In fact, in a 3.2-km/h (2-mph) wind, a release of cyclohexane above the 363 kg (800 lb) mol/h rate is required to reach the lower flammable limit of 1.3%. That occurs at only one point 107 m (350 ft) downwind. In a 16.1 km/h (10 mph) wind, releases of cyclohexane lower than 2,043 kg (4,500 lb) mol/h never produce flammable concentrations at the 6.1 m (20 ft) elevation.

Standards

The only added precaution that consensus standards take around facilities, capable of releasing large quantities of flammable gases or vapors, is to extend the classified area boundary to 30 m (100 ft) or more. It may seem, after reviewing Figures 3 and 4, that these boundaries should be extended several hundred feet. To do so, however, would result in the electrical classification of smaller plant sites in their entirety. It must be recognized that catastrophic releases of flammable liquids are exceptional incidents. Small releases, shown in Figures 1 and 2, may be expected more frequently and are usually dealt with in establishing the normal classified area boundary.

The new Sabine River Works standard provides additional protection by requiring a 30-m (100-ft) buffer zone around the high-risk hazardous area. The buffer zone is not a part of the classified area. However, special considerations, listed below, are given to electrical installations inside the zone.

The 30-m (100-ft) buffer dimension is not large enough to cover the full extent of the flammable range for large releases, Figures 3 and 4. However, the extended classified boundary plus the 30-m (100-ft) buffer zone will cover most of the area that could be subjected to the richest concentrations of gas or vapor. Special treatment of electrical installations inside the buffer zone also reduces the probability of gas or vapor cloud ignition without classifying extremely large areas.

The characteristic of heavy vapors remaining close to the ground in relatively calm or windy weather is shown in Figures 5 and 6. This characteristic allows system designers to provide clean air for pressurizing or ventilating electrical equipment rooms by locating the air intake ports at elevations above 6.1 m (20 ft). Around areas that contain large quantities of lighter-than-air gases, the air intake ports should be near ground level.

Construction and ventilation

Electrical equipment rooms, located inside a classified area, should be of solid wall vapor tight construction. Precast concrete is preferred. It is well known that exploding gas makes projectiles out of brick or cinder blocks. All wall, roof and floor openings in conduit, cable tray, ducts and pipe should be sealed to prevent air passage from the outside. Temporary holes in the walls should be carefully sealed at the end of each work day during construction periods. Door closers should be capable of closing doors unassisted against room pressure.

The ventilation system should maintain a minimum room pressure of 24.9 Pa (0.1 in. of water), relative to outside, will all room openings closed. Additionally, with all room openings open, the minimum velocity of air leaving the room should be 18 m/min (60fpm). Air should be ducted from outside the hazardous area. The duct intake port should be installed a minimum of 7.6 m (25 ft) above grade for areas that contain heavier-than-air flammables. The intake port should be located at ground level in areas that contain lighter-than-air flammables. Engineering judgment must be used in locating the intake port in areas that contain both heavier-than-air and lighter-than-air flammable vapors and gases, Figure 7.

Each electrical equipment room inside a hazardous area should be equipped with a continuous combustible gas monitoring system. One sensor should be located at the fresh air intake port, backed up by one sensor inside the pressurized room. Upon detection of 20% of the lower explosive limit, the monitor system should initiate an alarm in the central control room and an interlock to shut down the ventilation system. Personnel should refrain from opening the doors. When the room is depressured, the seal should not allow the gas or vapor concentration inside to exceed the lower explosive limit.

A differential pressure device should be installed to detect room positive pressure below 24.9 Pa (0.1 in. of water). The alarm, it actuates, may be time-delayed to a maximum of one minute to prevent nuisance alarms for normal room traffic.

An electrical equipment room, located outside of the classified area, but inside the 30–m (100-ft) buffer zone, should be constructed and sealed as if it were inside the classified area.

Room pressurization is not required. However, the ventilation system's fresh air-intake port should be at least 7.6 m (25 ft) above grade near areas with heavier-than-air flammables and at ground-level near areas with lighter-than-air flammables, Figure 8.

A continuous combustile gas monitoring system should be installed in the equipment room. One sensor should be installed at the air conditioner, one at the fresh air intake port, and one inside the room. The alarm and interlock should function with same as in the equipment room inside the hazardous area. With the building sealed, the possibility of a gas or vapor concentration above the lower explosive limit inside the building is remote.

No special treatment is recommended for electrical equipment rooms located outside a low-hazard area or more than 30 m (100 ft) from a high-hazard area.

Summary

To minimize the possibility of ignition of a large flammable gas release on a petrochemical plant, the greatest effort should go into the design, installation, and maintenance of the process facilities that contain them. The margin of safety can be increased, however, by preventing the accumulation of flammable concentrations inside standard electrical equipment rooms that are located near high-hazard potential areas. To accomplish this, techniques similar to those recommended for use inside hazardous areas may be applied.

Literature cited

1. *National Electrical Code,* Chapter 5, Article 500, Nat. Fire Protection Assn. No. 70 (1978).
2. Bodurtha, F. T., *Industrial Explosion Prevention and Protection,* McGraw-Hill, New York (1980).

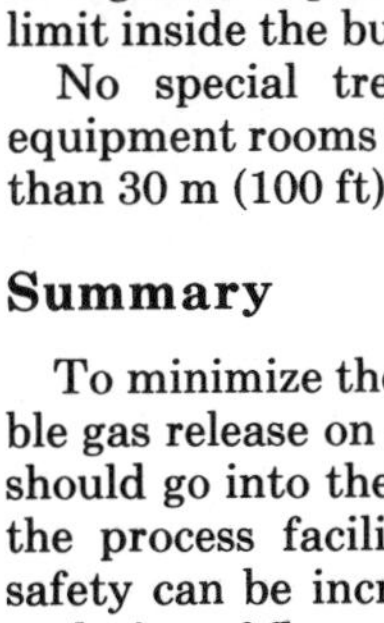

J. T. Woods, who earned his B.S.E.E. degree at Lamar Univ., is a staff design engineer at the Du Pont's Sabine River Works, Orange, Texas. For the past 17 years, he has worked in electrical generation and distribution design, operation, and maintenance at the plant.

DISCUSSION

MIKE FLESSNER, U.S. Coast Guard: Joe, I just have a comment about how you developed the hazard zones. I realize that you're an electrical engineer and you may not have had much to do with the vapor dispersion calculations. Some assumptions, which are not always valid, were made in using the Gaussian Models. They have certain deficiencies with respect to dense gases. For one thing they do not take into account any type of gravity spread phenomena that may occur when one has a release and the mixture density is significantly greater than air. A density of 2.9 times that of air would cause a gravity spread phenomena very close to the source. This spreading would then turn into an area source and one would have to utilize other types of equations.

The Europeans in particular have done a lot of work concerning the dispersion of dense gases, and Te-Riehle and Ooms, and more recently Colenbrander (I believe all of them were at one time or are currently at the Shell-Netherlands Labs) have worked out schemes of calculating this type of dispersion. I think that Americans should take note of what they are doing instead of continuing to use the Gaussian formulas for every case. We should perhaps compare the Gaussian solutions with some of the newer formulations which are numerical in nature but are not difficult to program on the computer and not terribly difficult to use.

J.T. WOODS: Your information is very well taken. I'd like to point out...that I was at a disadvantage on the timing of this paper. The reason is that the information involving the equation was done by two people in our home office, one of which has published a book since I wrote the paper. It covers the same information in greater detail. His name is Frank Bodurtha. He is becoming known as one of the leading experts in the field, even on an international scale. This book is now in the market. If you are interested, we can give you the ordering information, but your point is very well taken.

FLESSNER: This gentleman had a question and asked whether the misuse of the dispersion equations would change the conclusions. This is a difficult question to answer without conducting a quantitative comparison. There were some compensating factors that he mentioned; they used E stability in their equations which tend to provide conservative estimates of the dispersion. But, the 10 mile/h wind speed tends to disperse gases very rapidly. However, this relatively high wind speed may be justified because of the presence of machinery and the relatively higher turbulence levels.

We have been doing some tests on board tank vessels in the Houston area and we have been finding turbulence levels on the dock where there is an open, level area much lower than right on board the ship where there are a lot of piping systems. I think that those of us in the U.S. should examine alternative methods of calculating dispersion. I recommend looking at some of the European work and take note of the different ways to calculate dispersion instead of using the same old number crunching equations which are very easy to use but which in fact are not physically realistic for the situations we're trying to model.

WOODS: I would like to point out that none of the national standards that I am personally familiar with include vapor releases and gas releases when classifying an area, other than the diagrams which I did show you. Their classified areas are extremely small. Now I might say that Frank Bodurtha is not in complete agreement with our electrical people who are responsible for classifying our plants. So, we are still working on it. It is a problem.

W.B. HOWARD, Monsanto: I would like to reinforce a bit what Mr. Flessner has been talking about from an actual experience which we had in our company. We have published this in the past but I think possibly a number of people may not have realized what we have said or may not have heard it. Some years ago, we also in our cyclohexane oxidation plant had a major spill of about 37 tons; this occurred in about 20 minutes which is then equivalent to 100,000 kg/h flow rate. We had a vapor cloud which finally developed a length of about 600 meters and a width of about 350 meters.

This spill initiated in rather relatively stagnant air and so it built up, roughly, as a circle to begin with. That circle developed a diameter of about 130 meters and then the wind came up, a light wind, possibly like a couple of miles an hour, and the vapor cloud continued to expand upwind slightly as well as downwind. And I think this is very important to know that the cloud dispersion continued upwind, as well as downwind, and the final distance upwind was about 100 meters, the rest of the distance of course being downwind.

Now, as Mr. Flessner has mentioned, this is explainable on the basis of the density of the overall cloud. The molecular weight of cyclohexane itself is 84. This is appreciably greater than that of air, so that even when there is a mixture with air, we still wind up with a total mass having a molecular weight greater than air. This is equivalent to a static head, which in turn can equalize against a velocity head. There is nothing mysterious after you hear about this, but until you do, you think that all the published models are very nice, like Mr. Flessner said.

There is another factor, too, that contributes to density and which is also not immediately obvious. When a spill of this type comes from conditions where the liquid is well above its atmospheric boiling temperature, there is a lot of equilibrium flashing of liquid into vapor. This causes high shear which in turn creates very tiny droplets of the liquid which do not totally vaporize, and we get a fog. This fog in turn contributes additional density to the gas, air, fog mixture. So we wind up having a static head to equalize against the velocity head of the wind. Now there is another factor that often has not been mentioned in dispersion of gas-fog clouds. This is the effect of equipment in promoting turbulence. You indicated that some of your equipment items have fairly high elevations. Sometimes there are walls. All of these things have effects on the dispersion of gases and fogs into air.

I want to mention another thing; you spoke of your control room. We do essentially as you do with your control rooms, that is, we don't try to specifically pressurize them. We, of course, have fresh air coming in, and we normally have them air conditioned for protection of the instrumentation; it helps personnel comfort, of course. For minimizing entry of gases from spills into a control building, we have one or more emergency buttons which can shut down the air intake and all air flow through the building.

When we had the very, very bad spill that I mentioned, the air flow into the control room was not stopped, and the control room was immersed in the spill. People had to put on air-breathing equipment in order to stay in the control room to work. I did want to ask one question. Is your control room a particularly strong building or is it an ordinary nonstrength-design building? Do you know if that 2-lb basis is true for the roof as well as the wall?

WOODS: I believe that the strength is for 2 lb/in. We use precast concrete bolted together with steel framework. But we do not, as Flixboro recommended, install equipment on the roof as we did in the past, air conditioning or any other equipment.

HOWARD: The control rooms which we have been building in recent years have the roof as strong as the walls. For a 2-lb control room, our roof would also be 2-lb strength. We really have been going more nearly to what would be equivalent to the 3-lb basis. I was in one of our control rooms recently that has a smallest wall-to-wall distance of 60 ft, and yet it will stand that pressure.

WOODS: Thank you, I appreciate your comments. I will say that obviously there are a lot of variables that go into the solution of equations like this or setting up models and wind tunnel tests, various pressures and temperatures vs. just a spill on the ground with various wind conditions and weather stabilities. I will point out that this type of investigation was applied to the Port Hudson, Missouri, propane gas spill of some 77 tons a few years ago. It was ignited some 1,000 ft away, I believe, and this approach to that particular incident agreed quite well. Now that does not, of course, substantiate everything that we have said, but it is very good to know that it did agree in this case. We would have a flammable concentration at that distance with that weather condition.

SCOPE

Three Mile Island: The Chemical Engineering Aspects

Much of the problem could have been avoided or mitigated had there been a better recognition of elementary chemical and chemical engineering practices.

Alfred Schneider*, Georgia Institute of Technology, Atlanta, Ga.

A sequence of equipment malfunctions, operator errors, confusing communications, and the misunderstanding of certain chemical and physical reactions combined at Three Mile Island to escalate into a serious nuclear power plant accident.

The Three Mile Island Unit-2 (TMI-2) accident, which started in the early morning of March 28, 1979, has been investigated by many groups and considerable information has been released during the intervening year. The availability of a large store of direct experimental data as well as post-accident analytical results will certainly make this accident a classical case study. There were no human casualties, and the Nuclear Regulatory Commission determined that, in a legal sense, "this accident did not constitute an extraordinary nuclear occurrence." Nevertheless, the material damages and losses resulting from the accident are very high: estimates by the Kemeny Commission range from \$1 billion to \$1.86 billion, with over 60% of these amounts representing outlays for replacement power.

The protected nature of the accident and the unique role played by the news media generated great concern and anxiety throughout the country, particularly in the immediate vicinity of the power plant. Early conjecture that this accident may become a decisive factor in determining the future of nuclear energy may, retrospectively, have been exaggerated. However, the intervening delays in licensing actions for new receptors, several cancellations of projected nuclear power plants, and the generally cautious attitudes toward nuclear energy will certainly diminsh the rate by which the nuclear option will contribute to meet short-term energy demands.

A chemical engineer examining the accident's sequence of events will be impressed by the number of unit operations practiced in a nuclear power plant, the similarity of much of the equipment and control systems with those to be found in the chemical process industry, and the decisive role of certain chemical reactions in shaping the outcome of this accident. Indeed, a more rigorous implementation of "good" chemical engineering design and operating practices

**The author is Past Chairman of the Nuclear Engineering Division of AIChE.*

0149-3701/81/4361 \$02.00

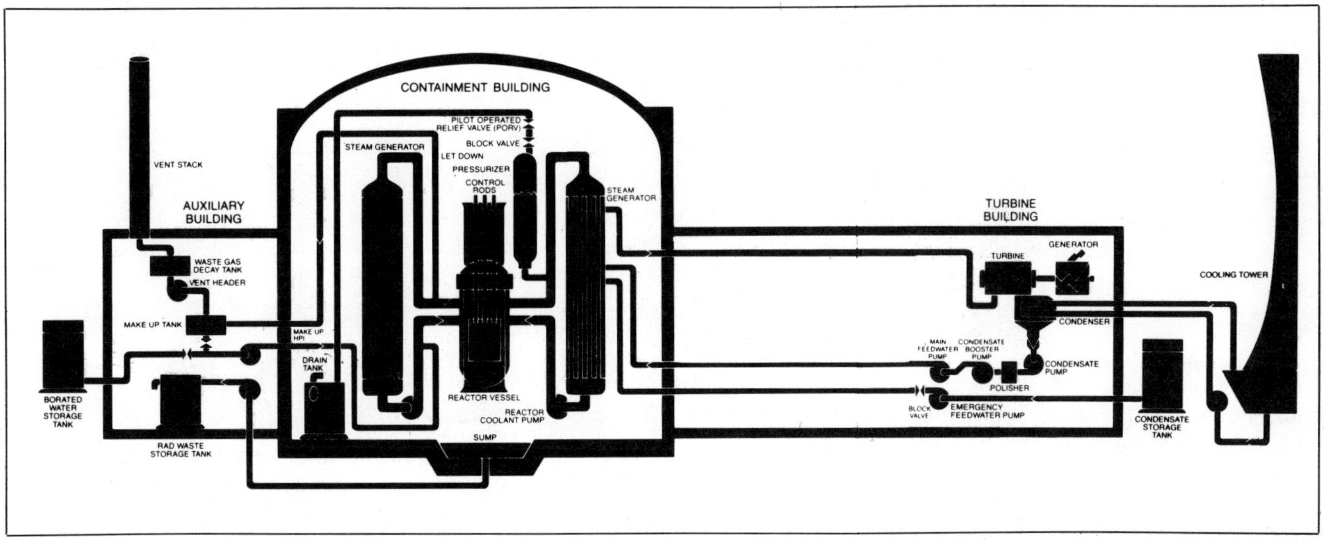

Figure 1. Schematic of the . . . . . . TMI-2 facility (5).

may have precluded the initiating event from happening. A better understanding of the elementary thermodynamics of water and of the chemical reaction of zirconium and steam could also have prevented the extensive damage to the reactor core and the subsequent release of volatile fission products.

The TMI-2 plant

TMI-2 is a pressurized water reactor (PWR) nominally generating 906 MW of electric power. During normal operation, the fission process which takes place in the core containing over 100,000 kg of 2–3% enriched UO_2 will generate over 2,500 MW of heat. The reactor is cooled by water circulating in the primary reactor coolant system. The heat is transferred in steam generators to the water of the secondary loop. The steam thus produced drives the turbine-generator set which produces electricity.

During normal operation, the reactor coolant is maintained at a pressure of 14,848 kPa gage (2,155 psig) and temperatures ranging from 290 to 318°C (554 to 604°F), conditions which preclude any boiling. In addition to the two water loops, there are many auxiliary systems to permit chemical control of the water by ion-exchange purification and by the addition of chemicals, removal of gaseous by-products, storage and decontamination of radioactive waste streams, and extensive containment of any potential leaks.

Figure 1 is a simplified diagram of the TMI-2 plant and Figure 2 shows a scale model of the nuclear steam system.

The accident

A detailed chronology of the accident is available from several references (*1, 2*).

An operational upset blocked the flow of condensate through the secondary side of the steam generators, thus depriving the reactor coolant system of its principal heat sink. The initial temperature increase caused opening of an automatic relief valve on top of the pressurizer. The latter is a component of the reactor coolant system which controls pressure in the primary loop; it is normally half-filled with water and half with steam. The subsequent loss of reactor coolant water and improper actions by the operating personnel, particularly their interference with automatic emergency systems, led to overheating of the core eventually resulting in extensive damage to the zirconium alloy tubes containing the UO_2 pellets. The structural failure of this barrier permitted radioactive fission products to enter the coolant water and a fraction of the more volatile species (noble gases and iodine) managed to bypass additional containment barriers and were released to the environment.

The condensate polishers

The condensate water is purified by passage of the entire stream through mixed ion-exchange resin columns to remove dissolved ions. The polishing system consists of eight parallel demineralizer vessels, with seven required for full-flow operation. A demineralizer removed from service is regenerated by transferring the resin by fluidization with water to the chemical regeneration system.

For about 11 hours prior to the onset of the accident, TMI-2 operating personnel had been trying to clear a resin block from the polisher outlet. The procedure called for injection of demineralized water and air into the line. The service air used for this purpose was cross-connected with the instrument air line, a practice not usually sanctioned in chemical plants. A single check valve was to prevent backup into the instrument air line, but post-accident inspection showed this valve to have stuck in the open position. The water is usually provided at pressures equal to or exceeding 1034 kPa gage (150 psig), while the service air is normally at 551 to 689 kPa gage (80–100 psig) pressure. Thus, a driving force was present for the infiltration of water into the instrument air lines.

The air operated polishing system outlet valves closed, stopping the condensate flow to the steam generators. A bypass line was provided in parallel with the bank of resin columns, but at this particular reactor it contained a manually operated valve which remained in the closed position. Considering the importance of maintaining feedwater flow to the steam generators, a manually operated bypass is certainly not adequate for this function.

This illustrates a fundamental difference in the design philosophies of chemical plant and nuclear reactor designers. The latter provide elaborate backup systems, such as emergency feedwater supplies and multiple emergency core cooling systems, while apparently downgrading the safety significance of certain auxiliary equipment, such as the outlying condensate demineralizers. Chemical plant designers, conditioned by a longer and occasionally painful learning experience, will customarily equip even auxiliary sections of a plant with independent safety provisions, rather than rely on backups within the main process complex to counteract transient upsets. In this case, flaws in the design and operation of a simple demineralizer system initiated the most serious nuclear power reactor accident in the United States.

The PVT relationship

A fundamental requirement for PWRs is that the temperature and pressure in the primary coolant are such that the water is always maintained below saturation conditions. To counteract any loss of coolant resulting, for example, from an accidental rupture of the primary system piping, a number of engineered safety features are provided to cool the core until most of the short-lived fission products will have decayed sufficiently to preclude damage by overheating.

The saturation pressure for the average coolant temperature of 304°C (580°F) is 9,136 kPa (1,326 psia), over 5,512 kPa (800 psi) below the normal operating pressure of 14,745 kPa (2,140 psia). As a result of the stuck pressure relief valve, undetected by the operators for nearly 2.5 hours, the primary system pressure reached saturation condition in about 6 minutes after the turbine trip, Figure 3. The decrease and eventual curtailment of reactor coolant flow at 1 hour and 40 minutes into the accident led to increases in core temperature, Figure 4; this, combined with the decreasing pressure, allowed boiling to occur in the core region.

The fate of TMI-2 was largely sealed during the period 1 hour 40 minutes (shutdown of the coolant pump) and 2 hours 22 minutes when the block valve in series with the sticking relief valve was closed. Boiling could have been prevented during this time period by using all available means to reduce coolant temperature. Subsequently, though means for pressure control were now available to the operator, boiling conditions were again allowed to occur when the pressure was deliberately reduced.

By this time, considerable hydrogen was present in the primary loop, as a result of the steam-Zircaloy clad reaction. Attempts to "collapse" steam voids in this system were unsuccessful because of the non-condensable hydrogen gas. In retrospect, the failure of the operators to associate the pressure-temperature indications for the reactor coolant with the resulting phase relations contributed in large measure to the core damage.

The zirconium-steam reaction

Zirconium alloys have been the preferred cladding material for water-cooled reactors because of their

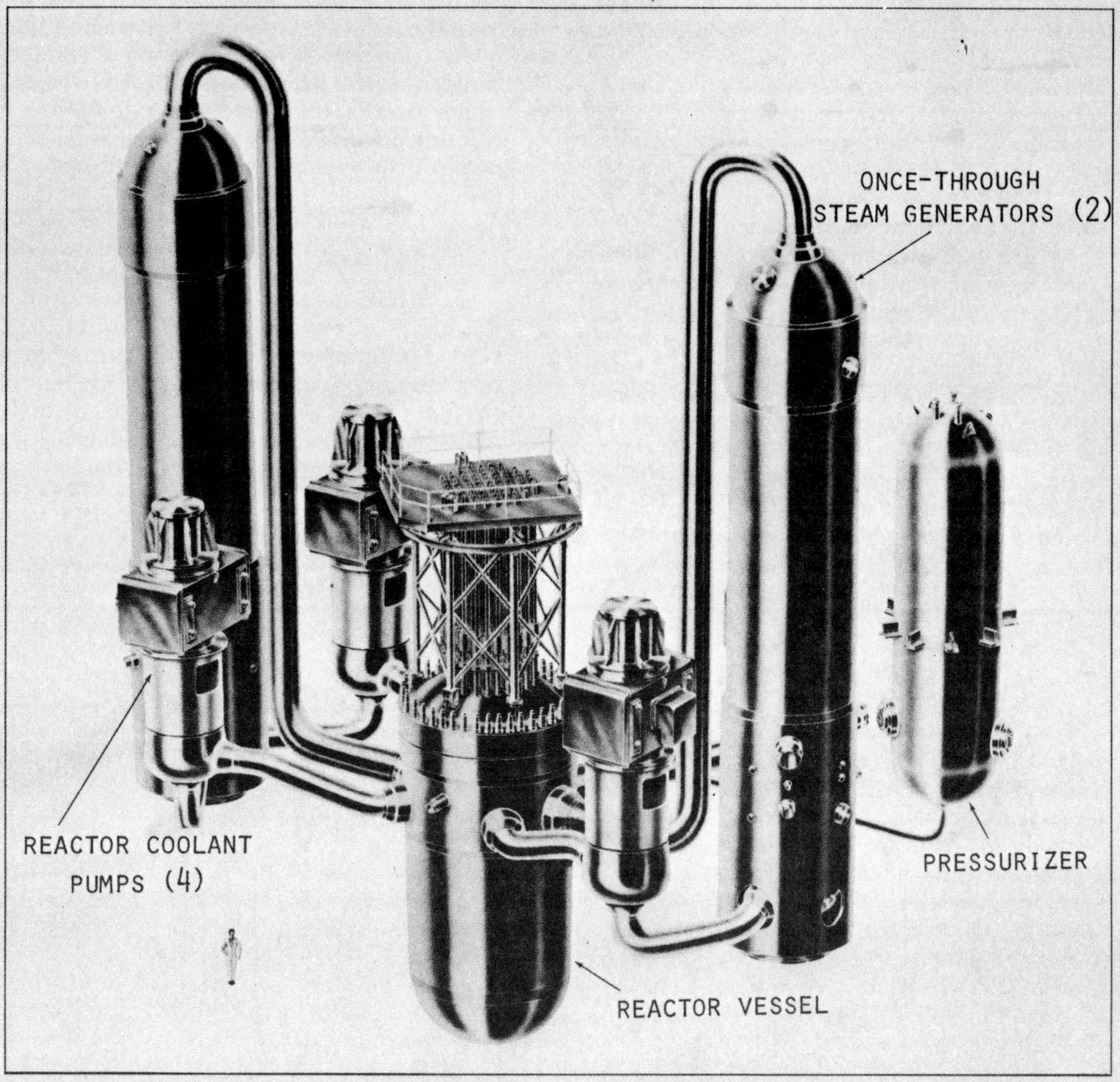

Figure 2. B&W nuclear system supply system (by permission).

superior nuclear and mechanical characteristics. The chemical reactivity of zirconium and its alloys has been a matter of some concern, but extensive studies showed that even under accident conditions mitigated by properly engineered safeguards, the integrity of the core would not be seriously impaired.

The clad material used in the TMI-2 reactor, Zircaloy-4, is a zirconium alloy with the following composition: Sn, 1.45%; Fe, 0.21%; Cr, 0.10%; and Zr, balance.

Analysis of the TMI-2 accident concluded that starting about 1 hour and 40 minutes after the turbine tripped, a portion of the core became uncovered. Cooling was then possible only by steam convection and by thermal radiation. Temperature estimates higher than 1,371°C (2,500°F) were deduced for zones in the proximity of several fuel assemblies. Zircaloy-4 undergoes a phase transformation between 782 and 954°C (1,440 and 1,750°F) which causes a sharp deterioration of its mechanical properties. The first consequence of the core uncovering was a splitting of the fuel tubes. It has been estimated that 90% of the fuel tubes failed.

At temperatures above 871°C (1,600°F), the following exothermic reaction becomes important:

$$Zr + 2H_2O \longrightarrow ZrO_2 + 2H_2 - 141 \text{ Cal}$$

The kinetics of this reaction, based on experimental data obtained between 982 and 1,510°C (1,800 and 2,750°F), is given by the Cathcart-Pawel equation (*3*):

$$\frac{d\tau}{dt} = \frac{0.01126}{\tau} \exp \frac{(-35{,}890)}{(1.987T)}$$

where

τ = thickness of oxide layer, cm

t = time, s

T = temperature, °K

The oxidation rate at 1,093°C (2,000°F) is such that the reaction heat release rate would have equaled the radioactive decay heat, which further increased the temperature and accelerated the oxidation rate.

Calculations based on an extrapolation of the Cathcart-Pawel equation showed that the 0.67-mm (26.5-mil) Zircaloy-4 cladding would require about 40 minutes at 1,649°C (3,000°F) and 12 minutes at 1,927°C (3,500°F) to be completely converted to oxide. The extensive oxidation of the cladding weakened the structural integrity of the core, and there is some indication that a restructuring of the core occurred 3 hours and 43 minutes into the accident.

The accurate determination of the damage must still await an inspection of the reactor internals and removal of the core. Indirectly, it is possible to estimate the amount of Zircaloy which had reacted from a material balance based on an accounting of the hydrogen produced. These calculations indicate that over 50% of the cladding was oxidized by reaction with steam. Failure of the fuel cladding permitted the release of substantial amounts of fission products from the fuel to the primary system and their subsequent transport to the Containment and Auxiliary Buildings. The hydrogen generated by the Zircaloy-steam reaction interfered with attempts to restart the reactor coolant pumps and caused great concern because of the explosive nature of hydrogen-air mixtures.

The hydrogen bubble crisis

Sixteen hours after onset of the accident, the crisis was practically over, since the operators had succeeded in repressurizing the system and in starting up a reactor coolant pump. Establishment of forced circulation now provided a stable mode of decay heat removal. Yet, evidence started to develop that an appreciable volume of hydrogen, an ominous proof of the extent of core damage, was present in the primary system. The response of the primary system to

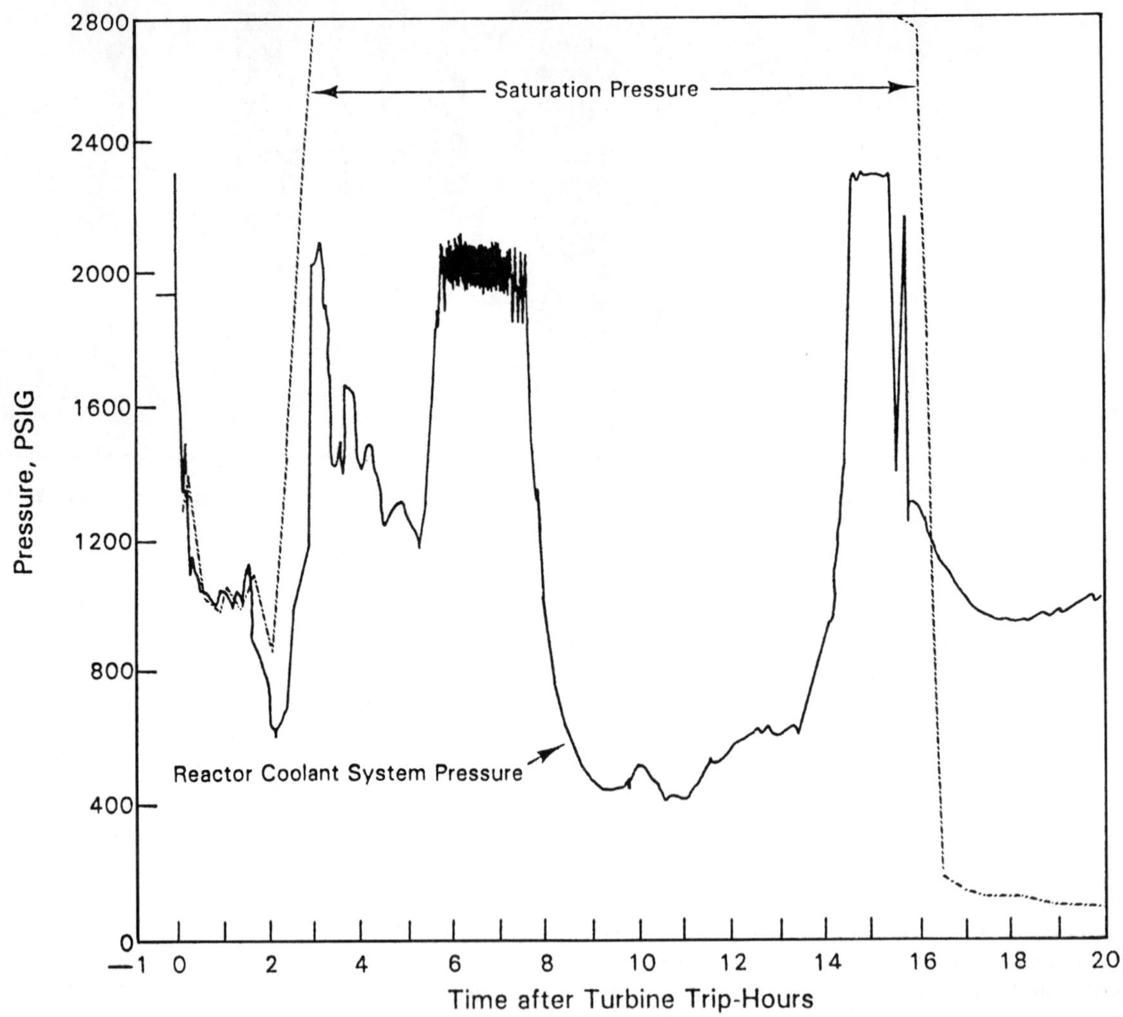

Figure 3. Reactor coolant system pressure and calculated saturation pressure (Source: Met-Ed).

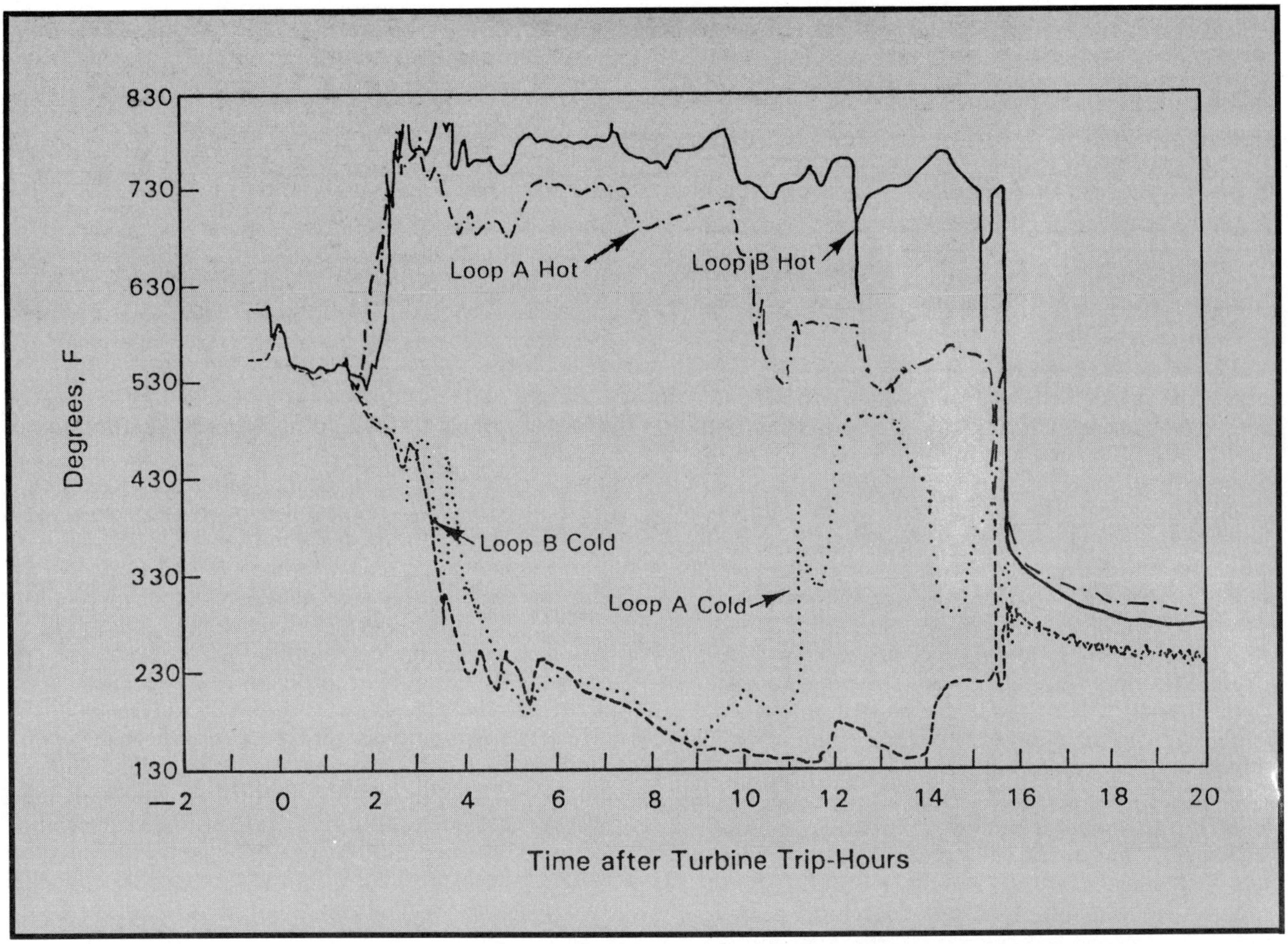

Figure 4. Reactor coolant loops temperatures (Source: Met-Ed).

pressure changes and anomalies in the thermal-hydraulic behavior could only be explained by the presence of a noncondensible gas.

A pressure pulse of approximately 193 kPa (28 psi) occurred at 1350 on March 28 in the containment building and this was soon recognized to have been an ignition of hydrogen mixed with Containment Building air. By March 30, the presence of "the hydrogen bubble" in the reactor vessel became a matter of great concern. It was estimated that, on March 31, the volume of the hydrogen bubble was 23–28 m^3 (800–1,000 ft^3) at the prevailing pressure of 6029 kPa (875 psia). If a correction is made for the solubility of hydrogen (by using Henry's Law) this volume reduces to 11–17 m^3 (400–600 ft^3).

Between March 30 and April 2, speculation by high-ranking officials of the Nuclear Regulatory Commission, that sufficient oxygen may be present in the "bubble" to result in an explosion, caused great excitement and concern. Plans were readied for the evacuation of the nearby population and frantic calls went out to radiation chemistry and detonation experts for an assessment of the hazard posed by the "bubble." Unfortunately, there was considerable delay before a plausible analysis of the situation was accepted and the "hydrogen bubble crisis" was officially declared to be over.

The only possible source of oxygen in the primary loop was the radiolysis of water:

$$H_2O \overset{\gamma}{\rightleftharpoons} H_2 + \frac{1}{2}O_2$$

This reaction was known to rapidly reach an equilibrium, if as little as 0.04 ppm of dissolved hydrogen is present in the water. Suppression of radiolysis is achieved in PWRs by adding about 1 ppm of hydrogen to the reactor coolant, a practice also followed at TMI. If the dissolved hydrogen level is continuously reduced by boiling or sparging, radiolysis of the water continues.

However, a material balance indicates that 20,000 volumes of steam would be formed by boiling for every volume of oxygen carried into the gas phase, an oxygen concentration far too low to result in an explosive mixture. Recognition of this situation led to the finding in the Rogovin Report that ". . . hydrogen never explodes in the reactor vessel; it blows up instead in the media" (*4*). The NRC was also criticized by the Kemeny Commission for failing to obtain proper professional opinions before allowing a nonexisting hazard to be escalated into a major crisis.

Release of gases and hydrogen

The extensive failure of the Zircaloy fuel tubes and the high temperature reached by the fuel pellets enabled a large fraction of the volatile fission products (xenon, krypton, and iodine) to be released into the primary coolant system.

Radiation monitors indicated that activity levels were rising rapidly in the Auxiliary Building. This implied that either reactor coolant was leaking into the Auxiliary Building where depressurization released the dissolved gases into the atmosphere, or that a communication existed between the atmospheres in the Containment and the Auxiliary or Fuel Handling Buildings. Several potential release paths were examined, and it was concluded that air transfer from the containment to the Auxiliary and Fuel Handling Buildings was not a major contributor to the fission gas release.

The most likely path for the transport of fission gases was through the reactor coolant letdown system. This system continually bleeds into the Auxiliary Building a portion of the reactor coolant water for purification by ion exchange and filtration and also for chemical adjustments, such as the addition of boric acid and hydrogen. The treated water is then returned to the primary loop or to the seals of the reactor coolant pumps. Under normal operation, water in the letdown system is cooled from 288 to 49°C (550 to 120°F) and depressurized from 14,848 to 138 kPa gage (2,155 to 20 psig). Since very little gas is normally dissolved in the reactor coolant, the combined cooling and depressurization do not release any significant amount of gas.

With the introduction of large amounts of fission gases into the reactor coolant, depressurization in the letdown system would lead to the release of radioactive fission gases into the Auxiliary Building atmosphere through tank vents, relief valves, or other openings. The gas was released to the environment then mostly through the Auxiliary Building ventilation system. Because of the greater solubility of iodine in water, the fraction released was considerably smaller than was the case for the noble gases.

The transport of the hydrogen produced in the reactor vessel to the Reactor Containment and Auxiliary Buildings was important for the following reasons: 1. the accumulation of hydrogen in the primary loop had created thermal-hydraulic problems and the scary "bubble"; 2. the release of hydrogen to the Containment Building had produced a flammable mixture; and 3. the pressurization of some off-gas systems, not designed for large volumes of gas, had caused component failures. At 10,631 kPa gage (1,543 psig) and 316°C (600°F), the hydrogen distribution between the gas and liquid phases is 67.5 and the coolant water can provide an effective means of transport for hydrogen to the Auxiliary Building, similar to the path postulated for the noble gases. However, most of the hydrogen in the Containment Building resulted from the coolant flow through the pressurizer relief valve. It is estimated that over 65% of the hydrogen generated was present in the containment prior to the ignition there.

Hydrogen control in Reactor Containment Buildings has been a matter of some concern. For PWRs, where normally there is no hydrogen production problem, a recombiner is provided for use in case of an accident. The design basis for the recombiner at TMI was the hydrogen produced by radiolysis in the water which would be released as a result of a loss of coolant accident. The capacity of the TMI-2 recombiner was a factor of 500, too small to cope with the much larger amount of hydrogen produced by the Zircaloy-steam reaction. The hydrogen concentration was reduced to 0.6% by June 1 through the operation of a larger recombiner installed after the accident.

Remedies

Even before the emergency at TMI was officially declared to have ended on April 9, many chemical engineers from the Electric Power Research Institute, Bechtel, Oak Ridge National Laboratory, and other

Figure 5. TMI-2 Radwaste processing systems (*6*).

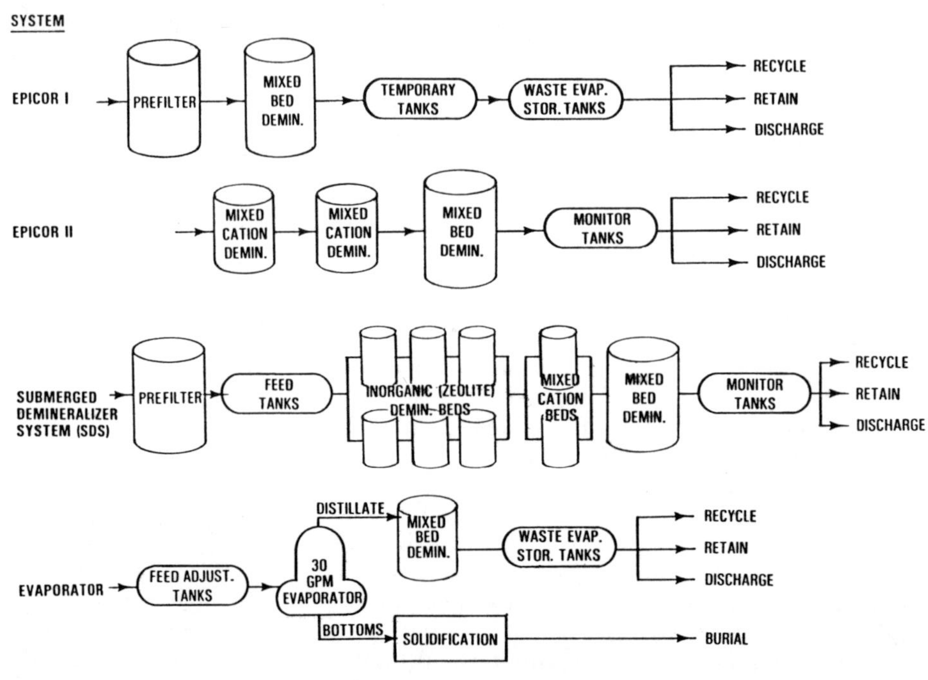

organizations arrived at the site of the stricken reactor to assist with problems connected primarily with contaminated air and water. Inventories of the contaminated water volumes and the radioactive nuclides no longer contained in the core showed that immediate actions will be required to minimize risks to the public and to permit the start of the recovery process.

The accident resulted in the release to the environment by April 27, 1980, of 13–17 curies of iodine and 2.4–13 million curies of noble gases. The resulting doses to the population of the surrounding area are not believed to pose any significant health hazards (*4*, *5*). Additional iodine releases were prevented by the installation of improved charcoal absorbers with a nominal decontamination factor capability of 1000.

Though the number of curies of noble gases released was large, these were mainly short-lived isotopes. The only remaining noble gas isotope, krypton-85, accumulated almost entirely in the Containment Building. Studies were made to trap this isotope by cryogenic or fluorocarbon absorption methods, but it was concluded that controlled venting of this gas would be preferable, considering the minimal potential health effects and the long delays which the alternate methods would entail.

During the accident and the succeeding months, a large amount of contaminated water was accumulated at TMI. Approximately 815,000 curies were dispersed in 4,169 m^3 (1.1 million gal) of water, as of June, 1980 (*6*). Over 2,464 m^3 (650,000 gal), the water in the primary loop and in the Containment Building, is designated as high-activity-level water and contains the bulk of radionuclides. About 1061 m^3 (280,000 gal) of intermediate-activity-level water (1–100 microcuries per mL) are contained in various tanks in the Auxiliary Building. It is estimated that an additional 26,530–34,110 m^3 (7–9 million gal) of contaminated water will be generated during the plant decontamination (*7*).

Several flowsheets, shown schematically in Figure 5, were developed for the water cleanup (*6*). The Epicor-1 system was used successfully for the processing of a large volume of low-activity-level water. The purity obtained with this system was considered sufficient for discharge to the Susquehanna River. However, a subsequent court ruling has prohibited any further discharge of water from the Three Mile Island to the river, regardless of the level of radionuclides contained.

The Epicor-2 system which also employs organic ion-exchange resins was designed for intermediate-activity-level water. Between October, 1979 and June, 1980, 1,251 m^3 (330,000 gal) were processed through this system with the removal of over 42,000 curies. The effluent contents of Cs and Sr, the principal radioactive contaminants, were below the maximum permissible concentrations specified in 10CFR20.

A shielded purification system (SDS) has been built to process high-activity-level water. Inorganic ion exchangers (zeolites) are used for the primary decontamination, since the high activity level could cause radiation damage to organic resins. This system has not yet been put into operation, but small scale tests were conducted at the Oak Ridge National Laboratory using actual high-activity-level water from TMI-2 (*8*).

The design of an evaporator system to process liquids resulting from the decontamination efforts was started. This is prompted by the concern that the liquid wastes produced during decontamination may contain a variety of chemicals and undissolved solids and not be amenable to purification by ion exchange.

The decontamination of buildings and equipment at TMI-2 is a formidable task. A variety of methods have been used, conventional (dry and wet vacuuming, wiping, and hosing) and more sophisticated ones (strippable coatings, Freon ultrasonic cleaning, Hydrolaser, etc.). Chemical procedures and reagents must be adapted for work in high-radiation areas and the compatibility of the reagents with many materials of construction must also be carefully verified.

In conclusion

The extensive analysis of the TMI-2 plant and of the accident which occurred there reveals the importance of several chemical processes and operations used at nuclear power plants. The initiating event, the unfavorable thermodynamic conditions for the reactor coolant during the accident, the Zirconium-water reaction which proved to be the most damaging occurrence, and the misunderstanding surrounding the "hydrogen bubble" could have been avoided or mitigated had there been a better recognition of elementary chemical and chemical engineering practices. Conversely, the chemical process industry could profit from the vast store of information now available about TMI-2, to improve safety standards, operator training, emergency procedures, instrumentation, and control philosophies in the design and operation of chemical plants.

Literature cited

1. Special issue *IEEE Spectrum, 16,* No. 11 (November 1979).
2. Nuclear Safety Analysis Center, "Analysis of Three Mile Island-Unit 2 Accident," Report NSAC-1 and NSAC-1 Supplement, Electric Power Research Institute, Palo Alto, Calif. (1979).
3. Pawel, R. E. et al., in "Zirconium in the Nuclear Industry," p. 119, ASTM Technical Publication 633, ASTM, Philadelphia (1976).
4. Rogovin, M. (Director), "Three Mile Island—A Report to the Commissioners and to the Public," U.S. Nuclear Regulatory Commission, Washington, D.C. (1980).
5. Kemeny, J. G., Chairman, "Report of the President's Commission on the Accident at Three Mile Island," Washington, D.C. (1979).
6. McGoey, R. V., "Processing TMI-2 Accident Radioactive Waste," Paper presented at 88th AIChE Natl. Mtg., Philadelphia (June 1980).
7. Brooksbank, R. E. and W. J. Armento, "Post-Accident Cleanup of Radioactivity at the Three Mile Island Nuclear Power Station," Report ORNL/TM-7081, Oak Ridge Natl. Lab. (Feb. 1980).
8. Collins, E. D., J. E. Bigelow, D. O. Campbell, L. J. King, and J. B. Knauer, "Flowsheet Development Studies for the Decontamination of High-Activity-Level Water at TMI-2," Paper presented at the 89th AIChE Natl. Mtg., Portland, Oreg. (Aug. 1980).

Schneider

FRP Duct Fire Incident

The quantity and continuite of combustibles were not readily apparent in this large building. But, there was ample fuel loading to sustain a severe fire. The significant continuity of combustibles developed when the duct burned products of combustion into the building.

D. W. Snow
Monsanto Co., St. Louis, MO.

This presentation describes a duct fire loss experienced in 1975. The process where the fire occurred is carried out in enclosures mainly of plastic construction although there is some minimal use of wood and metal. The enclosures measure approximately 35 x 12 x 7 feet high (11 x 4 x 2 m). The building involved is about 180 x 120 x 30 feet high (55 x 37 x 9 m). The building was not sprinklered.

High volumes of inert gas treated air are used in the process. The air is ducted to the enclosures under pressure and exhausted from the enclosures through FRP ducts by suction blowers. The air is scrubbed and then discharged to the atmosphere. The process enclosures are maintained under a slight negative pressure by the suction blowers located on top of the scrubbers.

The scrubbers, also of FRP construction, are located out-of-doors on the south side of the building. A schematic layout of the process lines and ducts are indicated in Figure 1.

There are four process lines arranged in parallel, lines A through D. A metal partition wall separates lines A and B from C and D. Exhaust ducts are oriented 90 degrees to the process enclosures and exit through the south building wall. Each process line includes 3 exhaust ducts; 2 of the ducts are 20 inches(508 mm) in diameter, the third is a 20 inch(508 mm) duct manifolded to a single 30 inch(762 mm) duct that serves all 4 lines. There are eight 20 inch (508 mm) ducts and one 30 inch (762 mm) duct exhausting to outside the building. The ducts are located overhead approximately 15 to 20 feet (4.5 to 6.0 m) above floor grade.

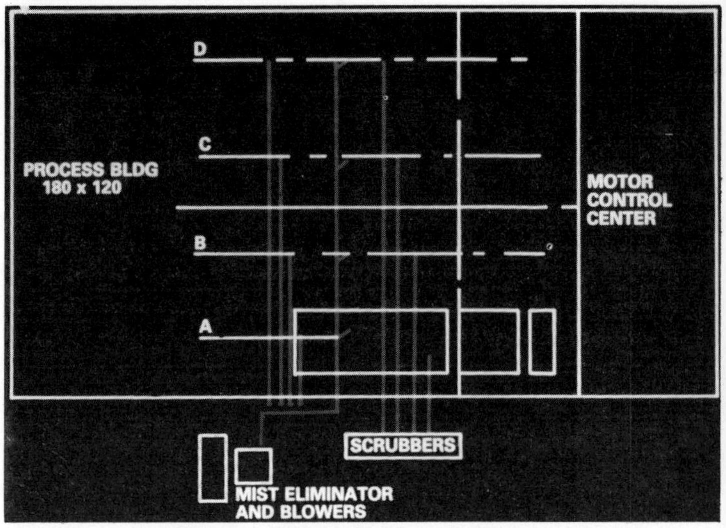

Figure 1. Process lines and ducts.

Duct construction was glass fiber reinforced polyester with no fire retardant additives. Air velocity through the ducts was about 1,000 feet per minute (5 m per second).

Building construction is metal on metal frame with extensive use of plastic panels having a flame spread rating of less than 25, in the roof. Sections of the process enclosures are illuminated and include clear plastic viewing panels in the walls.

At about 10:00 pm on the night of the fire an employe noted what appeared to be dimming of the lights in one of the enclosures in A line. He reported what was thought to be a lighting problem. Another employe checked the area and found the enclosure had turned black. Within a couple of minutes smoke was noted filling the area between lines A and B. The fire alarm was sounded, the plant fire brigade responded promptly and the building was evacuated.

About the time the brigade arrived, fire was noted outside at one of the scrubbers and the brigade was directed to the scrubber area. Hose lines were set up and water directed on the scrubber equipment. Fire fighters noted that the water was having little effect in reducing fire intensity.

Fire fighting was in progress about 15 to 20 minutes when brigade members saw fire inside the building through door windows and wall ventilators in the south wall.

Very shortly, thereafter, fire broke through the exhaust ducts and roof with flames shooting 50 to 70 feet (15 to 21 m) into the air, both horizontally and vertically.

Hoses were immediately directed into the building and the bulk of the fire was over in about 45 minutes, that is, 45 minutes from the time the fire was first noted outside at the scrubbers.

A view of the scrubbers and FRP duct located outside on the south side of the building is shown in Figure 2. Although not visible in the picture, gas flow is into the bottom of the scrubbers with exhaust out the top to the manifold header that is shown.

Figure 2. Outdoor scrubber and ducts.

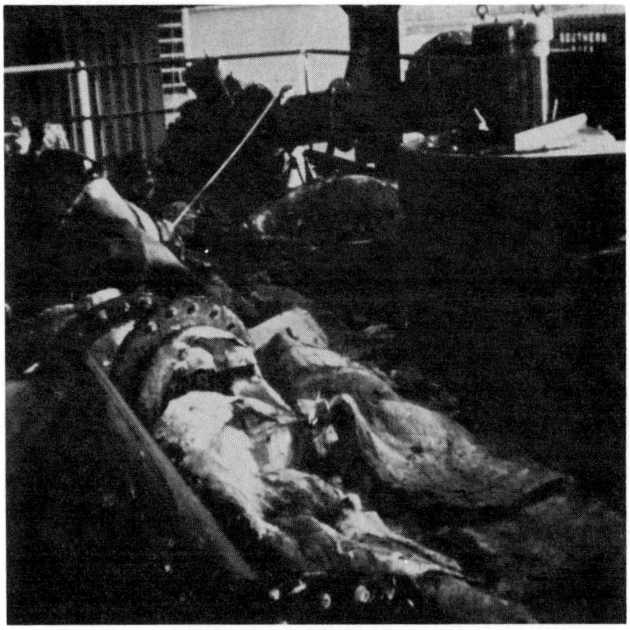

Figure 3. Scrubber deck.

Figure 3 is a view of the scrubber fan deck. Remains of some of the duct, only the glass fibers and metal joining rings, are shown in the foreground. Top sections of three of the scrubbers and two scrubber fans are visible in the background. The closest fan unit is still intact. The fan housing on the far end scrubber has partially burned away and the top mounted fan motor is tilted, partially falling into the scrubber.

The fan housing on the middle scrubber, barely visible, is completely burned away. The motor had fallen into the scrubber. This was the scrubber serving A line, the process line that sustained the most severe damage and where the fire originated.

Figure 4. Roof damage.

Some indication of the extent of building damage is illustrated in Figure 4. Damage was most severe over lines A and B but there was also roof damage over lines C and D.

Damage to cable trays, lighting, etc., was extensive throughout the building. Both aluminum and FRP cable trays were used. Little difference in damage was noted between the two types of trays for the same degree of cable damage.

In rebuilding the facility, automatic wet pipe sprinkler protection was provided throughout the building, inside the processing enclosures, and inside ducts having diameters greater than 12 inches. Also resin formulation for the new duct included fire retardant additives and the finished system has a flame spread rating of less than 25.

Automatic sprinkler protection for the outside ducts is shown in Figure 5. The sprinklers are supplied from the building wet pipe sprinkler system. The outdoor piping is insulated and electrically heat traced.

Figure 5. Duct sprinkler protection.

The sprinklers inside the ducts are spaced about 12 feet on centers. Sprinklers are also located inside the ducts at elbows as indicated in Figure 6.

Because of the corrosive nature of the air stream in the ducts special designs and materials of construction were needed for the short sections of pipe extending inside the ducts. This short pipe section was fabricated

Figure 6. Sprinkler piping at duct elbow.

of heavy wall machined PVC and designed to facilitate easy removal for examination. The pipe stub and sprinkler are inserted into the duct through special flanged duct nozzles.

The flange of the PVC pipe section is sandwiched between the duct nozzle flange and flange on the water supply pipe and the flange assembly then bolted together. One of the machined PVC pipes, pipe flange, and connected sprinkler is shown in Figure 7. The sprinklers are lead and wax coated. They are also covered with a lightweight polyethylene bag for added protection against corrosion.

Figure 7. Flanged PVC pipe and coated sprinkler.

SUMMARY

The exact origin of the fire and source of ignition within process line A was never definitely determined. Until nearly the end of the fire, burning was taking place totally with in the enclosures and ducts and completely inaccessible to manual firefighting efforts.

The quantity and continuity of combustibiles were not readily apparent in this large building. But obviously there was ample fuel loading to sustain a severe fire. The significant continuity of combustibles developed when the duct burned through venting the hot unburned products of combustion into the building. The ensuing large flash fire immediately involved cable trays, other utilities and roof panels throughout the building.

SNOW, D.W.

Plant Instruments: Which Ones Don't Work and Why

One reason why trips and control systems don't operate properly is that they violate the laws of physics; another is that they ignore human nature.

T. A. Kletz, Imperial Chemical Industries, Wilton, England

Modern processing plants depend greatly on instruments for their control and protection. Among many control and protective systems on our plants are a few that cannot do what we expect them to, for one of several reasons:

- They violate the laws of physics.
- They violate the principles underlying good control system design.
- They pay insufficient regard to human nature.

Laws of physics violated

A Sight-Glass with Limited Range. A sight-glass 1.2 m (4 ft) long was connected to vessel branches 0.6 m (2 ft) apart as shown in Figure 1. The level glass will indicate the correct level only when the liquid in the vessel is between the two branches. If the liquid level is below the lower branch, the liquid in the sight-glass is isolated and its level cannot fall. If the liquid level is above the upper branch, vapor will be trapped in the upper part of the sight-glass. As the level rises, this vapor will be compressed; if there is any noncondensible gas present, the pressure in the sight-glass will rise and the level will be depressed below its true value.

0149-3701/81/3792 $02.00 © 1981 AIChE

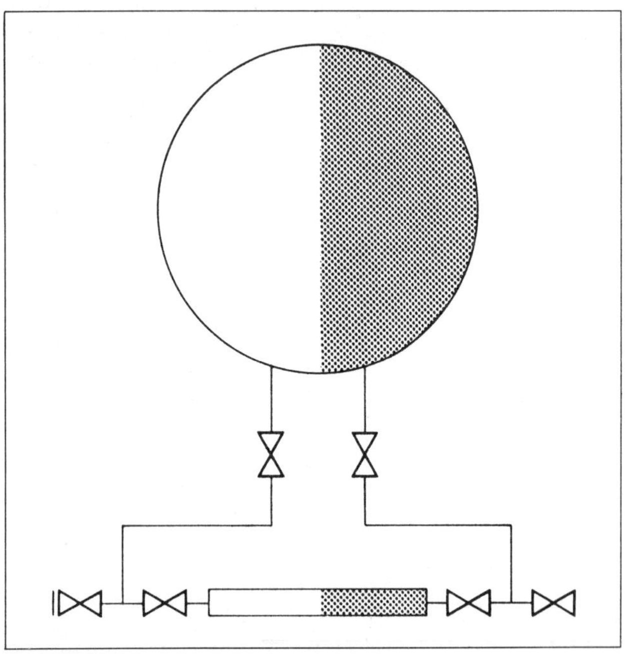

Figure 1. Sight glass with limited range.

An Explosion in a Nitric Acid Plant (1). Ammonia was vaporized, mixed with air and passed over a catalyst. The ammonia and air flows were measured and controlled so that the ammonia concentration was always below the explosive level, Figure 2. The vaporizer level controller was out of order and the level of ammonia was on hand control. The level got too high and droplets of ammonia were carried forward. The increased flow was not detected by the flow ratio controller and an explosion occurred.

All flow measurements are inaccurate when spray is present; the size of the error depends on the detailed design. If the spray increases the density of the gas by 50%, the flow of vapor and liquid could be 25% higher than the flowmeter reading.

Vapors and Noncondensible Gases Confused. The following has been discovered more than once during hazard and operability studies.

A vessel is kept half full of liquid. There is a level controller (not shown) and a high level trip isolates the inlet line if the level gets too high, Figure 3. If the high level trip fails, the relief valve will lift and discharge liquid to atmosphere. Thus, a high pressure trip was installed as well. If the space above the liquid contains nitrogen or other noncondensible gas, the system will work. As the level rises, the gas will be compressed and the pressure will gradually rise. However, the vessel contains liquefied gas. As the level rises, the vapor condenses and the pressure does *not* rise until the vessel is completely full of liquid. The pressure then rises too rapidly for the high pressure trip to prevent the relief valve lifting. The designer of the system probably did not understand the difference between a noncondensible gas, such as air or nitrogen, and a vapor.

Design principle violations

Problems of the Trip System, Figure 4. The pressure in the vessel is measured by the pressure transmitter (PT) and controlled by the pressure indicator controller (PIC), which adjusts the setting on the motor valve. If this control system fails to work and the pressure rises above the set point, the high pressure switch and trip (PSZ^{Hi}) operates to close the motor valve. At the same time, the high pressure alarm

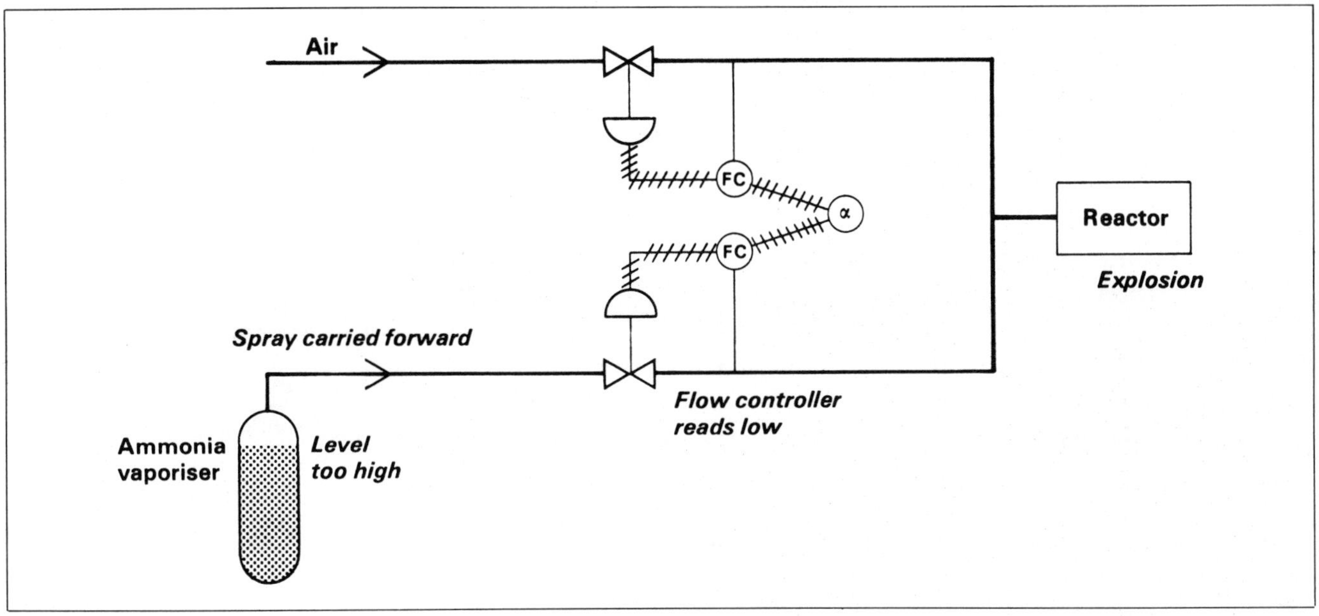

Figure 2. Nitric acid plant explosion.

(PA^{Hi}) operates. This trip system is almost useless.

The most likely causes of the pressure in the vessel getting too high are:

1. Failure or isolation of the PT. If this occurs the trip will not know there is a high pressure in the vessel.
2. Motor valve sticks open. In this case, the trip will know that there is a high pressure in the vessel and will send a signal to the motor valve, but the motor valve will not respond.
3. Failure of the PIC. In this case the trip will work.

The PIC is in the clean atmosphere of the control room while the other equipment is out on the plant. The PIC is thus more reliable than the other equipment. The trip will therefore operate on less than one-third of the occasions on which we want it to operate. Such a trip is not worth having. It may do more harm than good, as we may expect it to operate and not watch the pressure so closely.

The system shown in Figure 5 has a high reliability. The high pressure trip and alarm ($PSZA^{Hi}$) has an independent connection to the vessel and operates a separate motor valve. There is cross-connection to the control valve. A high pressure switch (PS^{Hi}) and prealarm (PA^{Hi}) give a warning that the pressure is approaching the trip setting and allow the operator to take action. This prealarm will operate, if the pressure rise is due to failure of the PIC or motor valve, but not if it is due to failure of the PT. If a high pressure occurs, the prealarm will operate on about two occasions out of three and the trip on almost all occasions.

Some designers may feel that the system shown in Figure 5 is too expensive. If so, they should omit the trip system altogether and let the bursting disk blow when the controller fails. The compromise system shown in Figure 4 is worse than no trip system at all.

Measure Directly the Property You Wish to Know. A packaged boiler was fitted with a trip to shut off the fuel when the air supply fails.

The flow of air was detected by measuring the voltage supplied to the electric motor which drives the air blower. If the air flow stops because the electricity supply to the motor has failed, or been isolated, the trip will work. However, air failure might be due to a broken or disengaged belt, to a broken or loose impeller, or to a plastic sheet over the fan inlet. In these cases, the trip will not work. The air flow should be measured directly and then the trip will work *whatever the cause of the air failure. Always try to directly measure the property one wishes to know, not some other property from which it can be inferred.*

An ethylene oxide plant tripped and a light on the panel told the operator that the oxygen valve had closed. As the plant was going to be restarted immediately, he did not

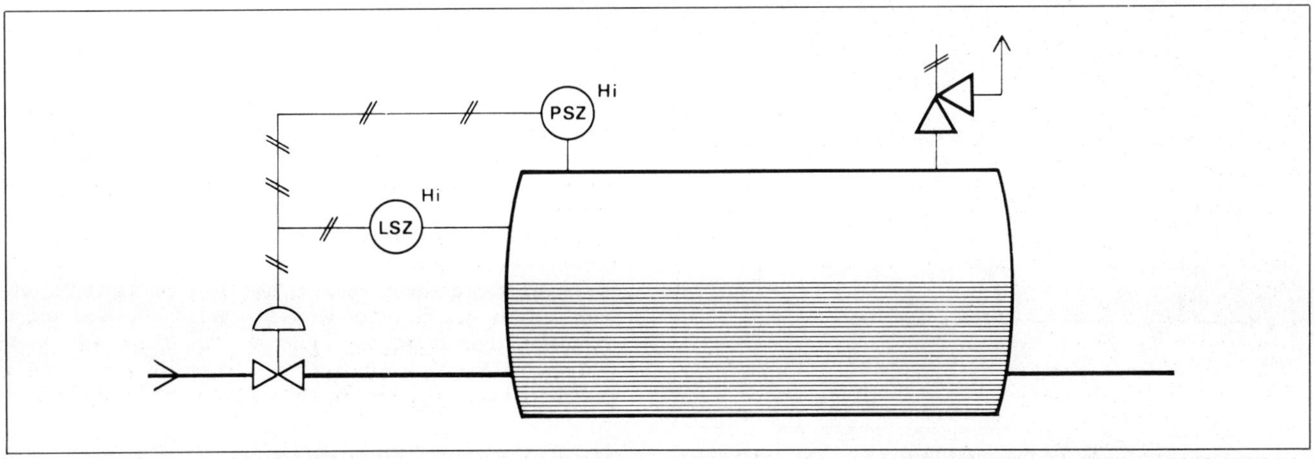

Figure 3. Vapors and non-condensibles are confused in this system.

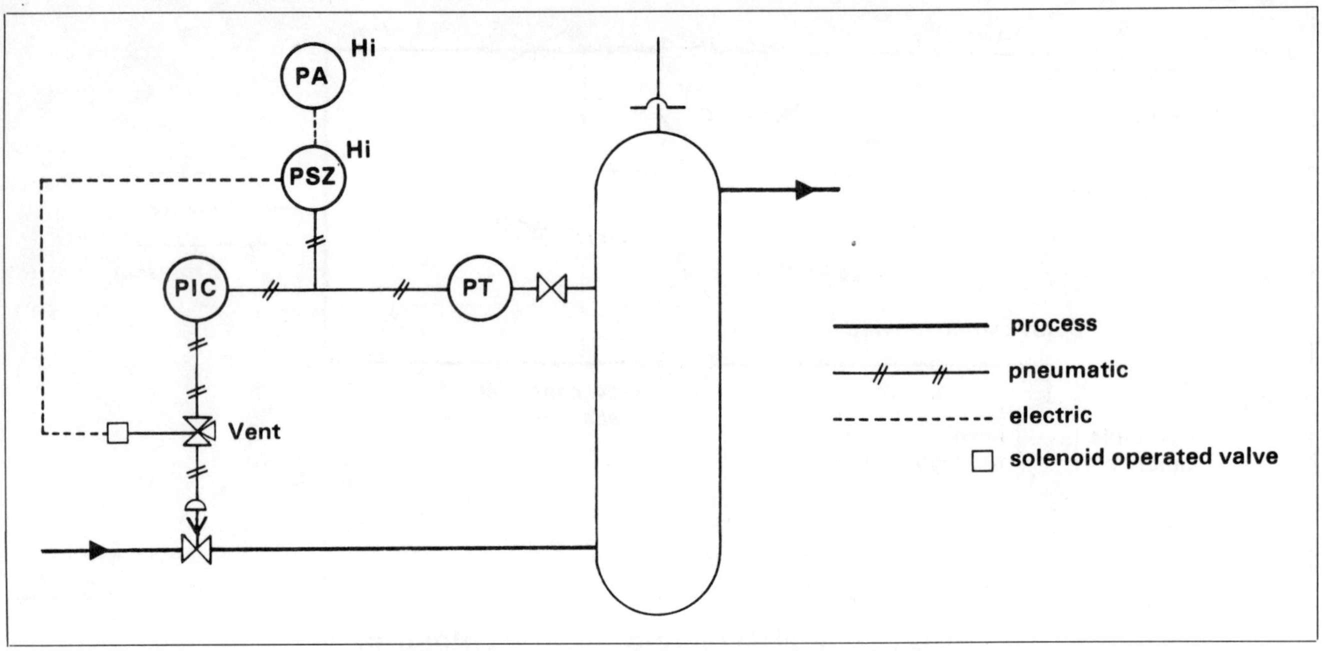

Figure 4. Poorly designed trip system.

close the hand valve as well. Before the plant could be restarted, it blew up. The oxygen valve had not closed and oxygen had continued to enter the plant, Figure 6.

The trip valve was closed by venting the air supply to the valve diaphragm, via a solenoid valve. The light on the panel merely said that the solenoid valve had been de-energized. Even though the solenoid valve is de-energized, the oxygen flow may continue because:

1. The solenoid valve fails to open.
2. The air fails to vent.
3. The trip valve fails to move.
4. With some designs of trip valve, the valve may not close even though the spindle moves.

Actually, the air failed to vent; the vent line on the air supply was choked by a wasp's nest (*2*).

The air line to a valve broke and the valve closed, isolating a diluent flow to a reactor and causing the reactor to overheat. Loss of diluent flow should have caused an alarm to sound. But, it did not as the loss of diluent feed was signalled by the electrical unit controlling the air supply to the valve and this was normal. The flow of diluent should have been measured directly.

At Three Mile Island nuclear power station in 1979, the relief valve on the primary cooling system lifted and failed to reseat. The operators did not understand what was happening. They thought there was too much water in the system, not too little, and they took action which made the problem worse. According to H. W. Lewis. (*3*)

> "The position of the main valve was not sensed directly; instead an indicator light on the main control panel signalled that the solenoid had been actuated and the position of the solenoid was considered to be indicative of the position of the valve."

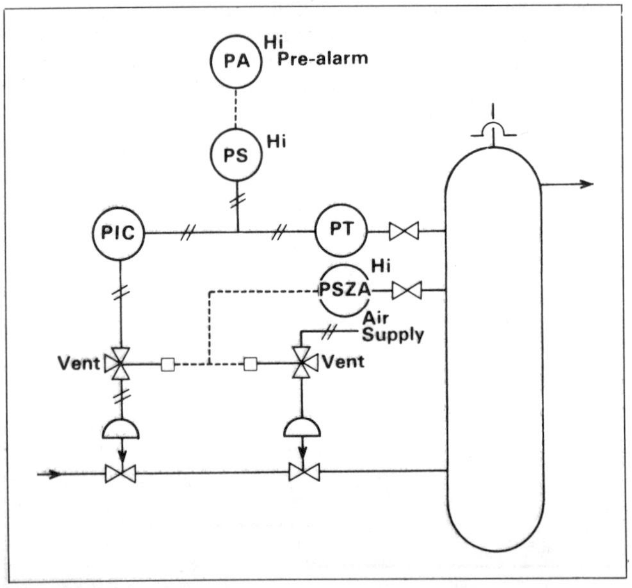

Figure 5. Highly reliable control system.

Combustible Gas Detectors. On more than one occasion, the atmosphere in a furnace has been tested with a combustible gas detector, no gas has been detected and a lighted poker has been inserted in the furnace (or a spark igniter operated). An explosion has then occurred.

The furnace fuel was an oil with a flash-point above atmospheric temperature. The furnace was hot and contained oil vapor. This was not detected by the combustible gas detector as the vapors condensed out in the sample tube.

In some combustible gas detectors the detector head is on the end of a lead and is placed at the point of test. However, the detector head is protected by a sintered metal disc and warm vapor will condense on this disc and may not be detected. Furnaces containing heavy oil vapor should be purged for a sufficient period of time to remove all oil vapor before they are relit.

On other occasions, welding has been permitted after a combustible gas detector gave a negative reading and an explosion then occurred. There was no flammable vapor present at the time of test, but traces of heavy oils or polymers were vaporized by the welding and ignited (*4*).

Human nature disregarded

The Trip That Is Too Easy to Disarm. Soon after a

Figure 6. Ethylene oxide plant control system.

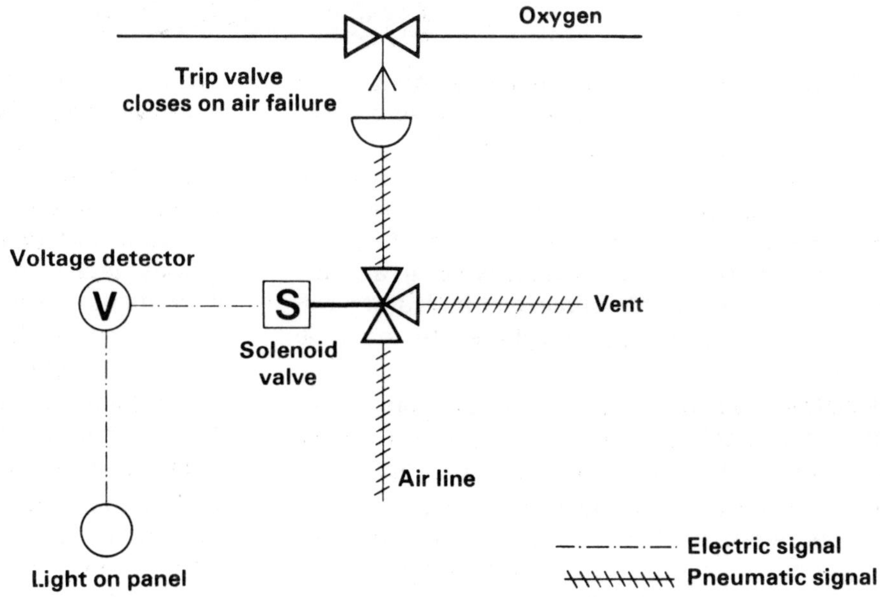

The light shows that the solenoid is de-energised, not that the oxygen flow has stopped

start-up, part of a plant was found to be too hot. Flanged joints were fuming. It was then found that the high temperature trip and temperature controller—a combined instrument—had been unplugged from its power supply.

It should be impossible to disarm a trip merely by pulling out a plug. The temptation to do so is too great. Trips should be disarmed only after proper authorization by responsible people, using a key, and the fact that they have been disarmed should be signalled in some way, for example by a light on the panel. Most trips should normally be designed to be operated when power is lost. Sometimes we do not want a trip to operate under these conditions. In these cases an alarm should sound when power is lost.

Trips and controllers should normally be independent, as the most common cause of a high temperature, pressure, or level is a fault in the temperature, pressure, or level control system. (Refer to the section on the *Problems of the Trip System.*)

Trips should be tested at start-up, if they have been worked on during the shut-down. On furnaces and compressors, they should be tested after all major shut-downs.

The Trip That Is Less Reliable Than a Man. A small tank was filled every day with sufficient raw material to last until the following day. The operator watched the level in the tank and switched off the filling pump when the tank was 90% full. This system was in use for several years before the inevitable happened and the operator allowed the tank to overfill.

The failure rate of the operator (about one in 1,000 occasions) was very low and would be achieved only in a situation in which stress and distraction are low. Nevertheless, it was decided to install a high level trip to shut the pump down automatically when the level reached a preset value. Surprisingly, the tank overflowed again after about a year.

When the trip was installed, it was assumed that: 1) the operator will occasionally forget to switch off the pump; 2) the trip will occasionally fail; and 3) the chance that both will occur on the same day is negligible. It did not work out like this, however. The operator decided to rely on the trip and stopped watching the level. The manager and supervisors knew this but felt pleased that manpower was being better utilized. A simple trip has a failure rate of about once every two years. Thus, a failure after a couple of years was almost inevitable. The trip was being used as a process controller.

The trip was tested every month, but this was a waste of time, because any failure was much more likely to be followed by a demand than a test. The options open to the manager, once the incident had occurred, were:

1. Remove the trip and accept a very occasional overfilling. Depending on the nature of the material, this might be an acceptable solution.
2. Install a process controller and a trip; the trip will be set at a slightly higher level and it will operate when the controller fails.

We have, rightly, a desire to compromise. Yet, there are occasions when extremes make sense but compromise does not. This is one such case (another was described in the section on the Trip System). Option 1 results in occasional spillages due to inevitable operator errors. Option 2 is expensive. So let us compromise on a simple trip. This, as we have seen, results in more frequent spillages than either of the extremes.

Literature cited

1. *Fertiliser News,* p. 35 (November, 1973).
2. Doyle, W. H., "Some Major, Instrument Connected, CPI Losses," paper presented at Chemical Process Industry Symposium, Philadelphia, Pa. (1972).
3. Lewis, H. W., *Scientific American,* **242,** No. 3, p. 33 (March, 1980).
4. Kletz, T. A., *J. of Hazardous Materials,* **2,** No. 1, p. 1 (1977).

T. Kletz, who has been with ICI since 1944, studied chemistry at Liverpool Univ. In 1968, he was appointed a technical safety adviser in the Oil Works with particular responsibility for process safety. He has also been appointed an industrial professor at Loughborough Univ. of Technology.

DISCUSSION

J.E. ROGERSON, Procter & Gamble: I would like to add one comment in connection with using combustible gas alarms. I have had a number of comments over the years to the effect that combustible gas alarms did not work when trying to detect a flammable vapor. I discovered that the analysts were trying to detect a flammable vapor in a nitrogen atmosphere. Combustible gas indicators will not work in a nitrogen atmosphere. Air must be present.

TREVOR KLETZ, ICI: That is very true. Some combustile gas detector manufacturers sell a device which mixes some air with the sample stream.

It might be worth adding a word or two on some of the other reasons why combustible gas detectors fail to work. The vapor can be absorbed by the sample tube and never reach the detectors. The sample tubes are usually of plastic and it's almost impossible to get a plastic that won't absorb something. Sometimes the tubing swells and blocks the sample tube. Some gas detectors are also poisoned by certain chemicals: silicones are one of the most common poisons. Every combustible gas detector should, therefore, be tested every time before it is used. We use a weak solution of isopropanol and water and this is kept in the control room. Sending the detectors away to the instrument section for a weekly test is not good enough because these instruments are not always fail-safe, and they should, therefore, be tested before every time they are used.

STAN SKINNER, Monsanto: Just one other point which I am always hammering home. The calibration will only work correctly for one gas; it doesn't detect everything independently.

KLETZ: Yes. However, in many cases, the detector is used as a "go, no-go" device and it doesn't matter if the reading is a bit in error.

To allow for the fact that a detector may be used on a different gas to what has been calibrated, we never allow welding to go ahead or anyone to enter a vessel if more than 20% of the lower explosive limit is detected. This gives a reasonable safety margin.

Control of Gas Detonations in Pipes

Use of passive detonation arrester devices to protect critical elements of the piping system is an appropriate safety measure.

M.F. Flessner
U.S. Coast
Washington, DC

R.A. Bjorklund
Jet Propulsion Laboratory
Institute of Technology
Pasadena, CA

INTRODUCTION

Within certain composition limits, combustible mixtures may burn in either of two modes: deflagration or detonation. Deflagration velocities are subsonic and the pressure change across the combustion front is small. Detonation velocities are supersonic, and a large step increase in pressure occurs across the combustion front. Under certain circumstances, a deflagration may accelerate and become a detonation.

The deflagration-to-detonation transition is a complex phenomenon and depends not only on the properties of the burned and unburned gases but also upon the conditions at the boundary of the combustible mixture. Wall roughness (1) and tube dimensions (2,3) both influence the distance required for development of detonation ("the run-up distance"). One other characteristic of this transition is the rapid pressure increase, which is of importance in considering safety precautions under conditions where detonations may develop. At the time of transition, pressure pulses in excess of the steady detonation pressure are often observed (4). Although these pulses are of very short duration, the pressure may be three to four times the steady detonation pressure. Other factors which are important to the deflagration-to-detonation transition are the fuel, the mixture composition, and the source of ignition.

There are many examples of industrial accidents involving detonations which have resulted in destruction of large amounts of capital equipment and loss of life. Several authors (5,6,7) have reviewed some of the major accidents and note that many of the detonations occurred in large processing equipment with hydrocarbon-air mixtures of marginal flammability. They present results of experiments designed to study detonation phenomena and compare the results with those of industrial accidents. Conditions which lead to detonative explosions are more prevalent today in large modern installations with long pipe runs, large vessels, and high-velocity vapor lines than those in the smaller refineries of the past.

Jacobs, et al (8) describes in detail the destruction of a large fluid hydroformer and associated equipment. Fragments of the reactor were spread over an area with a radius of 366 meters. Some of these fragments fell in two working tank fields. The impact on the tanks caused numerous fires which ultimately destroyed 63 tanks and approximately 1,270,000 bbl of crude and other products. Investigators were able to trace the propagation of the detonation through the piping system and

0149-3701/81/4775 $02.00

estimated the speed and pressure of the shock wave. The velocity of the wave was on the order of 1000-1200 m/s and pressures at some points reached 20700 to 28000 kN/M^2.

Sutherland and Wegert (9) reported details of an acetylene decomposition incident. Two plants separated by 8.8 km were involved and connected by 20.3 cm diameter pipe. In one plant, there was approximately 760 m of 15.2 cm diameter pipe, while in the other there was 1830 m of 20.3, 15.2, 12.7, 10.2 and 7.62 cm diameter pipe. A fire developed in the larger plant and resulted in an acetylene decomposition within the acetylene distribution network which rapidly developed into a detonation and propagated throughout the entire system external to the user and supplier operations--approximately 11.2 km of piping in 6 seconds. Fortunately, almost all the acetylene lines had been designed to withstand such a detonation and Linde hydraulic flash arresters had been installed in the acetylene producing unit supply line, in all user branch lines, and in the line to the flare stack in the smaller plant. Except for disruption of flows, little damage was done to either plant and the hydraulic arresters functioned as designed.

The U.S. Coast Guard regulates the Marine Transportation Industry in the U.S. and one of its responsibilities is to insure that the loading and unloading of flammable chemicals is carried out safely. Because of the imminent implementation of an air quality standard requiring the recovery of vapors emitted during cargo transfer operations of gasoline, the Coast Guard has had to examine the possibilities of fire and explosions in the piping system carrying the fuel/air mixtures. In particular, data concerning the adequacy of flame and detonation control devices was needed.

The purposes of this experimental study were: (1) the development of parametric data on the flame speed, flame acceleration, run-up distance to produce detonation, and pressure build-up due to detonation and (2) the demonstration of operational characteristics including pressure drop, flame quenching capability, and structural durability of several selected flame arrester devices before and after exposure to detonations in simulated vapor recovery piping systems. While the testing reported herein was performed specifically for the tanker-terminal problem, the results are applicable to many piping systems in which a fuel/air mixture is flowing and in which there is some potential source of ignition.

TEST APPARATUS

All testing was performed at the Jet Propulsion Laboratory (JPL) Edwards Test Station (ETS), a remote site in the Mojave Desert, California. The test facility flow system schematic diagram is shown in Figure 1. Table I gives a description of the symbols used in the diagram.

Air flow for testing was provided by a Spencer Model 80 x 10 SOH three stage centrifugal turbine compressor and was determined by measuring upstream total pressure and differential pressure loss across a Meriam laminar flow meter. The air was heated by compression up to 32°C above the ambient inlet temperature.

Liquid fuel was supplied from a 0.049 m^3 spherical aluminum pressure vessel, which was pressurized with gaseous nitrogen. Fuel was delivered through a 10 micron filter, turbine flowmeter, and remote metering valve for flow control. The fuel line proceeded through thirteen turns around an aluminum cylinder heated by four 2.4 kw electrical resistance elements. A fuel condenser loop was used to prevent the exhausting of vaporized fuel into the atmosphere during pre-test warm up and system stabilization. Fuel was injected upstream against the flow of air to improve the fuel to air distribution. A Komax mixer was installed downstream of the injector for additional turbulent mixing.

The detonation test piping was made from heavy walled stainless steel tubes that were bored, honed, and polished to a 15.32 cm internal diameter with a very smooth finish. There were eleven sections of pipe. Section 1 contained the hydrogen-air-spark igniter, two flame sensors, and one temperature sensor; Section 2 or flame stabilization section contained two flame sensors; Sections 3,4,5, each 4.57 m long, contained an expanded metal liner to generate turbulence in the propagating flame; Sections 6,7,8 or verification sections each contained two flame sensors and two pressure sensors. Section 8 had an additional pressure port for the test section upstream differential pressure measurement; Section 9 was the test section; Section 10 or witness section contained two flame sensors, two pressure sensors, one temperature sensor and one pressure port for the test section downstream differential pressure. The final section 11 was used to minimize light entering the witness section. Exit gas was sampled between the flanges of the witness section and the extension section and analyzed with an on-line Beckman Model 400 hydrocarbon analyzer instrument.

Steady-state data for the air system, fuel system, fuel vaporizer and condenser loop, fuel and air induction system, hydrocarbon gas analyzer and the pre-and post-test pressure loss measured across the test arrester was recorded on an Integrated Digital Acquisition and Controls System. Transient-state data was recorded on two high frequency FM tapes recorders and played back on an oscillograph. Flame velocities and peak pressures were manually scaled and calculated from the oscillograph traces.

Flame sensors used a DuMont type 6291, 10 stage multiplier phototube with a flat end window photocathode. High response pressure sensors were made up of PBC Piezotronic model 113AP quartz crystal and were flush-mounted opposite flame sensors in the verification and witness section, thus providing a second check on the detonation.

Commercial grade propane was used as the fuel in calibration tests while Indolene HO III clear gasoline was utilized in tests of devices.

TESTING AND RESULTS

Calibration Test

Initial calibration tests utilized commercial grade propane. Several interations of the test setup and ignition source were made prior to obtaining a stable detonation. The stable flame velocity flame obtained was 1830 m/s and the peak pressures ranged between 1000 to 5000 kN/m^2 with a stable pressure of approximately 2000 kN/m^2. The Indolene gasoline was substituted for the propane and the detonation obtained had a stable velocity of 1830 m/s and stable detonation pressure of 1630 kN/m^2.

In order to determine the point of deflagration to detonation, the apparatus was modified by removing the expanded metal liners from sections 4 and 5 and reinstalling them into instrumented sections, which contained the expanded metal liners. In the first series of tests, the initial mixture flow velocity was varied from 0 to 6.1 m/s with a constant gasoline/air equivalence ratio of 1.1. Figure 2 displays the results of those tests. The data indicated that the location of the transition point was about 15.4 m from the point of ignition and that the run-up length through the turbulent liner pipe was 11.2 m giving an L/D of 74. There appeared to be no discernible correlation between the initial flow velocity used in testing and the resulting run-up length. The stable detonation wave velocity downstream of the transition was about 1800 m/sec with a stable pressure of 1900 kN/m^2. Some peak pressures measured at the point of transition exceeded 4000 kN/m^2.

To determine the influence that the equivalence ratio has on the detonation run-up length, nine additional test firings were made at a constant initial velocity of 4.6 m/sec. The equivalance ratio was varied from 0.7 to 1.4 in steps of 0.1 for these tests. At an equivalence ratio of 0.7 the flowing gasoline/air mixture would not ignite. At an equivalence ratio of 0.8, the mixture ignited, but the flame did not accelerate to detonation within the length of the run-up distance or an L/D of 90. Stable detonations were obtained for all test conditions where the equivalence ratio ranged from 0.9 to 1.4. However, there was no obvious correlation between equivalence ratio and run-up length. An L/D of 74 appeared to hold for all conditions tested above an equivalence ratio of 0.8.

A series of air flow tests were made to measure the pressure loss through the 15.2 cm diameter run-up piping with varying lengths of turbulent liners and without the liners. The measured data were used in the Darcy equation to determine the pipe wall friction factor. The Moody diagram (10) was used to find the relative roughness of the pipe. The lined pipe was found to have a relative roughness of 0.010 to 0.015, which is comparable to riveted steel pipe. The unlined pipe had a relative roughness less than 10^{-6}, which is less than any commercial grade smooth pipe.

Device Tests

The following devices were tested for detonation quench ability: pipe tees and rupture discs, two types of spiral-wound crimped ribbon arresters, metal foam arresters, two types of water-trap arresters, and a vertical packed bed. The following paragraphs discuss each device in more detail. The results are summarized in Table II.

Pipe Tees and Rupture Discs.
Initial tests were conducted on an arrangement which caused the detonation wave to make two 90° turns before proceeding along the length of pipe. Blowout discs inline with the detonation wave were installed along with a restrictor orifice plate downstream. Both rupture discs were blown out in all test firings and the detonation passed through the assembly with little reduction in pressure and velocity. There was no correlation between the magnitude of the pressure loss and the rupture disc pressure rating, which was varied between 345 and 4137 kN/m^2. Similar results were obtained when the rupture discs were installed non-inline.

Shand and Jurs Spiral Wound, Crimped Aluminum Ribbon Arrester Test Assembly.
The spiral wound, crimped aluminum ribbon core element was removed from a Shand and Jurs commercial flame arrester housing model No. 94/305/16/11. This arrester had a hydraulic diameter of 0.114 cm and an arrester length of 15.2 cm. The core element was installed into a high-pressure housing made from an extra strong 30.5 cm diameter pipe 15.2 cm long and was supported on each end of the housing by a mounting ring with retainer grid. The arrester test assembly is shown in Figure 3.

The Shand and Jurs arrester test configuration was subjected to nine stable detonations. There was no evidence in the witness section that detonation flame passed through the arrester. A low level peak pressure pulse of approximately 290 kN/m^2 was measured at the downstream side in the witness section. Changing the rupture disc rating and, in a later test, replacing the rupture disc with a blank flange, had no effect on the quenching capability of the arrester. In the final test, the arrester was removed and the two flanges bolted together. The detonation wave passed through the test section, verifying the functional capability of the flame sensors in the witness section.

The pre- and post-test measured pressure drop across the arrester test assembly averaged 0.152 kN/m^2 at an average flow rate of 4.68 m/s. Post-test inspection of the arrester showed a slight indentation of the retainer grid on the downstream side of the core element. There was a separation of the crimped ribbon layers near the outer perimeter of the core at the upstream interface with the mounting ring.

Amal Spiral Wound, Crimped Stainless Steel Ribbon Arrester Test Assembly.
The Amal spiral wound, crimped stainless steel ribbon arrester core element, model No. 188/905/15/18/CN, was installed in a high pressure housing 25.4 cm diameter by 3.8 cm long. This arrester had a hydraulic diameter of 0.038 cm and a length of 3.8 cm. It was installed in the test facility similarly to the Shand and Jurs Arrester. This Amal arrester test assembly was successful in stopping the detonation flame in only the first of four test firings. Post-test inspection of the core element showed no evidence of damage which could have caused the failures.

Whessoe Foamed Metal Arrester.
The Retimet 80-grade foamed nickel-chrome alloy core element was removed from a prototype Whessoe arrester and installed into a high pressure housing 25.4 cm diameter by 1.3 cm long. It was then placed into the test assembly as in the case of the spiral crimped ribbon of Figure 5. The arrester was successful in stopping the detonation flame in three tests. Pre-and post- pressure drop measured across the arrester test assembly was 0.241 kN/m^2 at an average flow velocity of 4.74 m/s. After the third test, the measured pressure loss dropped to 0.110 kN/m^2. Post-test inspection revealed that a section of the Retimet foamed metal core element had been extruded through the downstream retainer grid. The

80-grade foamed metal was replaced with a 45-grade foamed metal element and reinstalled. An additional test firing was made but this arrester failed.

Water Trap Arrester Test Assembly.

The water trap arrester test assembly is illustrated in Figure 4. The arrester vessel contains two 15.2 cm diameter flanged side ports, one of which contained a pipe elbow inside the vessel with a 30.5 to 15.2 cm diameter concentric pipe reducer directed towards the bottom. The lower end of the reducer had three rows of 2.54 cm diameter holes drilled through the wall so that the assembly functioned as a bell mouth diffuser. There was no flame penetration of the water trap arrester in five tests and there was no measurable peak pressure pulse passing through the witness section. However, the water level in the arrester was reduced by 0.64 to 1.91 cm during each test and was replaced before the next test firing. The loss could be related to the length of air flow run-time before each test firing. There was no evidence of water in the downstream piping or at the shock tube exit. Pre-and post-test pressure loss measured across the water trap arrester test assembly averaged 1.655 kN/m^2 at an average flow velocity of 4.66 m/s.

In the sixth test, all the water was drained from the arrester vessel before firing. The detonation passed through the arrester test configuration.

Vertical Bed of Aluminum Ballast Rings Arrester Test Assembly.

The vertical bed arrester test assembly (figure 5) utilized the water trap vessel for the lower housing. The vessel was disconnected from the covered inlet pipe elbow and reversed 180°, exchanging the inlet and outlet ports. A blind flange was used to close off the unused outlet port. The vessel cover was removed and a 45.5 cm diameter flange pipe spool section was placed on top. A support grid ring covered with heavy wire mesh was installed into the housing assembly just above the inlet port. The internal volume between the inlet port and the outlet port, a distance of 63.5 cm, was filled with 2.45 cm diameter by 2.54 cm long aluminum ballast rings (pall rings). A second wire mesh grid ring was installed on the top of the rings and held securely in place.

Five tests were made with the vertical bed arrester. The detonation flame was arrested in all tests and there were no measurable peak pressure pulses passing through the witness section. Pre-and post-test pressure loss measured across the arrester averaged 0.05 kN/m^2. Post test inspection revealed compaction of the bed and some distortion in the pall rings.

Linde Hydraulic Back-Pressure Valve Arrester Test Assembly.

The Linde Hydraulic back-pressure valve was a standard commercial unit model No. H-20 which had to be modified by removing a 7.6 cm diameter one-way check valve on the inlet piping and the enlargement of the inlet port to 10.2 cm. It was then connected to the 15.2 cm diameter pipe tees with reducer elbows as shown in Figure 6. The arrester was filled with water to a depth of 15.2 cm.

Five test firings were made on the Linde arrester configuration. The detonation flame was quenched in all cases and there was no measurable peak pressure pulse passing through the witness section. The water level loss ranged from 0.63 to 4.45 cm per test depending on the length of air flow run-in time. No visible quantity of water was detected in the downstream piping or at the exit of the shock tube. Pre- and post-test pressure loss across the arrester averaged 2.027 kN/m^2 at an average flow velocity of 4.65 m/s. Further tests with the water completely drained produced flame-through to the witness section.

Arrester Evaluation Tests Using Downstream Ignition.

Because the water trap and Linde arresters had significant differences in the geometric arrangements between inlet and outlet configurations, it was desirable to test them with downstream ignition so that the detonation wave would enter the arrester through the arrester outlet. The shock tube was reversed for downstream ignition. The detonation wave had the same terminal velocity and peak pressure pulse although the mixture flow rate was slower (1.5 m/s) and propagation was against the mixture flow. The results of the tests with the two arresters were the same as in the experiments reported above. Both devices sucessfully quenched the flame and detonation pressure wave.

Parametric Tests of Two Arresters

From the above series of tests, two arresters were selected for further testing which included attempts of optimization of each arrester design. Gasoline was used as the fuel in all these tests. Of the arresters which successfully quenched the detonation, the Shand and Jurs crimped ribbon and the vertical packed bed arresters were judged to be the most practical for field use. Both were made of off-the-shelf components and required a minimum amount of maintenance. Because of the problem of maintaining the water level in the water trap and Linde arresters these two were considered to be significantly less reliable than the Shand and Jurs or the packed bed arresters. The metal foam arrester was not chosen because of the question of its structural capability to sustain multiple detonations.

Shand and Jurs Crimped Ribbon Arrester

Tests were conducted with various arrester lengths in the case of the Shand and Jurs Arrester in order to investigate the relationship between arrester length and the downstream pressure pulse. The existence of this low-level pressure pulse passing through the arrester caused some concern as to the possibility of a reignition occuring in the combustible fuel and air mixture in the piping at some distance beyond the arrester. Therefore, tests were carried out with arrester lengths of 15.2 cm, 20.3 cm, and 30.5 cm with resulting downstream pressures of 344 kN/m^2, 320 kN/m^2, and 260 kN/m^2. If one extrapolates the apparent linear relationship between arrester core length and downstream peak pressure pulse to zero pulse pressure, the arrester core length would be 58.5 cm. It was impractical to evaluate this result by further testing since the hardware was not commercially available. The 20.3 cm long Shand and Jurs Arrester was selected for further testing.

This arrester was put through a battery of tests which were designed to more closely simulate an actual plant scenario. The arrester was connected inline with the shock tube and the arrester quenched the detonation. Next, a 13.4 m extension of piping was placed between the witness section and the verification section. This was done in order to explore the possibility of reignition occuring at some distance downstream of the arrester. After a detonation was initated in the tube and quenched by the arrester, the fuel/air mixture was allowed to continue to flow for up to 2 minutes. No reignitions occured. A summary of the tests is given in Table III.

Vertical Bed of Aluminum Ballast Rings Arrester Test Assembly.

Efforts were made to decrease the size of the vertical packed bed arrester and to check the significance of the packing size. The arrester was tested with four bed diameters, three bed depths, three ballast ring sizes, and increasing rupture disc pressure rating from 690 kN/m^2 to a blind flange condition. The diameters tested were 43.2 cm, 33.7 cm, 30.5 cm, and 25.4 cm. The bed depths tested were 63.5 cm, 45.7 cm, and 22.9 cm. The arrester failed when the bed depth was decreased to 22.9 cm (diameter = 43.2 cm). When the arrester was packed with the 3.81. and 5.08 cm diameter Pall rings, it failed. The final arrester configuration which quenched all detonations had a diameter of 25.4 cm and a bed depth 45.7 cm, and utilized Ballast Rings of 2.54 cm diameter and 2.54 cm length as the packing.

This last arrester assembly was subjected to the same continuous flow tests which were carried out on the Shand and Jurs Arresters. As in the continuous flow tests, the detonation flame was arrested on all test firings and there was no evidence of reignition in the witness or extension sections. No lingering flame was observed in the test run-up section and the pressure pulse was observed to decay in the extension section. A summary of the tests is given in Table IV.

DISCUSSION

The characteristics of the detonation used in these tests compare quite well with those produced by other investigators. The run-up distance with an L/D of 74 agrees reasonably well with empirical correlations (11, 12). Detonation tests performed with propane/air mixtures in a 150 mm diameter shock tube (13) produced a detonation run-up L/D of approximately 80 which is very close to 74 of this study. In addition, the flame velocity of the stable detonation (approximately 1800 m/s) obtained with the gasoline/air mixture was similar to those of other studies. The study of detonation in propane/air mixtures cited above (13) showed steady velocities of 1850 - 2000m/sec. Another study performed with 'petrol' produced detonations with velocities of approximately 1750 m/sec (14).

The results of the device testing paralleled those of other investigations with a few exceptions. In our tests four types of arresters were successful: the spiral wound crimped metal; the foamed nickel-chrome metal; vertical packed bed of aluminum Ballasts rings; and the water trap or hydraulic back-pressure valve.

Howard (15), using a 10.18 cm shock tube with a 90% hydrogen-10 % methane mixture, tested Amal arresters of similar configuration to the the one tested in this study. Detonation velocities were on the order of 2000 m/sec. Howard's study showed that some of the detonations were quenched while others were not. These results agree with those of our study.

Howard (15) also tested a type of water seal arrester, but obtained different results from ours. The initial shock wave was quenched but reignition phenomena occurred and there was a secondary explosion in the apparatus. One explanation the author offered for the failures included the displacement of significant amounts of the water from the arrester with passage of the detonation. Another of his explanations was that the flame front may have heated the internal apparatus above the autoignition temperature of the hydrogen gas.

Our experience with gasoline/air mixtures has been that the water trap functions quite successfully. Since the Linde Arrester has performed so well in an acetylene plant accident (9), perhaps the failures cited above were due to a design flaw. One negative feature of the water trap is the loss of the hydraulic fluid due to vaporization and mass transfer to the flowing fuel/air mixture. Unless some means is found to continuously replenish this fluid, the arrester after some period of time would be ineffective. It is felt that in the marine terminal, the requirement for replenishment of the liquid would be neglected and that the arrester might provide a false sense of safety.

Experience of other investigators with the packed bed arrester seems to support the results of this study. Broschka and Will (16) set up detonations in a butane/air mixture within a 7.63 cm diameter pipe. A 2.5 meter expansion of the tube to a 15.2 cm diameter pipe was packed with 0.95 cm diameter pall rings. Detonations were arrested in all cases. Other tests with acetylene were carried out by Miller and Penny (17). They used a tower 25.4 cm in diameter, 1.22 m high, packed with 2.54 cm diameter raschig rings, and irrigated with water. All detonations were arrested within 46 cm of dry packing and within a few centimeters of irrigated packing. (Although the irrigation of the arrester seems to provide some additional factor of safety, this measure would require additional piping for the water supply and would probably increase pressure drop across the arrester).

The results of the testing of the foamed metal arresters supported earlier tests carried out by Davies et al (18) representing Dunlop Ltd, and Barton, et al (19) of the Ministry of Defence, United Kingdom. These studies have shown that the device quenches detonations in high pressure acetylene and low pressure oxy-acetylene lines. Our findings with gasoline/air mixtures demonstrated that the 80 grade foam would stop detonations but that the 45 grade would not. The major drawback to this type of arrester is that it is succeptible to structural damage and that it may fail in an accident scenario involving multiple detonations.

RECOMMENDATIONS

The first step in designing a piping system which will carry fuel/air mixtures is to insure that the mixture is not within the flammable range. However, to depend on this measure alone as an insurance of safe operation is impractical. Use of passive detonation arrester devices to protect critical elements of the piping system is an appropriate safety measure. Of the successful devices tested in this study, the Shand and Jurs crimped ribbon arrester and the packed bed arresters have been judged to be the most practical and reliable. Each is a passive device and has a low pressure drop associated with gas flow through it. It is interesting to note that these devices are similar in concept to those specified as appropriate detonation arresters in West Germany (20).

It is emphasized that the data reported is strictly applicable to gasoline/air mixtures in 15.2 cm diameter pipe. It is felt however that the results may be extended to other saturated hydrocarbon fuels and to piping systems less than 15.2 cm diameter. They may even be scaled up in a limited range to large sizes, 20.3 cm or 25.4 cm diameter with a reasonable degree of confidence. With much larger pipe size, 30.5 cm to 61.0 cm diameter, it would be highly desirable to obtain additional data concerning the deflagration to detonation transition and the required arrester sizing, configuration, and durability for the quenching of a detonation.

LITERATURE CITED

(1) Lewis, B., and von Elbe, G., Combustion, Flames and Explosion of Gases, Academic Press, Inc., New York, p. 547,(1961).

(2) Urtiew, P.A. and Oppenheim, A.K., "Onset of Detonation," Comb. and Flame 9, pp. 405-407 (1965).

(3) Urtiew, P.A., and Oppenheim, A.J., "Transverse Flame-Shock Interactions in an Explosive Gas," Proc. Roy. Soc Lond. A 304, pp.379-385 (1968).

(4) Craven, A.D., and Grieg, T.R., "The Development of Detonation Over Pressures in Pipelines," I|Chem E. Symposium Series 25, pp. 41-50, (1968)

(5) Ginsburg, I., and Bulkley, W.L., "Hydrocarbon-Air Detonations .... Industrial Aspects," Chem. Eng Progress 59, pp. 82-86, (1963).

(6) Jacobs, R.B., "Occurrence and Nature of Hydrocarbon Detonations," Processing American Petroleum Institute 39, pp. 15-20, (1959).

(7) Pipkin, O.A. "Detonation - Old Processes Are Not Immune," IBID., pp. 21-33, (1959).

(8) Jacobs, R.B., et al, "Destruction of a Large Refining Unit by Caseous Detonation," Chem. Eng. Prog. 53, pp. 565-573 (1957).

(9) Sutherland, M|E| and Wegert, H.W., "An Acetylene Decomposition Incident," Chem. Eng. Prog. 69 pp. 48-51,(1973).

(10) Moody, L.F. "Friction Factors for Pipe Flow,"American Society of Mechanical Engineers Transactions, 66, no. 8, pp. 671-684, (1944).

(11) Bollinger, L.E., et al, "Experimental Measurements and Theoretical Analysis of Detonation Induction Distances, " American Rocket Society J. 31, pp. 588-595, (1961).

(12) Schelkin, K.I. and Troshin, Ya. K., Gasdynamics of Combustion, (B.W. Kuvshinoff and L. Holtschlag, transl) Mono Book Co., Baltimore, pp. 174-179, (1965)

(13) Holzapfel, G., and Schoen, G. "Fur Ausbilding von Gasdetonationen in Rohren", Chemie-Ing. - Techn., 37, pp 493-497, (1965).

(14) Fraser, R.F. "Detonation Velocities in Liquid Fuel Vapors with Air or Oxygen at 100°C and Atmospheric Pressure.", 7th Inte rnational Symposium on Combustion pp. 783-788 , (1958).

(15) Howard, Walter B. "Flame Arresters for H igh-Hydrogen Content Fuel-Air Mixtures," 79th National Mtg, AICHE, (1975).

(16) Broschke, G.L., and Will, R.G. A Study o f Flame Arresters in Piping Systems, Project 3721, Amoco Oil Company, Naperville, Ill., (1 975).

(17) Miller, S.A. and Penny, E. "Hazards in H andling Acetylene in Chemical Processes," Sym posium on Chemical Process Hazards Inst. Chem . Engrs. pp. 92-94 (1960).

(18) Davies, J.P. et al, "Performance of the Metallic Foams as Flame Arresters," Dunlop Lt d.

(19) Barber, R.R. et al. "The use of Reticula ted Metal Foam as as Flashback Arrester Eleme nts," Ministry of Defense, FMz Branch, RARDE, London.

(20) Schon, G. et al, Flammendurchschlagsiche re Einrichtungen, PTB Mitteilungen Deutscher Eichverlag GMBH .33 Braunschweig (1968).

FLESSNER, M.

BJORKLUND, R.A.

Table I. Symbols and Description for Flow System Schematic Diagram.

Symbol	Description
	Manual Globe Valve
	Electric solenoid operated valve
	Electric motor operated valve
	Electric motor operated ball valve
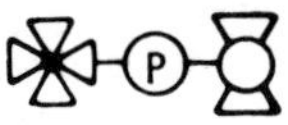	Air piston operated ball valve
	One-way flow check valve
	Pressure relief safety valve
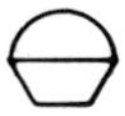	Dome pressure regulator valve
	Manual set pressure regulator valve
	Electric motor operated pressure Regulator valve (dome loader)
	Pressure rupture-disc assembly
	Pressure gage
	Voltmeter transducer
	Ammeter transducer
	Temperature transducer
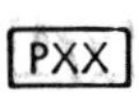	Pressure transducer
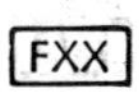	Flame sensor transducer
	Flowmeter transducer

Table II. Summary of Detonation Flame Arrester Test Results

Ignition Location and Type of Arrester	Number of Detonations Sustained	Detonations Quenched		Air Velocity (averaged)		Pressure Loss (averaged)	
		Yes	No	m/s	(ft/s)	kN/m^2	(psid)
A. Upstream Ignition							
1. Two pipe tees, rupture discs, and orifice plate	4		4	4.68	(15.35)	0.200	(0.029)
2. Pipe tee, pipe elbow, rupture disc, and orifice plate	4		4	4.71	(15.46)	0.214	(0.031)
3. Shand and Jurs spiral wound, crimped aluminum ribbon	9	9		4.68	(15.35)	0.152	(0.022)
4. Amal spiral wound, crimped stainless steel ribbon	4	1	3	4.69	(15.39)	0.428	(0.062)
5. Whessoe 80-grade Retimet foamed metal	3	3		4.74	(15.54)	0.241	(0.035)
6. Whessoe 45-grade Retimet foamed metal	1		1	4.78	(15.68)	0.130	(0.019)
7. Water trap with water	5	5		4.66	(15.29)	1.655	(0.240)
8. Water trap without water	1		1	4.84	(15.88)	0.028	(0.004)
9. Vertical stacked bed of aluminum Ballast rings	5	5		4.64	(15.23)	0.048	(0.007)
10. Linde hydraulic back-pressure valve with water	6	6		4.65	(15.27)	2.027	(0.294)
11. Linde hydraulic back-pressure valve without water	3		3	4.63	(15.20)	0.407	(0.059)
B. Downstream Ignition							
1. Linde hydraulic back-pressure valve with water	6	6		1.61	(5.28)	1.917	(0.278)
2. Water trap with water	5	5		1.52	(5.0)	1.393	(0.202)

Table III. Summary of Parametric Test Results for the Shand and Jurs Spiral Wound, Crimped Stainless Steel Ribbon Arrester Assemblies

Arrester Core Size: Diameter, cm (in)	Arrester Core Size: Length, cm (in)	Inlet Pipe Connection Configuration	Rupture Disc Pressure Rating, kN/m² (psid)	Number of Detonations Sustained: Yes	Number of Detonations Sustained: No	Run-Up Pipe Air Velocity, (averaged) m/s (ft/s)	Arrester Pressure Loss, (averaged) kN/m² (psid)	Downstream Pressure Pulse, (averaged) kN/m² (psid)
30.5 (12)	15.2 (6)	Indirect	690 (100)	4	1	4.78 (15.68)	0.151 (0.022)	344 (49.9)
30.5 (12)	20.3 (8)	Indirect	690 (100)	6	0	4.58 (15.89)	0.152 (0.022)	320 (46.4)
30.5 (12)	30.5 (12)	Indirect	690 (100)	6	0	4.48 (14.70)	0.230 (0.033)	260 (37.7)
30.5 (12)	20.3 (8)	Direct	690 (100)	1	0	4.35 (14.27)	0.134 (0.019)	318 (46.1)
30.5 (12)	20.3 (8)	Direct	2068 (300)	1	0	4.38 (14.37)	0.138 (0.020)	334 (48.4)
30.5 (12)	20.3 (8)	Direct	4137 (600)	1	0	4.55 (14.93)	0.138 (0.020)	325 (47.1)
30.5 (12)	20.3 (8)	Direct	Blanked-off	3	0	4.46 (14.63)	0.138 (0.020)	323 (46.8)

Table IV. Summary of Parametric Test Results for the Vertical Bed of Aluminum Ballast Ring Arrester Assemblies

Arrester Bed Size: Diameter, cm (in)	Arrester Bed Size: Depth, cm (in)	Ballast Ring Size, (dia. & lg.) cm (in)	Rupture Disc Pressure Rating, kN/m² (psid)	Number of Detonations Sustained: Yes	Number of Detonations Sustained: No	Run-Up Pipe Air Velocity, (averaged) m/s (ft/s)	Arrester Pressure Loss, (averaged) kN/m² (psid)	Downstream Pressure Pulse, (averaged) kN/m² (psid)
43.2 (17)	63.5 (25)	2.45 (1.0)	690 (100)	6	0	4.52 (14.82)	0.075 (0.011)	181 (26.3)
43.2 (17)	45.7 (18)	2.45 (1.0)	690 (100)	6	0	4.62 (15.14)	0.053 (0.008)	197 (28.6)
43.2 (17)	22.9 (9)	2.45 (1.0)	690 (100)	4	2	4.59 (15.04)	0.052 (0.007)	236 (34.2)
43.2 (17)	63.5 (25)	3.81 (1.5)	690 (100)	2	2	4.73 (15.52)	0.067 (0.010)	290 (42.1)
43.2 (17)	63.5 (25)	5.08 (2.0)	690 (100)	1	2	4.68 (15.35)	0.056 (0.008)	304 (44.1)
33.7 (13.3)	63.5 (25)	2.45 (1.0)	690 (100)	6	0	4.55 (14.93)	0.131 (0.019)	149 (21.6)
30.5 (12)	63.5 (25)	2.45 (1.0)	690 (100)	6	0	4.51 (14.78)	0.147 (0.021)	142 (20.6)
25.4 (10)	63.5 (25)	2.45 (1.0)	690 (100)	7	0	4.62 (15.14)	0.278 (0.040)	132 (19.1)
25.4 (10)	45.7 (18)	2.45 (1.0)	690 (100)	1	0	4.55 (14.93)	0.162 (0.023)	142 (20.6)
25.4 (10)	45.7 (18)	2.45 (1.0)	2068 (300)	1	0	4.53 (14.86)	0.197 (0.028)	141 (20.4)
25.4 (10)	45.7 (18)	2.45 (1.0)	4137 (600)	1	0	4.60 (15.09)	0.197 (0.28)	141 (20.4)
25.4 (10)	45.7 (18)	2.45 (1.0)	Blanked-off	3	0	4.57 (14.99)	0.220 (0.032)	164 (23.8)

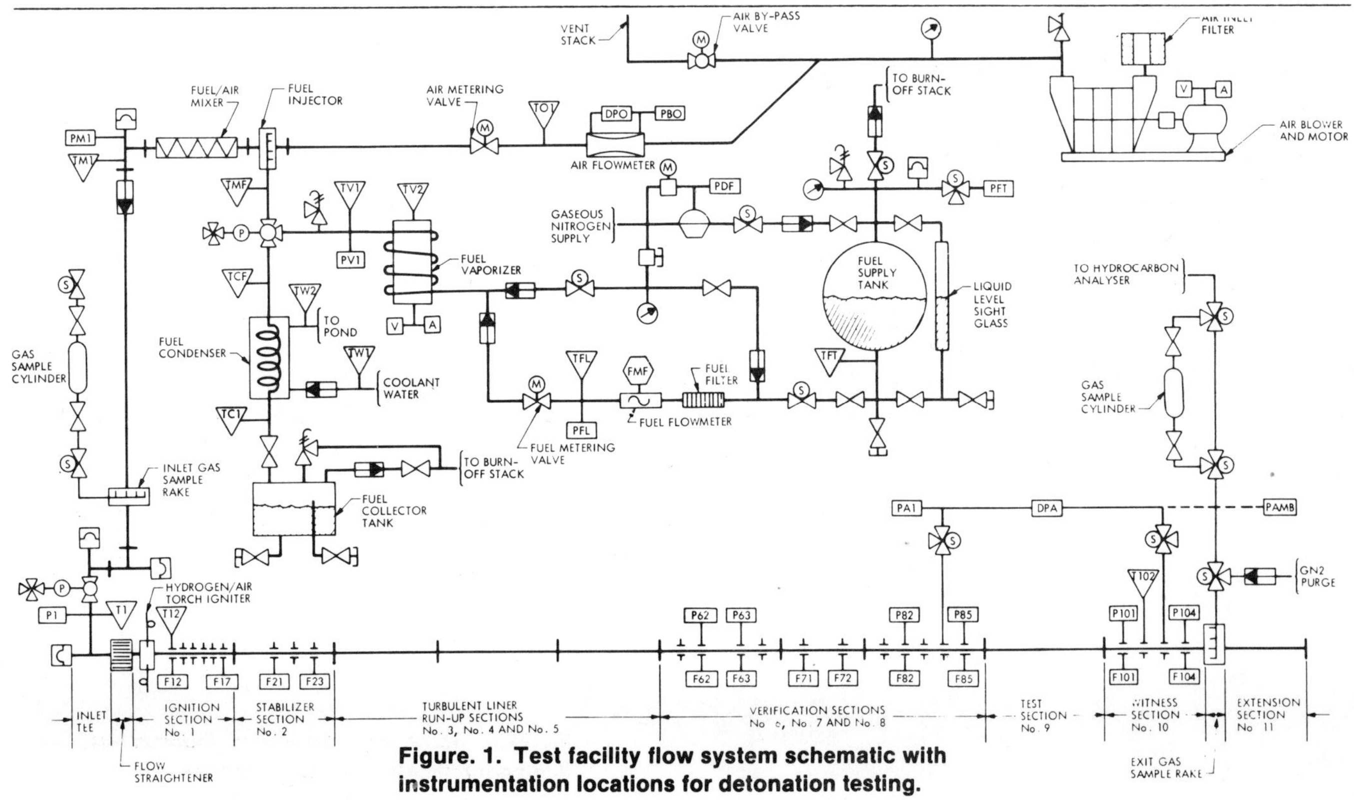

Figure. 1. Test facility flow system schematic with instrumentation locations for detonation testing.

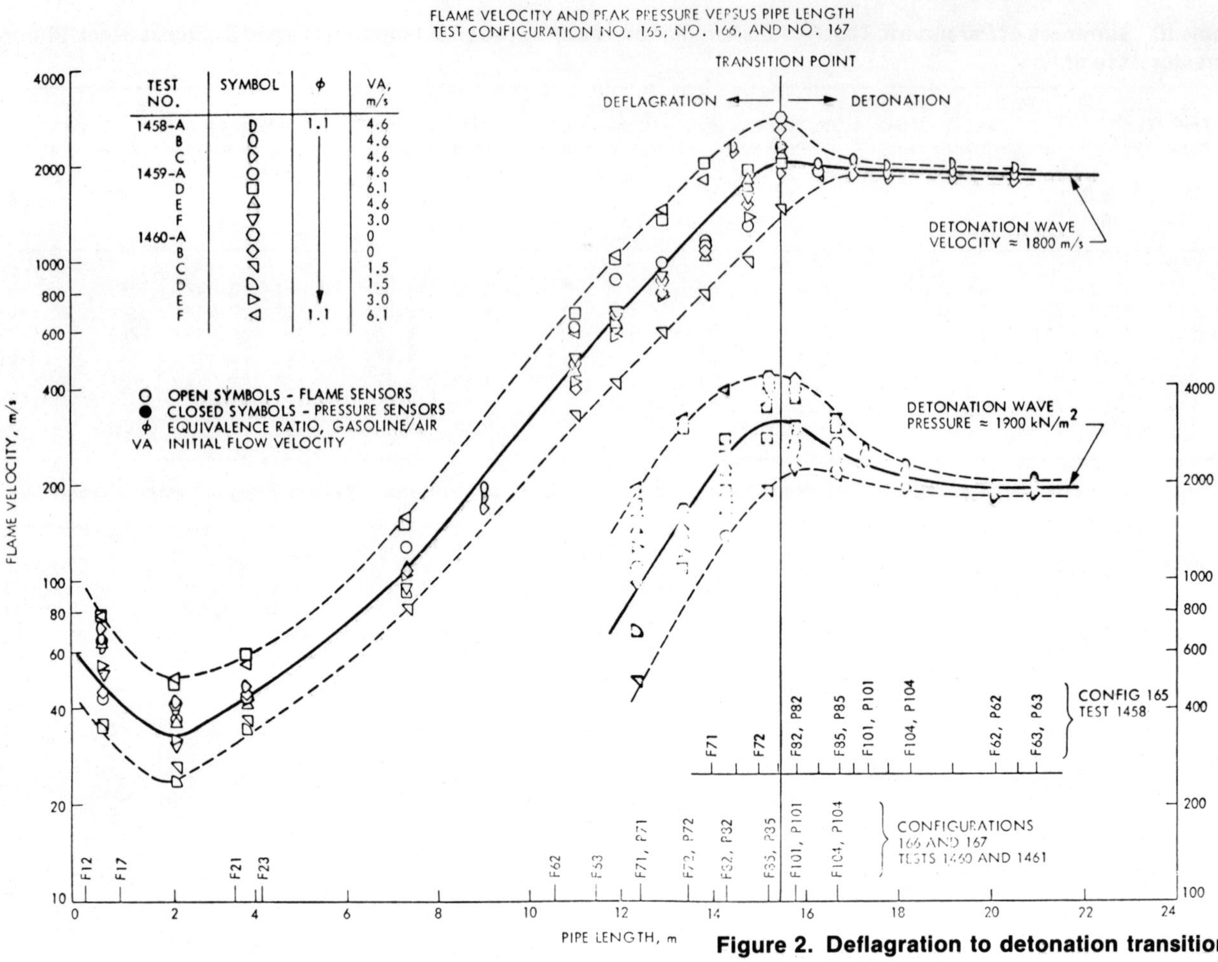

Figure 2. Deflagration to detonation transition for gasoline air mixtures.

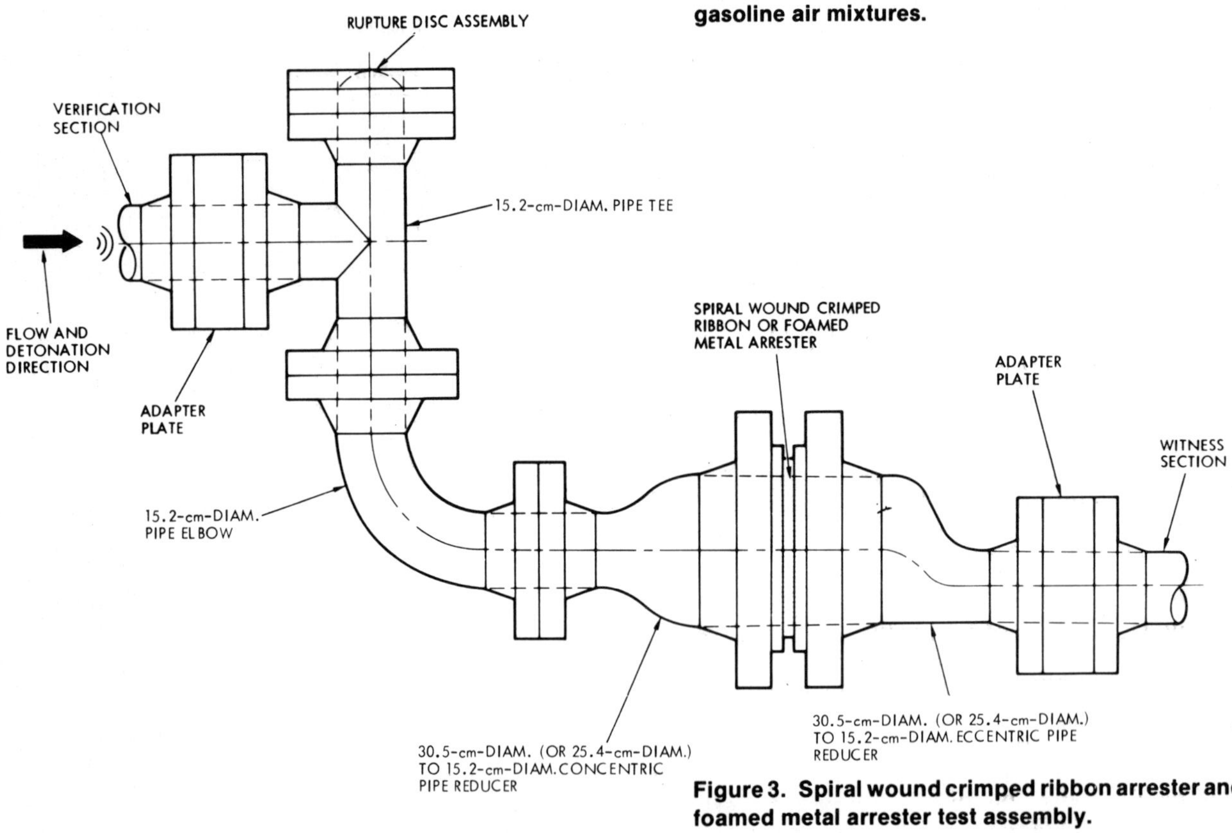

Figure 3. Spiral wound crimped ribbon arrester and foamed metal arrester test assembly.

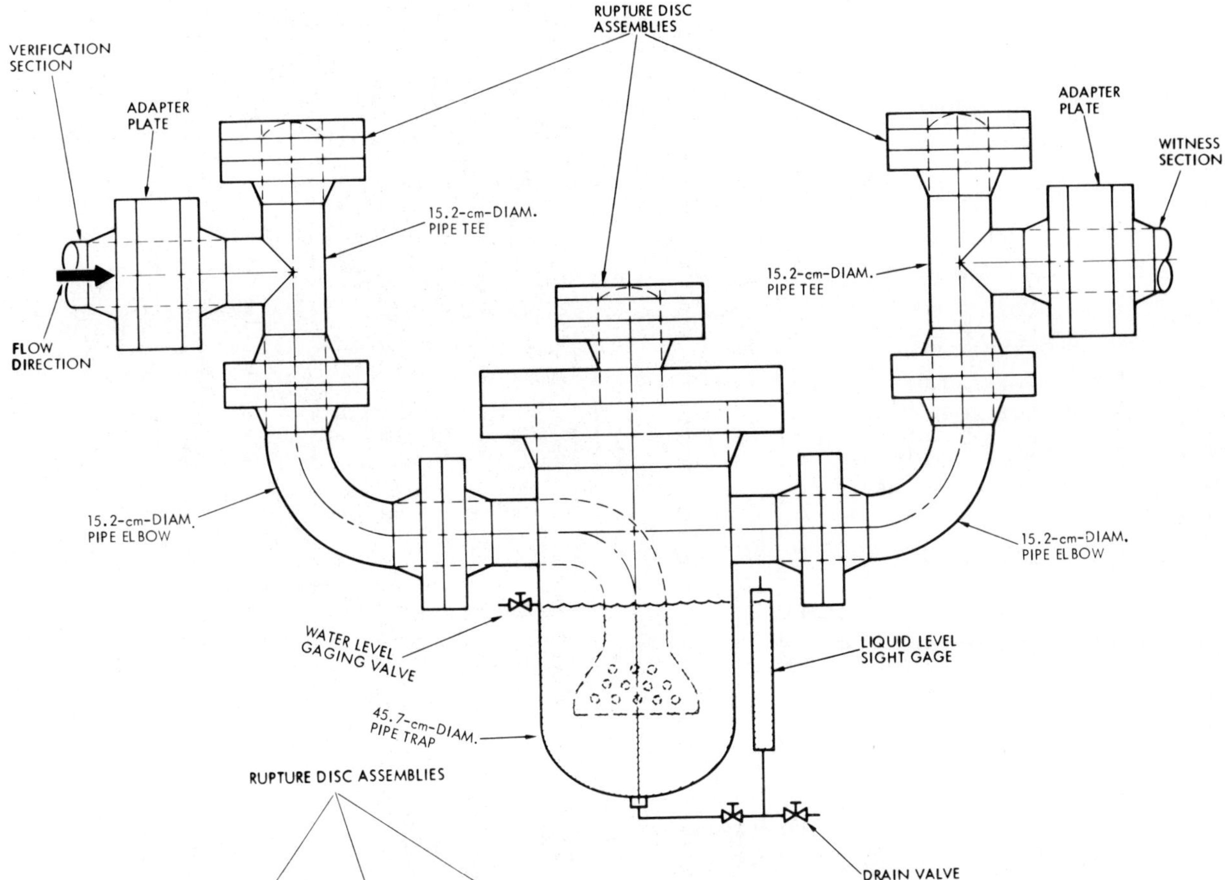

Figure 4. Water trap arrester test assembly.

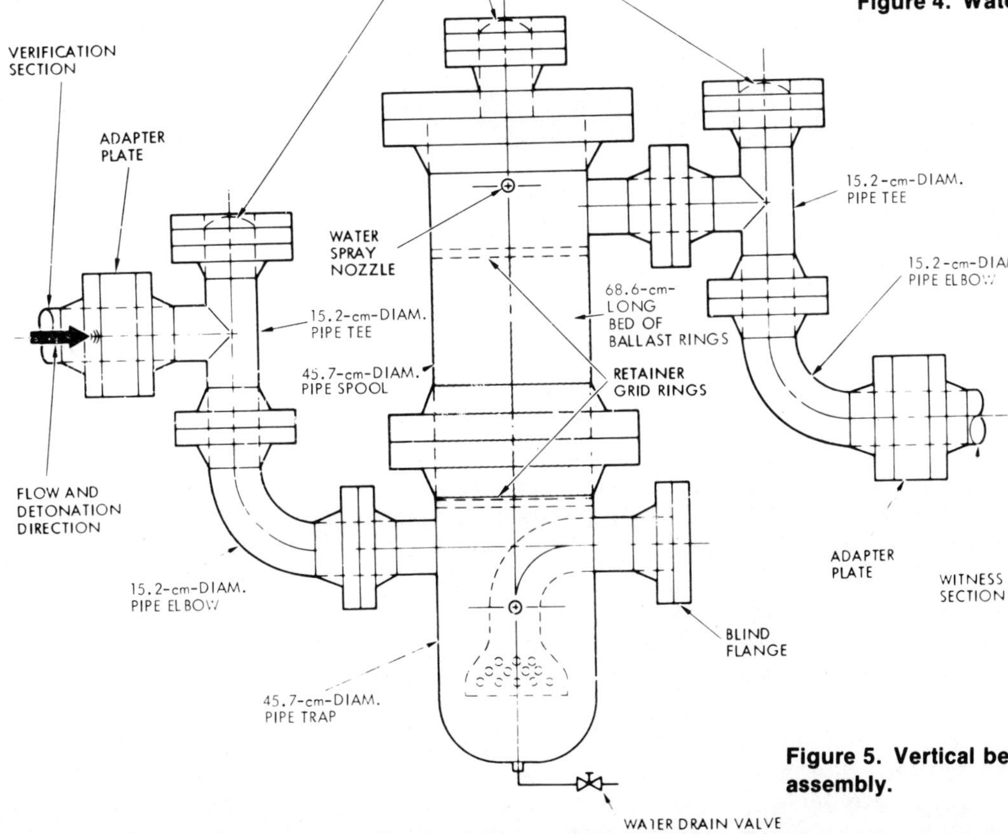

Figure 5. Vertical bed of ballast rings arrester test assembly.

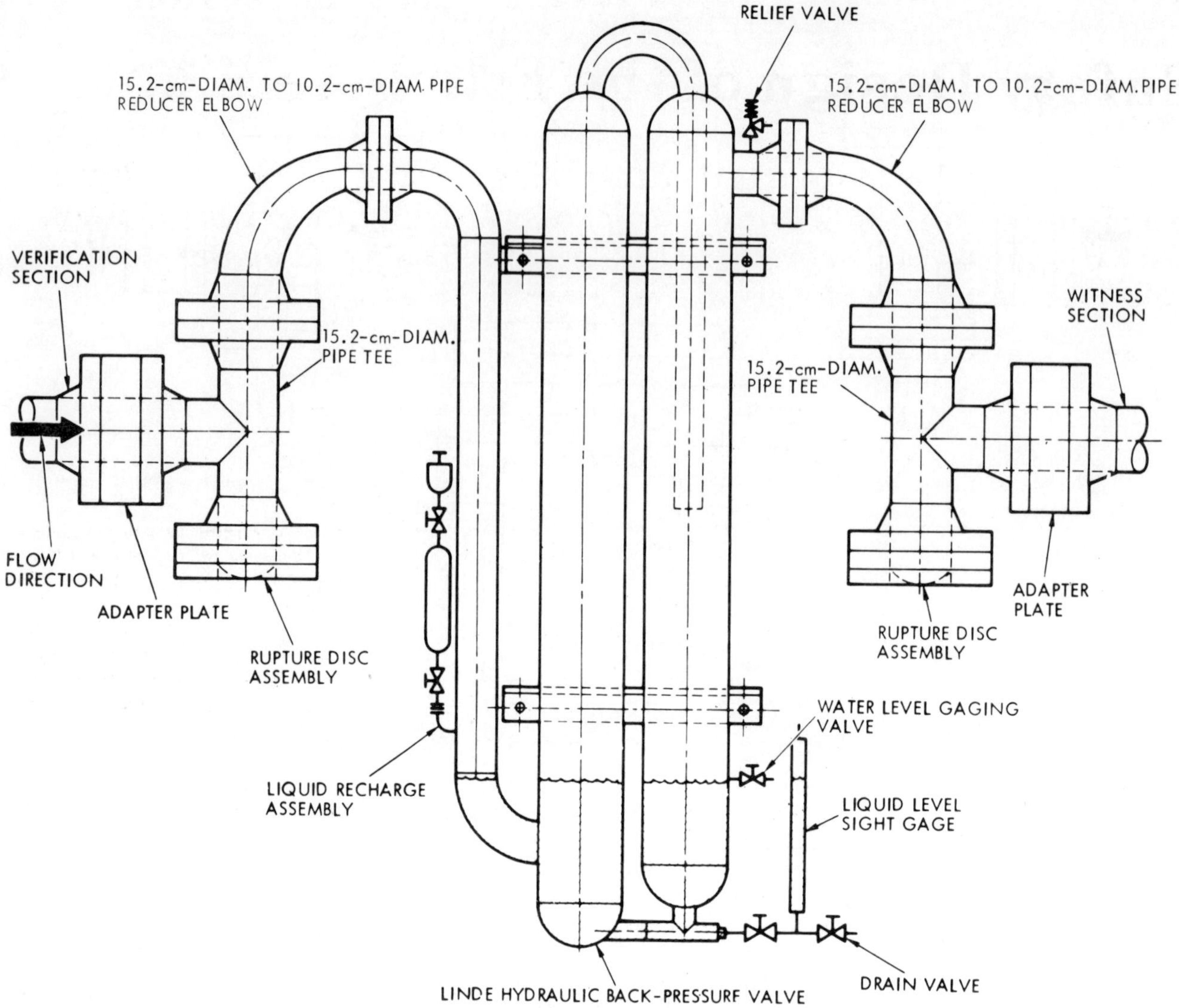

Figure 6. Linde hydraulic back-pressure valve arrester test.

DISCUSSION

W.B. HOWARD, Monsanto: First, I would like to compliment the author on giving us some good, new, "hard number" safety information. It isn't easy to get new data, and we are very grateful for this. I did have a question. On the test that you did with slow velocity flames, deflagrations, have you done those under conditions which would allow pressure to develop in the piping system before the flame reached the face of the arrester with your 30 mesh screen? What I am wondering about is the pressure effect which can develop with the deflagration.

M. FLESSNER, USCG: No, the specific tests that I alluded to have been carried out to simulate a flame propagating in the open environment, and actually what we had was a large conduit pipe and it had a very flimsy plastic bag on each end sealing the compartment. The plastic seals have been specifically designed to produce what we would call flashback flames and a flashback flame in our definition is one that is propogating completely unconfined; this particular housing ruptured at about 0.1 16/in^2. In earlier work, we did test the screens installed within the pipe, and I do not recommend that screens be placed anywhere except at the end of a pipe. I would not, even with methane, place screens of any dimension anywhere but at the end of a pipe. I think as flashback flame arresters the screens are fine. Note, however, that all these devices are safe for saturated hydrocarbons, but not for ethylene or acetylene or other unsaturated hydrocarbons.

Safety Design of the Ethylene Plant

To eliminate the potential safety hazards, a plant handling hazardous materials should be properly designed with various safety devices. However, the same emphasis should be put on the perfection of humans as on plant design.

T. Bøckmann, G. H. Ingebrigtsen and T. Håkstad
Norsk Hydro a. s., Oslo 2, Norway

Prior to the discovery of oil and gas in the Norwegian part of the North Sea, the petrochemical industry in Norway was limited. When the Norwegian Government in 1974 obtained an option for supply of NGL to Norway, a new basis for such industry was created.

Subsequently, a large petrochemical complex has been built in Bamble, South Norway. Except for the HD polyethylene plant, which was put into operation in 1979, the other plants (an ethylene plant, a chlorine plant, a VCM plant, a LD polyethylene plant, and a polypropylene plant) were all completed and in production by 1978.

0149-3701/81/3807 $02.00 © 1981 AIChE

The safety designs reported here refer to the ethylene plant which has been in operation since 1977. The plant has a capacity of 300,000 ton/yr ethylene and 70,000 ton/yr propylene.

Since the safety philosophy for such plant involves all stages of the project, it is difficult to cover all subjects. Therefore, some specific topics, basically to the plant safety design, are being discussed here to convey some of our thinking on this complicated subject.

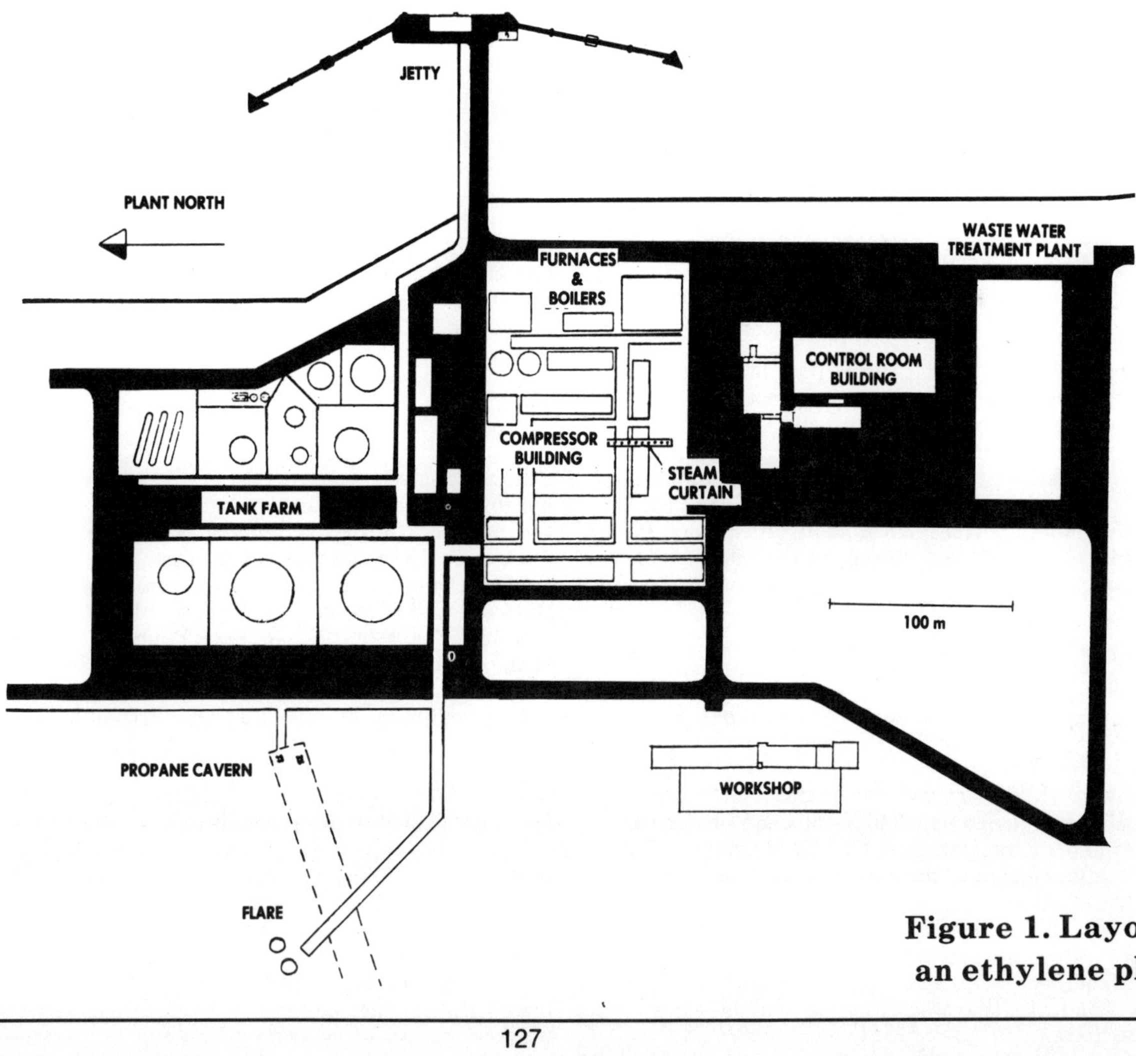

Figure 1. Layout of an ethylene plant.

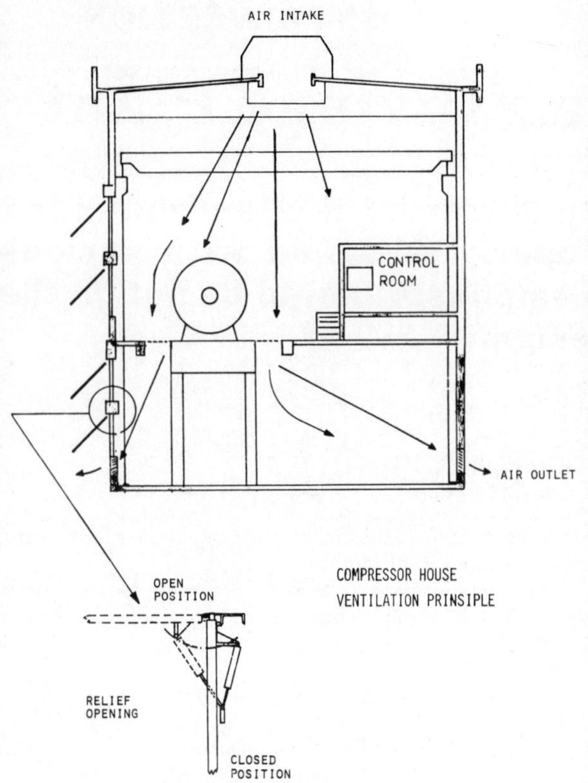

Figure 2. Ventilation principle.

Some of the questions we raised at the beginning of the planning and design phase were:

- To what extent could experience from accidents or "near accidents" in other plants be incorporated in the design of our plant to avoid similar accidents?
- When accidents happen, what are the main reasons for injury and death to people inside and outside such plants, and how could we reduce such risks?
- How could the environmental requirements for the plant be met without jeopardizing the desired safety aspects?

Norwegian law required the authorities to approve the plant with respect to safety. The existing national code of safety was, however, hardly adequate and suitable due to previous lack of such industry in Norway. Therefore, the government authorities, and the designers and operators of the plant jointly worked out a new national code of safety for oil refineries and petrochemical plants in Norway. Part of this work is thus reflected in some of the safety designs chosen for our plant.

Safety strategy for plant layout

The plant consists of a process area, tankage area, flare area, maintenance area, waste water treatment plant, and a jetty, Figure 1. Available land with suitable ground quality was limited; but to obtain an adequate spacing between the various plants, an area of 230,000 m^2 (57 acres) was allocated for the ethylene plant. Hazardous equipment such as furnaces, boilers, reactors, and blowdown drums, are all located on an open side of the process area. The pyrolysis furnaces and auxiliary boilers are located on the east side facing the sea. The reactors and blowdown systems are located on the opposite side together with the fractionation columns. This was done with due consideration to the prevailing winds in the area, running mainly along on north-south axis.

Safety considerations were paramount in selecting the location of important buildings such as control rooms, laboratory, offices, etc., as several reports on explosions in other similar type of plants concluded that collapsed buildings and glass splinters were one of the main reasons for injuries and death to people.

The obvious approach to that problem was to look into the design and location of buildings in which a number of people most likely could be present. Typical was the control room building in which the laboratory, switchroom, emergency power unit, and offices were located, attached to the control room.

The strategy was to locate this building on the periphery of the plant, not between the process and tankage area which for practical and economical reasons could be favorable. We decided to locate it south of the process area, 30 m from any hazardous equipment. This was also a minimum distance required by the authorities. Only the instruments for the tankage and the jetty were located in a small control room building between the process and tankage area. All essential instruments were, however, also in the main control room so that the tankage control did not require permanent manning.

The main compressor building (to be discused further later on), being a closed building 55 m long and 20 m high, is located in the middle of the process area. This building is thus a physical hindrance for the movement of a potential hydrocarbon gas cloud from the fractionation section to the furnace section where ignition can occur.

To extend this barrier, we have installed a steam curtain on a concrete wall in the longitudinal direction south of the compressor building. This steam curtain has to be actuated from outside the control room, so that the operator can assure himself that people are not hurt by the steam.

The tankage area located north of the process area was arranged so that the distance between the large cold tanks and the furnace section should be as great as practically possible. The distance between the ethylene tank and the furnaces is 140 m, an even greater distance than the insurance companies recommended (minimum of 100 m). We should like to have had an even greater distance.

The flare area is located on a hill top 45 m above the plant. The high flare stack is 90 m high. The location was chosen both from an environmental point of view as well as for safety reasons. The underground storage for propane, being a rock cavern of 100,000 m^3, is 150 m underneath the flare area.

Implementation by hazard analysis

To establish the design criteria for the various buildings, a hazard analysis (consequence analysis) was carried out. It approximates the consequences of a major gas leakage in the cold section of the process plant with a subsequent ignition in the hot section of the plant.

After evaluation of this analysis, it was decided to design the front wall of the control room building for an explosion pressure of 60 kPa (0.6 bar, 9 psig) and the remaining part of the building for a pressure of 30 kPa (0.3 bar, 4.5 psig). The tankage control room had to be designed for 60 kPa (0.6 bar) all over. To meet these requirements, buildings of monolithic concrete had to be built. Furthermore, special consideration had to be taken to avoid weak sections of the building which normally are windows, doors, ventilation openings, cable and pipe breakthroughs, etc.

Due to these severe design requirements for the front wall of the main building and for the tankage control building, windows could not be certainly installed. The exit doors from the control rooms were designed for the same pressure as the front wall.

The only parts of the main building with windows installed are the laboratory and office sections with windows facing away from the plant. An extensive test program was carried out to find an acceptable window design: we concluded that polycarbonate glass supported in special steel frames gave the best results. These windows proved to withstand an explosion pressure of 30 kPa (0.3 bar). These windows were not only installed in the buildings in the plant (main building and control room inside the compressor building), but also in the buildings in the near vicinity of the plant.

To avoid any gas leakage into the building, an overpressure ventilation system was installed. All ventilation openings are safeguarded with automatically closing shutters to avoid any shock wave propagation into the building. Even the room for the emergency power drive, which is a diesel engine, has air intake and exhaust outlet secured with such shutters.

Gas detectors are installed in the air intakes which close automatically in the event of a gas detection alarm. In such case, the overpressure is maintained by closed circuit circulation.

The heating source for the building is low-pressure steam from the process plant. As a result of the explosion in the control room building of the German VEBA ethylene plant in 1976, we decided not to heat directly with LP steam. This explosion was caused by hydrocarbon leakage into the building from the polluted LP steam system. Also in our plant it is possible for the LP steam to be polluted by hydrocarbons through leakages or maloperation similar to what happened in the VEBA accident. A secondary water circulation was, therefore, used for heating the building. The heat exchanger, in which the secondary water system is being heated by condensing LP steam, is located outside the building. By ensuring that the secondary water pressure is always higher than that of the steam, the possibilities for such accidents should be eliminated.

Compressor buildings

There are two compressor buildings in the plant: the main one located in the process area and another in the tankage area. In the main compressor building, there are three large centrifugal compressors and four screw compressors. These together are so noisy that it would be impossible to meet the noise limitation of 40 dB(A), 1000 m from the plant. This noise limitation was required by the environmental authorities.

We, therefore, decided to close the building instead of providing separate insulation on each machine, Figure 2. This closing could: 1. avoid winterizing; 2. make maintenance and operation easier during the wintertime; and 3. make the building serve as a physical barrier between the fractionation and furnace section of the process plant, as already mentioned.

However, this building introduced a potential safety hazard, if hydrocarbon gas leaked into the closed space. Our approach to this problem can be summarized as follows:

- If leakage occurs, how much gas could be released?
- If an explosion occurs, how could an acceptable pressure inside the building be obtained?
- How should the ventilation system be designed to secure proper ventilation and to avoid hydrocarbon gas pockets inside the building?

The building as such was then designed for an explosion pressure of 15 kPa (0.15 bar). Pressure relief openings had thus to be installed. The authorities required a venting area of 0.05 m^2/m^3 corresponding to 928 m^2, which is 35% of total wall area and 25% higher than the venting area necessary according to the NFPA code.

The dynamic explosion pressure, based on the assumption that the entire volume of the building was filled with a gas/air mixture and with all relief openings released, was thereafter calculated to approximately 50% of the design pressure for the building.

However, to reduce the possibility of collection of gas in the building of subsequent risk for such explosion, a ventilation system with air intakes on the roof and outlets at the ground floor level was designed. As leakages most likely occur on the ground floor level, the gas would thus be pressed downwards by the air flow. This will reduce the possibilities for hazardous gas/air mixtures in the upper part of the building. The air flow from the first floor to the ground floor is being pressed through narrow openings running around the three large compressors, thus obtaining an air curtain effect around each of the compressors.

The ventilation system was designed for 10 and 20 air exchanges per hour during normal and emergency operation respectively. The emergency operation can be initiated by any one of the 22 hydrocarbons gas detectors located in the building, and the max ventilation capacity will be obtained

Figure 3. Reactor shutdown system.

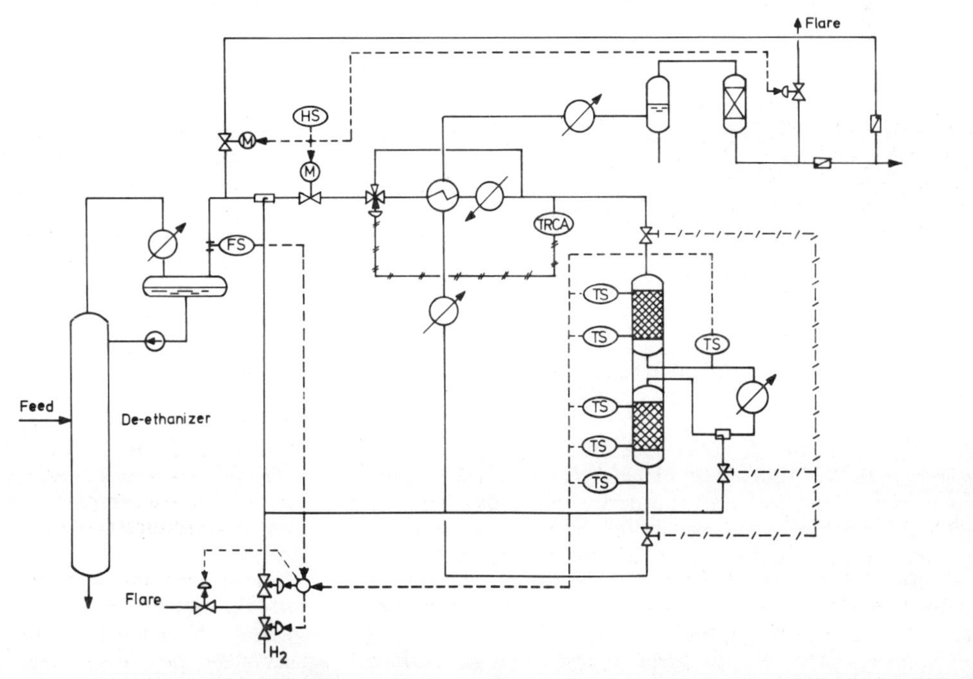

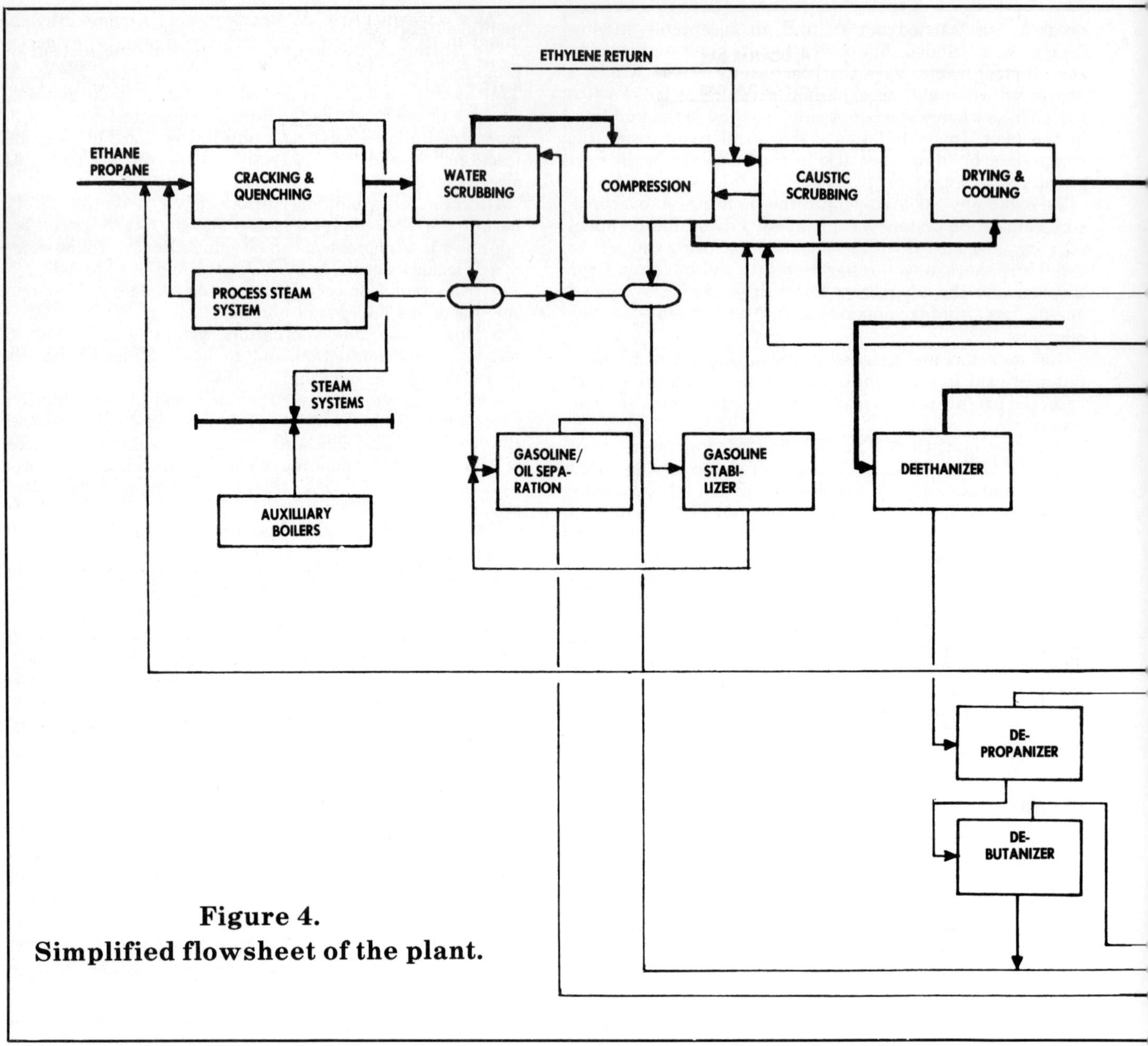

Figure 4.
Simplified flowsheet of the plant.

6 s after a detector has initiated the emergency ventilation.

The fire protection system inside the compressor building consists of a water spray system on the machines and a high expansion foam system at the ground floor level. Both systems can be started by push buttons at the entrances to the building.

Safety measures

All safety valves in the plant except those installed on the storage tanks are connected to a blowdown system which again is divided in a hot and cold system. The cold liquids are vaporized in a large water bath heat exchanger prior to discharge to the flare. The blowdown and flare system, which is designed for total shutdown of the plant due to power failure, has a capacity of 400 ton/h. Furthermore, a ground flare is installed for environmental reasons and a separate high flare for all the storage tanks, as no such direct emission of hydrocarbon to the atmosphere was permitted by the authorities.

To minimize the quantities of hydrocarbon gas leaking from a damaged pipe or component, remote-operated, isolating and venting valves are installed on systems with large volumes of liquid hydrocarbons. Large fractionation columns with surge and reflux vessels, storage tanks, and long feedstock lines are examples on systems which thus can be remote-isolated and vented in the event of leakages.

To be able to detect leakages quickly, 140 *gas detectors* are installed in the plant. Acoustic and visual alarm is given in the main control room at 30% of LEL. In general, the detectors were located where leakages most likely occurs, as for instance at pumps sealings. There are 70 detectors in the process area, 42 in the tankage area, 22 in the compressor building, three on the jetty, and three in ventilation air intakes. All detectors are tested every third month.

As all detectors have individual alarms, the leak source can be traced rapidly. With a quick action, this can certainly reduce the quantities of gas leaking to the atmosphere.

So far, we have not experienced leakages of such magnitude that several detectors have been activated by the same source, but we have had several alarms due to minor leakages in mechanical seals and flanges.

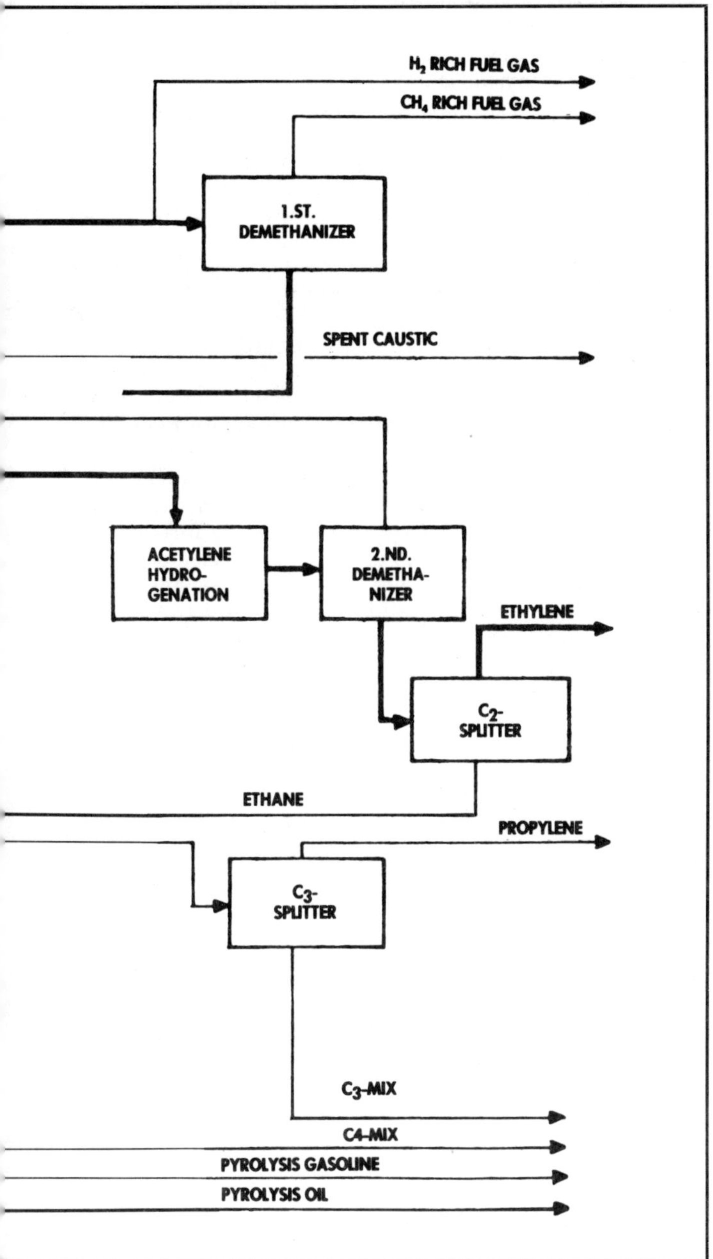

As a consequence of the accident in a Japanese ethylene plant in 1973, where a temperature runaway in the *acetylene hydrogenation reactors* resulted in a disaster, we carried out a comprehensive evaluation of the design of this system. Refer to Figure 3.

The plant was designed for tail end hydrogenation of the C_2 stream which contains from 0.5 to 1.5% acetylene, Figure 4. Each reactor has two beds in series with hydrogen supply upstream the reactor and between the two beds. If, for any reason, a runaway is being initiated, the too high temperature will be detected by at least one of the six temperature switches on the reactor vessel. These switches close two independent shutoff valves in the hydrogen line to the reactors. The line between these two valves is simultaneously vented to flare in order to avoid any leakage of hydrogen into the reactor system through the shutoff valves.

The hydrogen supply will also be stopped in the same way by too low C_2 stream flow in the reactor, or simply by actuating handswitches located in the plant or in the main control room.

Furthermore, the valve in the hydrogen supply line is interlocked with the isolating blockvalves up and downstream the reactor to make sure that the reactor cannot be filled with hydrogen only.

The reactor system can be isolated and depressurized by actuating handswitches in the plant or in the control room. The valve in the feed line is then being closed while the bypass line is opened. Furthermore, the vent line from the reactor system is being opened by the same signal.

With this shutdown system, we feel that even with considerable maloperation or malfunction of the plant, the reactors are properly protected against temperature runaway.

Shutdown and alarm systems

Due to the strict authority regulations on explosion-proof equipment, it was decided to use pneumatic instruments in the plant. For the shutdown and alarm system, we selected a high signal level electronic system. In the original design, some 30 important shutdown inputs were implemented as 2 out of 3 (2/3) voting systems. To improve the reliability and safety in these systems and to avoid erratic shutdowns of the plant, we have converted about 25 more shutdown inputs to 2/3 systems. These 2/3 systems are used for such

Figure 5. Propane cavern shutdown system.

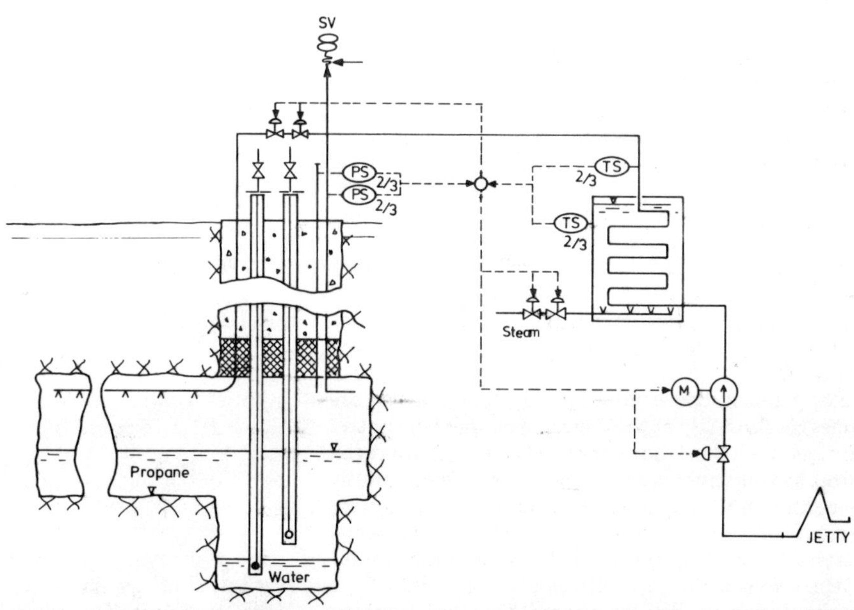

instances as suction pressure switch on big compressors; as high level switch on storage tanks, on quench tower liquid level; and as pressure and temperature switch on the large underground cavern for propane.

We also put considerable emphasis in finding the correct fail-safe position on all control and remote-operated valves.

The *fire protection system* was designed for a major fire in the tankage area where 1800 m^3/h (7,920 gal/min) fire water is required. All storage tanks have water spray system. All oil and gasoline tanks have in addition foam injection. There are in total 13 water monitors in the plant, of which five can be remote-operated from the main control room via remote-operated television cameras. All water spray systems are also remote-operated from the control room. The control rooms are installed with automatic gas systems.

The *underground cavern* for propane feedstock has a capacity of 100,000 m^3 of liquid propane, Figure 5. As the cavern is pressurized, it is located so deep that the ground water pressure is higher than the cavern pressure.

We selected this type of storage partly due to limited land availability and partly due to safety reasons, as such quantities of propane in an above ground storage would increase the potential hazard considerably.

As such caverns are excavated in rocks, there will always be a certain risk for leakages through the rock and up to the surface. This can occur, if the ground water level is not maintained during excavation due to bad rock conditions or due to overpressure during operation. As we have operated our cavern since 1977 without any trace of leakages so far, we suppose that such overpressure during operation is the only way we can destroy the cavern.

To ensure the integrity of the ground water system, we have installed an adjustable water injection system in the rocks above the cavern.

Such leakages will not probably create any immediate hazard to safety, but will be an economical disaster if the cavern has to be given up due to such leakages. Unwanted overpressure can occur when:

- LPG feedstock with too high vapor pressure is introduced.
- Overfilling the cavern.
- Overheating the feed stream to the cavern.

Except for the first possibility, such maloperations and/or instrument failures can be controlled by stopping mass and heat flow to the cavern, and if necessary by starting venting. Why not install safety valves large enough to protect the cavern for the worst possible case of maloperation?

The point is that the feed heater upstream the cavern is designed to heat light fully refrigerated propane to 10°C at full discharge rate (500 ton/h), and there are safety valves installed which protect the cavern against maloperation and/or instrument failures in such a case. It is, however, possible to unload semirefrigerated tankers, and if all available heat is added to this feed stream, the required venting rate from the cavern would be enormous. We have, therefore, installed a comprehensive shut down system which we believe will give adequate protection.

In connection with the cavern gas phase two independent 2/3 voting pressure switch systems are installed. On the feed heater, which is a water bath heat exchanger, there is one 2/3 voting temperature switch system on the propane outlet stream and one 2/3 voting temperature switch system on the water side of the heater. Furthermore, there are handswitches on the jetty, at the cavern, and in both control rooms. All these switches shut three independent valves in the feed line, two independent valves in the steam supply line to the heater and finally the feed pump.

To learn more about the behavior of the cavern during maloperation of the complicated thermodynamical conditions in the cavern, we established a comprehensive computer simulation model. Compared to later operating experience, the model has proved to be rather good. The model indicated that a serious maloperation could last for several hours under normal initial conditions, and more than one hour at the most unfavorable initial condition before the pressure increased above 100 kPa (1 bar, 14.7 psi). Although the human element in design of the safety system was not considered, it must be an improbable negligence of the operators if the cavern pressure is allowed to rise that much as a consequence of total breakdown of the shutdown system.

Human aspects

Eliminating the potential safety hazards in a properly designed plant by adding various safety devices can be of minor value. The inadequate or neglected operation and maintenance procedures are probably the greatest potential safety hazards. Experience shows that human error during operation and maintenance seems to be the main reason for accidents in such plants. Modification of the original design of the plant might hopefully improve the performance as intended, but might also well increase the risk for accidents due to unconsidered reasons. Equally important is the performance of the modification work, and the control and documentation of the completed modification.

It is thus important to remember that in spite of all safety considerations built into a plant handling hazardous materials, the plant operators also have to be well qualified, disciplined, and trained. The operating staff in our plant was trained more than two years prior to startup and the operators, approximately one year prior to startup. We have, furthermore, introduced systems and procedures for modification work in the plant which, due to all required checking routines, are rigid and bureaucratic. The possibility for accidents due to incorrect modifications is considerably reduced.

A general problem is, however, that operators without being aware of it always tend to find the most convenient way of working. This may increase the risk for accidents, unless the same emphasis is put on the perfection of humans as it is on the plant design. #

T. Bøckmann is production manager of the I/S Noretyl ethylene plant. A chemical engineering graduate of the Technical Univ. of Norway, he previously worked in the petrochemical group at Norsk Hydro Research Center.

T. Hakstad is manager, production department in the Norsk Hydro Petrochemical Div. A graduate of the Technical Univ. of Norway with a degree in mechanical engineering, he was the deputy project manager for the Noretyl ethylene plant.

G. H. Ingebrigtsen is manager, project service, for Norsk Hydro Engineering. A graduate of Loughborough College of Technology, United Kingdom, he had been project manager for Norethyl eythlene plant prior to his current position.

DISCUSSION

A.A. CARONA, Mobile R & D Corp.: In one slide, you showed the compressive building; there were two white storage tanks between the compressor building and your boilers and compressors. Were those tanks storing flammable liquids?

HÅKSTAD: No. Those are tanks for boiler feed-water.

W.B. HOWARD, Monsanto: Since this plant was designed by Linde, I assume that you have some cold boxes. Is this correct? Have you encountered any problem in the cold boxes with either acetylenes or nitrogen oxides?

HÅKSTAD: Yes (we have cold boxes). We don't have such problems.

J.E. ROGERSON, Procter & Gamble: You said that you had 140 gas detectors located in various places in the plant; this might be a wonderful source of data for determining the frequency of hydrocarbon leaks and also the duration of the leak. This would be extremely valuable data that could be used in determining the classification of hazardous areas for electrical design purposes.

HÅKSTAD: Well, I agree with you. We are trying to collect the information as well as we can, but the amount of data is not large enough to make statistics out of them.

ROGERSON: Is this likely to be ever put in the literature?

HÅKSTAD: Yes. I think we will do that when we have sufficient data.

CHARLIE SOLOMON, Esso Chemical, Europe: I would like to comment on two aspects of your philosophy on designing control houses. First, you base the blast resistance on a theoretical explosion and then design the control house to withstand the specific blast. In Exxon, we take a different approach; we design all control houses for the same potential blast, namely to withstand a blast equivalent to 1 ton of TNT at 100 ft. However, at the design blast there will be no permanent deformation of the structure. Consequently, the building would not suffer permanent deformation or collapse until many times the design blast.

This brings up the second point. The slide which you showed seems to be of a two-story building which, on our design basis would have a much lower resistance to total collapse than a single-story structure. It has been estimated that a single-story building designed to resist 1-ton TNT equivalent blast at 100 ft would not start to collapse until about 16-ton TNT equivalent. However, a two-story building, designed to the same standards, would start to collapse at perhaps a 4-ton TNT equivalent blast, unless special collapse resistance is included in the design.

HÅKSTAD: I have to make sure that I understand what you mean. You mean a building with two floors?

SOLOMON: Yes.

HÅKSTAD: I think that the angle from which the picture has been taken has misled you. It's true that the control room is higher than the laboratory which is located just beside it; but there is not a second floor in the building.

SOLOMON: It is just a high building, isn't it?

HÅKSTAD: It's approximately four to five meters high.

SOLOMON: It probably was just the angle of the camera. O.K. fine.

W.H. DOYLE, Consultant: In connection with the propane cavern, have you considered the probability, the possibility rather of gas hydrate formation and plugging up piping instrument lines and so forth with gas hydrates?

HÅKSTAD: Yes, that is a potential problem. At the design stage of the cavern, we determined the minimum temperature based on the criteria that hydrate formation should be avoided. The temperature and thereby the pressure must, however, not exceed the maximum design pressure for the cavern.

Hydrate formation has, however, not created any problem for us so far. We have experienced unstable operation of the submerged pumps once or twice which may have been caused by hydrate formation.

BILL WOOD, Consultant: There is an as-yet-unpublished account of an incident in a cavern storing butane where the pressure was inadvertently increased because of becoming liquid full, a failure in control. You didn't mention it, I think, but, the real (actual) seal around the cavern is the water, the ice sheath that forms around cold material. This instantaneous overpressure caused that ice shieled to be breached and the gas came out, not a very great amount of it, and it went up and reached the frost line. This happened in the winter time, traveled across the frost line and found a convenient warm wall and went through and blew up a house. It is necessary to take extreme precautions to prevent any such liquid filling of these things or overpressuring in any way. One other point I wondered about; I saw the elevated nozzle that you said could be directed to any part of the plant. I hope that is operated from the ground, because I wouldn't want to be up there on that post in a fire.

HÅKSTAD: Well to answer the second question first, it's correct that the monitors can be operated from the ground and it can also be operated from the control room. Regarding the first question you raised, I believe there must be a misunderstanding as the temperature in the cavern is not below the freezing point. The temperature is between 8 and 10°C in the cavern and the pressure will then be the corresponding equilibrium pressure. Water is leaking continuously into the cavern. If the cavern pressure is higher than the pressure of the surrounding ground water, the gas will enter into the rock fissures and break the water seal. Such fissures will probably never be filled up with water again.

Figure 1. 1,000-m^3 propane storage facility.

Safe Blowoff Conditions For Storage Vessels

It is often simpler and safer to have a safety valve on a pressure vessel vent directly into the atmosphere than to connect it to a flare system.

N. H. Gerardu, DSM Corp., McGeleen, The Netherlands

Pressurized vessels for the storage of liquefied light hydrocarbons should be adequately protected against high temperatures and pressures. To prevent unduly high-pressure build-up, such vessels are provided with safety valves. If direct blowoff from the safety valve on top of the vessel through a short vertical vent pipe is used, the possible effects of this blowoff should be quantified and the vessel should be protected from these (secondary) effects both from the pressure wave generated (if the escaping gas should ignite), and from the secondary radiation effects.

This article focuses on the secondary effects, with the primary radiation intensity at the foot of the vessel being calculated by the API equations. Most calculations are made for a 1000-m^3 spherical propane storage vessel with two mechanically linked safety valves. For geometrically different vessel, the calculations differ slightly, but not essentially.

Determination of line diameter

At the basis of all calculations is the maximum capacity of the safety valve, i.e., the maximum exit rate of the vapor (propane in this particular case). This is determined by the following factors:

- the installed valve diameter;
- the maximum pressure build-up beyond the valve setting;
- omission of the safety factors used in calculating the

0149-3701/81/3808 $02.00 © 1981 AIChE

valve area to be installed.

At sufficiently high exit speeds, the size of the gas cloud formed (i.e., the size of the cloud formed by gas in concentrations between the upper and lower explosion limits) no longer depends on the exit flow rate, but is a function of the line diameter only. It is, therefore, recommendable to take the line diameter as small as possible, considering the allowable back pressure of the safety valve. This allowable back pressure depends on the type of valve used. As a rule, however, it is 10–15% of the set pressure for an unbalanced safety valve.

Two situations should be distinguished in the calculation of the exit velocity, the conditions at the end of the line, and the back pressure behind the safety valve. 1. The exit velocity at the end of the line (V_E) is lower than the velocity of sound (V_{SE}). In other words, the pressure at the end of the line (P_E) equals atmospheric pressure: 2. $V_E = V_{SE}$, in other words, P_E is higher than atmospheric pressure.

We can now, for either case, calculate the conditions at the end of the line, assuming the gas to expand adiabatically from the pressure before the safety valve, P_0, to the final pressure, P_E, with the enthalpy being given by: $h_E = h_0 - \frac{1}{2} V_E^2$. For instance:

Max. valve capacity, $G_{max} = 16.67$ kg/s
Temp. in vessel when safety valve opens, $T_0 = 59°C$
Line diameter
Ø 250 mm nom.
$P_E = 1$ a bar
$T_E = 15°C$
$V_{SE} = 247$ m/s
$V_E = 165$ m/s
Ø 200 mm nom.
$P_E = 1.03$ a bar
$T_E = 5°C$
$V_{SE} = V_E = 244$ m/s

On the basis of the final conditions thus calculated and the geometry of the line, it is now possible to calculate the pressure immediately behind the safety valve, assuming adiabatic, compressible flow (*1*). If the back pressure thus calculated is lower than the allowable value, the choice of line is justified; otherwise, a larger line diameter will have to be used.

Determination of pressure wave

The gas cloud, forming when there is no wind and where

Figure 2. 500-m^3 C_4 bullets.

density differences between the vented gas and air are relatively small, can best be described by the formulas of Hess *et al.* (2, 3).

$$V = \frac{\pi a^3 d^3}{486} \text{ and } V_g = 1.5 \times \frac{CO1}{100} \times V \qquad \text{and } L = a \times d.$$

where

V = volume of the gas cloud in concentrations within the explosive limits (m^3)
V_g = gas content of the cloud
$a = \frac{CO}{CO1} \times 5.3 \times \sqrt{\frac{M_I T_I}{M_E T_E}}$
$C\phi$ = volumetric concentration of the explosive component at the vent opening (%)
$CO1$ = lower explosive limit (%)
M_I = molecular weight of air (g/mol)
M_E = molecular weight of gas (g/mol)
T_I = temperature of air (K)
T_E = temperature of gas (K)
d = line diameter, if $P_E = 1$ a bar
d = line diameter ψ $\sqrt{P_E}$, if $P_E > 1$ kPa (bar)
L = length of radius to $CO1$

If there is wind, the gas cloud will be deflected and the volume of explosive gas in the cloud will decrease considerably. However, our further calculation will be based on no-wind conditions.

To calculate the energy released upon explosion of the gas cloud formed we use the following formula, valid for explosions of unconfined gas clouds:

$$E = V_g \times \rho_g \times C \times C^* \times h_c$$

where

E = energy released (J)
$V_g \times \rho_g$ = quantity of explosive gas (kg)
C = correction factor for a non-stoichiometric mixture (0.3)
C^* = mechanical efficiency of the explosion (0.3)
h_c = heat of combusion of the gas (J/kg)

Conversion to TNT equivalents results in: $E' = 0.22 \times 10^{-6} \times E$ kg TNT.

The damage radii are now calculated by the following relation, partially based on work by Brasie and Simpson (4).

$$R = C'' (E')^{1/3}$$

where

R = distance to explosion center (m)
C'' = constant depending on the overpressure at distance R ($m/kg^{1/3}$)

ΔP (kP, bar)	C'' ($m/kg^{1/3}$)	
0.01	70	Window damage (~10% of panes)
0.03	25	Window damage causing injury
0.1	10	Reparable damage to buildings and house fronts
0.3	5	Heavy damage to buildings and damage to process plant

The explosion center should be thought to be in a place where the concentration is between the lower explosion limit and the stoichiometric composition. The above method of calculating damage radii is suitable for explosions of unconfined gas clouds. It is questionable, however, whether a long and narrow cloud—as will form in the case we are considering—can be regarded as a point source.

A comparison with similar explosions (aerial decomposition experiments, for instance) shows that the above approach leads to conservative results. However, this method is used, as there is no better method of calculation available at the moment.

For the example referred to we get:

ΔP (bar)	R (m)
0.01	220
0.03	80
0.1	30
0.3	15

Prevention of liquid escape

As appears from the example, no allowance was made in the dimensioning of the blow-off line for the blowoff of liquid through the safety valve, with or without vapor. Theoretically, liquid may escape through the safety valve as a result of: 1) overfilling of the vessel by the feed pumps; 2) entrainment of liquid when vapor is being blown off; and 3) thermal expansion of the liquid and gas hold-up.

In spite of the fact that the capacity of the safety valve is sufficient to handle the liquid flow, it is very undesirable to have liquid issued through this valve, in view of the direct discharge of the line into the atmosphere. To restrict the risk of overfilling to a minimum, the vessel is provided with two independent level-control devices acting on the feed pump. About entrainment and the chance of its occurrence, little has so far been published. In dimensioning safety valves for storage vessels, the possibility of liquid entrainment is ignored.

If due attention is paid to limiting the maximum degree of filling of a tank, it is realistic to assume that no entrainment will occur. In wintry conditions (storage temperature, 0°C), for instance, the maximum degree of filling with C_3 should not exceed 80–85% of the vessel volume (for spherical vessels). In the summer (storage temperature, 25°C), a figure of 85–90% should be adhered to. For C_4, the values are 85–90% in winter and 90–95% in summer (assuming the temperature at which the safety valve opens to be 55°C. These figures should be increased to account for the gas hold-up, but this will make a difference of only 1–2% for liquified light hydrocarbons.

Protection of storage vessel

In comparison with the hazards that may be caused by overpressure in a storage vessel, the consequences of unduly high temperatures may be even more far reaching, the more so as causes and safety measures cannot always be quantified. In the case of a fire near a storage vessel, two sources of radiation may be distinguished: 1) heat radiation from the burning pool of liquid; and 2) heat radiation from the burning gas cloud over the vessel.

Both heat sources cause direct danger for the storage vessel, in that the heat radiation may result in unduly high metal temperatures, or the secondary heat radiation may cause the exit rate to become so high that the vapor flow can no longer be handled by the safety valve.

As mentioned before, the primary radiation from the burning pool of liquid is calculated by the API formulas. However, as normally such provisions are made in the area under the tank that the fire-extinguishing water and the burning medium will flow off to a tank pit situated elsewhere, an estimate will have to be made of the secondary irradiation of the vessel—and especially of the gas-filled top.

The following formula holds for hydrocarbons irradiated by the burning liquid pool under no-wind conditions (5):

$$q = 213 \times r^{-2} \times D^{1.72} \text{ kW/m}^2$$

where

r = distance between tank and tank pit (m)

D = diameter of tank pit (m)

$$1 < r/D < 3$$

For calculating the irradiation by the burning gas cloud, we assume the length of the flame to equal the length of the gas cloud to the point where the concentration equals the lower explosion limit (L). We can now represent the flame as a point source at a distance x_m from the vessel. The value of x_m is defined by (6):

$$x_m = \sqrt{H(H + L)}$$

where

H = length of blowoff line to vessel

At a distance r' of x_m, the irradiation under no-wind conditions (q') is:

$$q' = \frac{3.82 \times 10^{-6} \times h_c \times G_{\max} \times \sqrt{M_E}}{(r')^2} \quad \mathrm{kW/m^2}$$

Integration of the two radiation intensities q and q' for the whole of the area radiated makes it possible to calculate the total amount of heat irradiated. For the complete calculation refer to (6).

In the case under consideration—with a 1000-m^3 propane storage vessel—the total secondary radiation flux incident on the spherical vessel ($q + q'$) is 50 kW/m^2 (tank pit, D = 12 m, is at 25 m from the vessel). In calculating the amount of heat effectively contributing to raising the surface temperature, a correction should be made for the emissivity of the spherical surface. Depending on the nature of the material of which the sphere is made and the condition of the vessel, the emissivity will have a value of between 0.5 and 1.0. The effective radiation will, therefore, be about 25–50 kW/m^2.

If the effective radiation flux is 25 kW/m^2, the surface temperature will be 400–500°C for the gas-filled top, a value considerably higher than the maximum admissible temperature at the pressure prevailing in the vessel, when the safety valve opens (about 250°C). It will be clear, then, that the vessel should be protected from such high temperatures. As already mentioned, the wind certainly influences the gas cloud and hence the flame, even at high exit velocities. If there is a wind, the flame will not be situated straight over the vessel and comes closer to the vessel. The radiation will then be even higher than the value calculated for no-wind conditions.

A number of protective measures offer more or less effective protection (7 and other references), but hardly any measure is sufficient if used alone. Complete or partial *insulation* of the vessel would be at first sight seem an adequate precaution. However, this solution is less attractive (high cost), since every vessel must be externally inspected at regular intervals. Also, the possibility of corrosion under the insulation should not be underestimated. These problems may be overcome by the recently developed sublimating heat coating.

Placing the vessel on a *sloping* plane and locating the tank pit at a distance from the vessel will reduce the primary irradiation. But, protection against the secondary radiation from the burning tank pit will have to be provided. The use of a *depressurizing valve* which keeps the pressure in the vessel so low as to allow higher metal temperatures is not to be recommended for vessels whose safety valves open direct into the atmosphere (no flare connection).

To avoid instrument-air failures, the safety valve would need to have its own air reservoir. However, as in the situation we are considering, the valve must also blow off directly into the atmosphere. A dangerous situation may arise, if the valve is accidentally opened during normal operation or if during a fire the valve opens too soon. In either case, a gas cloud of considerable size may form if the pressure in the vessel is low. This makes it clear that the use of a depressurizing valve for vessels not conected to a flare is not recommendable.

A means of protection, that is very common for storage vessels and is used also by DSM, is a *system of sprinklers* cooling the vessel in the event of a fire and limiting the amount of heat influx. This method of protection is highly effective, provided the following conditions are satisfied.

- The sprinkler should be actuated with sufficient rapidity (i.e., automatically).
- Capacity and water distribution should be adequate. As to capacity, DSM keeps to a value of 10 $L/m^2/min$ for a total secondary radiation intensity of up to 60 kW/m^2 (8).
- The water line system should be laid underground up to a point immediately near the vessel. The aboveground section should be zinc-coated and should always be empty when not in use.
- The water grid should be kept pressurized from two ponds by two independently driven pumps.

For extra protection of the storage vessel in the place where the heat radiation is maximum and where the hazard from high metal temperatures is greatest, i.e., on top of the spherical vessel where the nozzles are located, DSM decided to install an insulated *heat shield* in this place, at a distance of about 2.5 m from the vessel (diam. about 5 m for a 1000-m^3 propane storage vessel).

This shield also prevents damage to the instruments on top of the vessel as a result of radiation, so that during a fire-process data keep being transmitted to the control room, Figures 1 and 2.

The location of *hydrants* near the vessel and the use of an effective warning system enabling the fire brigade to reach the fire shortly after it has broken out are additional expedients in keeping a vessel cool in the event of a fire.

In conclusion

In many cases, it is simpler and safer to have a safety valve on a pressure vessel vent directly into the atmosphere than to connect it to a flare system. However, due attention should be paid to the design of the blowoff line. Also, the consequences of a possible gas cloud explosion should be quantified and the possible danger to the surroundings should be evaluated. Measures should be taken to limit the secondary heat radiation to the storage vessel. Stress is laid on the use of a reliable sprinkler system and a heat shield, as well as a tank pit for extinguishing water and burning hydrocarbons, located at a distance from the vessel. #

Literature cited

1. Powley, M. B., *Canadian J. Chem. Eng.*, **36,** 241 (1958).
2. Hess, K., and R. Stickel, *Chemie-Ing. Techn.*, **42,** 467 (1970).
3. Hess, K., W. Leuckel, and A. Stoeckel, *Chemie Ing. Techn.*, **45,** 323 (1973).
4. Brasie, W., and D. Simpson, *Symp. Loss Prevention in the Process Ind.* (1968).
5. Hoftijzer, G., *CTI-TNO-report*, Ref. No. 76-04069 (April, 1976).
6. Kovacs, F., and G. Honti, *Loss Prevention*, **1** (1974).
7. Kletz, F., *Hydroc. Process.* (Aug., 1977).
8. van Eijndhoven, P., J. Nieuwenhuizen, and R. Wally, *Loss Prevention*, AIChE (1974).

N. H. Geradu, senior process engineer with DSM Engineering, in Geleen, The Netherlands, earned his engineering degree at Eindhoven Technological. He joined the firm in 1975 and has been involved in process designs for chemical plants.

DISCUSSION

RUSSELL HILL, Oxirane International: I think you've done an extremely valuable survey article here which will be of great use to engineers designing these systems. I have one comment and one questio. When doing the diffusion calculations for a jet discharge from a safety valve, it is important to recognize that an oversize spring-operated safety valve has three positions when it discharges: wide open, closed and 25% open, depending on the degree of oversize of the valve. It may spend a significant portion of the time discharging at a rate that is only one-quarter of the design rate; this can have a significant influence on your diffusion calculations because the change in velocity of discharge. The question: I have personally never seen the concept of heat shields before and I know we're not applying them in our plant in Rotterdam. Is this becoming accepted practice in the Netherlands now?

GERARDU: We did it for the first time on these three bullets and one spherical vessel. But, we didn't yet do it on the other vessels, because it is very difficult to install a heat shield on an exisitng bullet because of all the welding involved. Besides, when we decided to install the heat shield, the sublimating protective coating wasn't yet available in Holland. If it were, we might have decided to use it rather than the heat shield. When you design a safety valve which proves to be too big, it will open and release much more than it ought to. As a result, the valve will alternately open and close.

HILL: I beg to disagree with you. I have checked this out with many safety valve manufacturers and some of them to their surprise after they have done experimental work have come to the conclusion that a valve reaches its set pressure and, if it is properly popping, it goes wide open. When the pressure starts to decrease in the vessel, the valve will close fairly rapidly, if it is oversized to about 25% of its normal opening and it will just rest there until the pressure drops down to the normal closing pressure.

KLETZ, ICI: Where we have relief valves discharging to atmosphere to meet the point raised by Russell Hill, we use pilot-operated relief valves. These are designed so they are either fully open or fully closed and you do not get the 25% open effect to which he referred. If you have an ordinary relief valve discharging to atmosphere, you can, as Russell Hill said, get the situation where the valve is not putting out the vapor with much veloctiy and it is liable to drift down to ground level. Have you given any thought to pilot-operated relief valves?

GERARDU: No, we didn't have pilot-operated relief valves on this vessel.

KLETZ: Second point is a comment rather than a question. I hear what you say about the relative merits of insulation and water cooling and so on. We, as a general policy, do fit insulation. The disadvantages that you mention are quite true; it does have weight and this has to be allowed for; bits do have to be removed for inspection. Nevertheless, the insulation is there as an immediate barrier to heat input and it does not have to be commissioned. Water has to be turned on. Whatever situation has led to the leak may have caused injuries to the operators or it may have damaged the water lines and it may not be possible to commission the water immediately. The insulation is there all the time. Experience has shown that vessels exposed to fire can burst in as little as ten minutes; the metal has softened and the vessel has burst. Ten minutes is not a lot of time to get your water turned on especially in a situation where all hell is breaking loose and the operators have everything else to do. Therefore, I feel the insulation is worthwhile.

Because we put insulation on, we don't use so much water; some of the money we spend on insulation we get back in that way. I don't think you said what your water rate is but presumably you use the standard 0.2 gpm/ft^2. When installing insulation, we reduce the water rate to about a third of that figure.

GERARDU: On the first part. The problem is that the authorities want the vessel inspected once every two or three years and they couldn't guarantee us that we would not have to take off the whole insulation. So, this would be a rather high cost that would come back every two or three years. The other part, the sprinklers that we use are actuated automatically by temperature sensors, without the intervention of an operator.

KLETZ: Well, I agree with you that if you have to remove the insulation every two or three years, this is an impossible situation. You couldn't possibly do that and perhaps this is another example to add to the collection which some of us are building up of authorities making us do things which we think are actually unsafe. The authorities in effect make you leave off the insulation, maybe, you have a less safe situation than if you had it on and inspected it less often.

There are a number of cases occurring here and there throughout the world of authorities doing this sort of thing. I was pleased to see that in one respect your tank is better than another LPG tank whose pictures I have seen in Holland where the authorities apparently insisted on a concrete wall around the tank to protect it from missiles. In my view, this was a waste of time and money. All this concrete interferes with the ventilation and a small leak could build up within the confined area to an explosive concentration; in a more normal situation, it would disperse.

So, does this mean that the Dutch authorities have changed their minds and no longer want these concrete walls around tanks?

GERARDU: The concrete walls are quite a different matter in that we place concrete walls around atmospheric storage tanks but not around pressure vessels.

KLETZ: I know of one. It's in the Rijnmond area which, of course, is a different authority than what you deal with. Incidentally, I have seen a letter from the authorities in Holland that says they can agree to radio graphic inspection of the sublimating heat insulation once every two or three years. So, in this case, you don't have to remove the whole insulation.

It is perhaps worth mentioning that at the 1973 meeting Ruby Feldman presented a paper describing the sublimating coatings, *Loss Prevention Technical Manual* (Vol. 8). At the time, they were not available commercially; it has taken some time for them to come on the market. But, they are now being marketed and appear to provide a very effective lightweight form of heat insulation.

IAN SWIFT, Union Carbide: I noticed in your paper on which I congratulate you that you said you would ignore the effective entrainment in sizing your relief capacities; this is in the case of the vessel becoming liquid filled. Did you also consider the possibility of two-phase relief in the event of fire exposure to the storage tank?

GERARDU: No, I didn't. The reason was when you lower the maximum filling degree of the vessel while taking into account expansion of the liquid and gas hold-up, the chances of liquid pull-out will be minimum.

SWIFT: I was referring particularly to the case where your cylindrical vessels which are raised off the ground were heated from below over practically all of the surface area in the fire situation.

GERARDU: As you can see from the slides, the cylindrical vessels are placed on sloping ground and all liquid will directly flow away.

SWIFT: Thank you.

Risks in the Cyclohexane Oxidation

Cyclohexane oxidation can attain the level of safety required for average chemical organic processes.

S. Ciborowski
Institute of Industrial Chemistry, Warsaw, Poland
A. Krzysztoforski
Nitrogen Works Tarnow, Poland

After Flixborough disaster in 1974 cyclohexane oxidation process gained the opinion of an extremely dangerous one. T. A. Kletz compared it to a lion kept in a strong cage /1/. The problem is very essential one taking into account that nearly 100 plants of that type are working all over the world with some 4 million tons of cyclohexane being processed per year, and there is no economical alternative to this process.
We hope that some of our experience could be of interest for the bigger number of readers.We feel to be authorised to express our opinion, because there are in Poland in operation two commercial plants for cyclohexane oxidation based on our own process elaborated in cooperation of Instytut Chemii Przemysłowej /Institute of Industrial Chemistry/ in Warsaw with Zakłady Azotowe /Nitrogen Works/ in Tarnów. This process was worked out in laboratory and pilot plant scale,a commercial plant of a capacity 25.000 t of cyclohexanone per year was put into operation in 1974, the second one, 50.000 t/y - in 1976.
In 1977 a contract was signed for a 80.000 t/y cyclohexanone plant in one line being built in Czechoslovakia. Some data on this process were published /2/.
We would like to argue that cyclohexane oxidation process not necessarily should be more dangerous that many other chemical processes considered as acceptable from the point of view of safety requirements.

SOURCES OF DANGER IN CYCLOHEXANE OXIDATION

There are potential three main ones:

1/ Large amount of superheated liquid heaving about 150°C at boiling temperature 80°C.This is however not specific for cyclohexane oxidation.Let us mention Halcon propylene oxide process /oxidation of tert-butylene to hydroperoxide/,toluene oxidation in liquid phase as well as many processes including separation and storage of C_3 - C_4 fraction.
2/ Possibility of formation of explosive mixtures.Such possibility however does exist in all oxidation processes in liquid, as well as in vapour phase.
3/ Possibility of ignition of organics in oxidation reactor.

This may happen in presence of large amount of deposits which make heat

0149-3701/81/3809 $02.00

removal difficult.
All above mentioned factors were in fact sources of heavy accidents in industrial plants. In Flixborough large amount of superheated cyclohexane expanded through broken tube, formed explosive cloud which exploded. In another plant explosive mixture exploded in oxidation reactor. There are also known cases of cyclohexane ignition inside oxidation reactor, investigation showed deposits as a cause.
Knowing the potential sources of danger we can undertake measures which would neutralize that danger. They are:

- decreasing the amount of superheated cyclohexane in the plant
- avoiding the possibility of explosive mixtures formation
- avoiding the deposite formation

It should be emphasized that no chemical process can be absolutely safe. In the report concerning Flixborough disaster /3/ we read what we we consider principal for the philosophy of risk in industrial activity:"hazard analysis recognises that hazard cannot be entirely eliminated and that it is necessary to concentrate resources on those risks which exceed a specific value", and further on:"No plant can be made absolutely safe any more than a car, aeroplane or home can be absolutely safe".

SAFE CYCLOHEXANE OXIDATION PROCESS

We feel very strongly that the cyclohexane oxidation process may be markedly improved with an increase in operation safety, and the risk cut down to such a level, which would be acceptable and which is inevitable at every industrial activity.
To prove that point of view we will compare the process solutions of the Flixborough plant /Fig.1/ according to the desciription included in the report /3/, with the process design of our commercial plants of cyclohexane oxidation /Fig.2/.
According to the first process design the oxidation step was operated under the pressure of 0.86 MPa in a series of six reactors with the liquid reaction mixture /mainly cyclohexane/ flowing from one reactor to another one through pipe connection of 71 cm fitted up with expansion bellows. After passing the reaction system the pressure of the reaction mixture was increased further on to 1.5 MPa and the mixture was flowing through the series of four mixers and three separators /two of them, S-2538 and S-2539 are shown on the drawing, not to scale/, where a caustic soda solution was added to neutralize the product.
This step was followed by the recovery of the main part of cyclohexane in the series of three columns and then by the purification of the remained liquid. According to the second process design the oxidation step is operating at a similar pressure but with a single multisection reactor of a special design. Next to the reactor there is a decompression of the product closely to atmospheric pressure and the unreacted cyclohexane is distilled off and recycled into the main process stream. Subsequently cyclohexanol and cyclohexanone are distilled off from the nonvolatile by-products /while at the same time the hydrolysis of cyclohexyl esters is carried out and cyclohexanol is recovered/ and the gaseous top product enters the column scrubbed with caustic soda solution, where the neutralization and saponification of the volatile acids and esters /4/ at a pressure of 0.4 MPa takes place. Vapours of cyclohexanol and cyclohexanone leaving the system, are only to be purified by distillation. It is obvious that the nature of process is similar in both cases and in both methods the reaction proceeds through the superheated liquid state. But it is simple to make following statements on the process design presented in Fig.2:

- the residence time of the liquid under reaction pressure is limited to the very minimum and all the subsequent operations are performed under a markedly lower pressure than the reaction pressure itself
- due to the special reactor design there are no pipe connections on the reactor liquid side between individual stages /the collapse of

such a connection was the main cause of the Flixborough disaster/
-due to neutralization and saponification performed in the vapour phase the significant decrease in superheated dangerous media content in the installation takes place, this is precisely what T.A.Kletz suggests:"........to develep more gaseous phase processes......", and "Plant inventory also might be more limited........".
Therefore it is quite clear, that the process design of the cyclohexane oxidation method itself includes the possibility of an important increase in the safety of the operation.
The technological comparison of the processes is not the subject of this paper, but we would like to mention, that the process design we are using here, presents also advatages of another kind, assuring the possible lowest consumption of caustic soda /of about 40 kg/t of cyclohexanone/, significant reducing of arduous alkaline effluents therefore, and easy utilisation of by-products /heavy acid residues/.

WATER - POISON OR REMEDY?

R.King described his point of view /5/ that the presence of water in the reaction system of cyclohexane oxidation in Flixborough led to some increase in pressure, which caused the bursting of an uncorrectly fitted by-pass and consequently the disaster.
It cannot be ignored that water and cyclohexane form a heteroazeotrop with a higher vapour pressure than the pressure of either component, but to our mind with a correct design and operation of the plant the probability of pressure increase due to that fact may be practically brought to null. When searching after the Flixborough disaster for efficient means to combat the dangers of flashing liquids, we found unexpectedly that if a stream of cold water is fed in a specific way into the system containing hot cyclohexane under pressure, there is a rapid cooling down of the system even without an instantaneous incresase in pressure.
The equipment for method developed /6/ is under construction for the Slovak plant and is going to be incorporated into both Polish plants of cyclohexane oxidation.
It will allow, in case of a leakage in the reaction system, to cool down its content within 1 to 2 minutes below the boiling point of cyclohexane under atmospheric pressure, thus stopping the cyclohexane flashing and preventing the explosive "cloud" to be formed.
The advantage of the method is that the oxidation reactor not necessarily has to be larger to provide additional space for water.
In cyclohexane oxidation the volume of the aerated liquid is 75-80% of the total reactor volume. Since hold-up is about 20% of the liquid volume the not-aerated liquid occupies only 60-65% of the reactor. This volume decreases to 55% after cyclohexane has been cooled from 150-160°C to 70°C. We therefore have space for 0.8 cu m of water per 1 cu m of cyclohexane /at 70°C/ while 0.7 is sufficient.
We carried on pilot plant experiments on various systems of putting water into cyclohexane oxidation reactor. In strictly definite conditions no pressure increase was observed, the time of the operation was 1-2 minutes, the water introduced was 70 vol.% of that of cyclohexane.

DEPOSITS - THE REAL DANGER

Deposits are real danger in cyclohexane oxidation. They may be cause of "dry" spots formation in oxidation reactor with no possibility of reaction heat removal because of the lack of liquid which can evaporate.
The ignition temperature of cyclohexane is not very far from the reaction temperature.
At deficiency of liquid cyclohexane in the area of air inlet the open fire may arise. The only measure which can be undertaken is avoiding deposits formation.
It is very complicated and difficult task. Nevertheless it is fully possible, as we found in our pilot plant /2.000 t/y/ and in our two

industrial plants.In the first one no deposits were formed during six years operation, in the others - no deposits were formed since their putting into operation,ie for 5 and 3 years.This problem was solved by applying some technological measures.
It should be mentioned that in some technologies the usual procedure is cleaning cyclohexane oxidation reactors every 3 months.

SUMMARY

Our experience shows that cyclohexane oxidation can be made safe enough to attain the level of safety specific for average chemical organic processes.
The sources of danger have been well identified,the technological measures which have to be undertaken have been worked out.Several accidents which took place in the past in industrial plants were not related to the specificity of this process but had a technological background and their causes can be fully eliminated in the future.

LITERATURE CITED

1. More safety urged in process designs,Chemical Engineering News, March 31,1975,p.11
2. Poland details new cyclohexanone from benzene process.Europ.Chem. News,May 4,1973,p.28.
3. The Flixborough disaster - Report of the Court of Inquiry.HM Stationerv Office.1975
4. Ciborowski,S. et al. Polish patent 70457,British patent 1 323 372, Belgian patent 763 736,French patent 2099 050, to Institute of Industrial Chemisty /March 11, 1970/.
5. Flixborough - role of water reexamined, Process Engineering,Sept. 1975,p.69
6. Ciborowski,S.et al.,Polish patent 96763,Swiss patent 596112,U.S. patent 4071561,British patent 1490909,Czech.patent 188255, Belg.patent 833968,Indian patent 441685,Iranian patent 15505,to Tarnow Nitrogen Works /October 10,1974/.

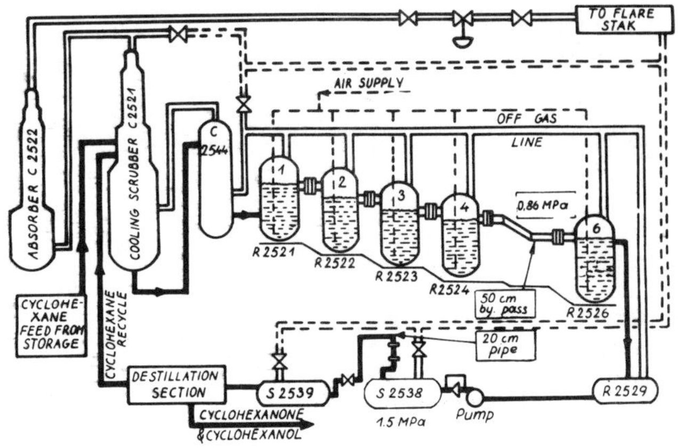

Figure 1. Simplified flow diagram of the cyclohexane oxidation plant-report (3).

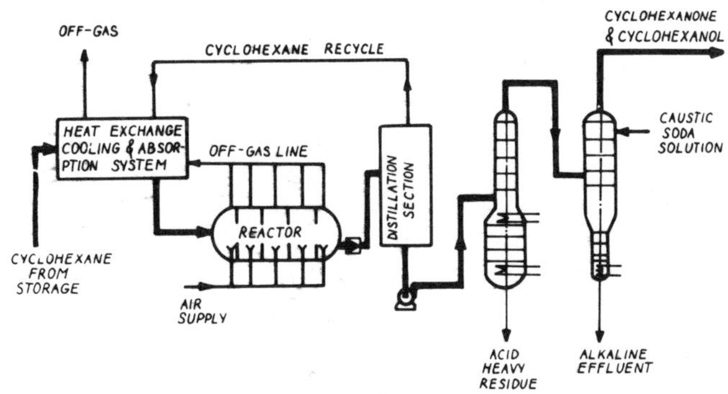

Figure 2. Simplified flow diagram of the Polish cyclohexane oxidation plants.

CIBOROWSKI, S.

KRZYSZTOFORSKI, A.

DISCUSSION

TREVOR KLETZ, ICI: I would like to congratulate Mr. Ciborowki on a most interesting paper and on his excellent delivery. I would like to ask two questions. Could you apply your water quenching procedure to an existing plant with several reactors in series?

CIBOROWSKI: This is possible, but, of course, is more complicated than in the case of one rig. But, it is possible.

KLETZ: What are the procedures and methods you use to give freedom from deposits?

CIBOROWSKI: That is a very complicated problem. You know, I can tell you that in the beginning of our pilot plant operation, we had to remove deposits every six weeks. We took some technological measures which are rather not to be disclosed. Perhaps, we had some luck to find the proper way.

KLETZ: Thank you.

Blast Wave from Deflagrative Explosions: an Accoustic Approach

Simple acoustic source theory has been applied to determine the maximum overpressure obtainable by the deflagration of nonspherical clouds. In three dimensions, over-pressure is generated not by the rate of energy addition but by the first-tune derivative of the rate. Therefore, deflagrative combustion of edge-ignited clouds produces markedly less overpressure than central, spherical ignition.

R.A. Strehlow
University of Illinois at Urbana-champaign, Urbana, IL.

The determination of the blast wave produced in free space by the nonspherical combustion of a nonspherical vapor cloud has not been examined in any detail, either analytically, or numerically. The problem is important because at the present time spherical theory is being used to estimate the overpressure for clouds of arbitrary shape and no one knows how conservative this approach is. This paper applies simple acoustic source theory to the deflagrative combustion of a few simple free cloud shapes to determine a first cut estimate of the maximum overpressure that one could expect as a function of cloud size and shape.

THE SIMPLE ACOUSTIC SOURCE

Stokes[1] first showed that a simple source of mass at a point $\dot{m}(t)$ (Units: Kg/sec) generates a sound wave in three dimensions which has an overpressure which is proportional to $\ddot{m}(t)$. Specifically, as Lighthill[2] states,

$$P - P_o = \ddot{m}\left(t - \frac{r}{a_o}\right) / 4\pi r \tag{1}$$

where $t - \frac{r}{a_o}$ replaces t because the wave is propagating away from the source region at the velocity of sound, a_o. Lighthill also states that at some distance from the source the sound will appear to emanate from a point (i.e., the wave will be very close to spherical) if the quantity $(\omega r/a_o)^2 \ll 1$. Here, ω is the circular frequency of the source of sound and r is a characteristic radius of the source region. Finally, for a source which has a finite duration $\int \ddot{m}(t)dt = 0$, and therefore $\int (P-P_o)dt = 0$ in the wave. This means that the acoustic pulse from a source of finite duration must have a negative phase impulse equal to the positive phase impulse. This should be contrasted to the pulse emitted by a finite duration source in a strictly one-dimensional channel. In this case the acoustic overpressure is proportional to $\dot{m}(t)$ and the overpressure in the pulse is always positive everywhere.[1]

Now consider the deflagrative combustion of an unconfined cloud of arbitrary shape. For simple source acoustic behavior the effective rate of mass addition $\dot{m}(t)$ may be replaced by an effective rate of volume addition multiplied by the initial gas density, that is, $\dot{m}(t) = \rho_o \dot{v}(t)$.

However, for deflagrative combustion

$$\dot{m}(t) = \rho_o \dot{v}(t) = \rho_o \left\{\frac{V_b - V_u}{V_u}\right\} \frac{d}{dt}(S_u(t) \cdot A_f(t)) \tag{2}$$

where $S_u(t)$ is the effective normal burning

0149-3701/81/3810 $02.00 © 1981 AIChE

velocity, $A_f(t)$ is the effective frontal area of the flame and the ratio $(V_b-V_u)/V_u$ is the new volume produced per unit volume of gas burned at constant pressure. Using the notation of Strehlow et al.[3] $(V_b-V_u)/V_u = \hat{q}/\gamma_0$. Here, $\hat{q}$ is an effective dimensionless heat addition. Its relation to the $\tilde{q}$ of Strehlow et al. is given in Appendix A. Therefore, using the definition of the velocity of sound in an ideal gas and combining Eqs. (1) and (2) yields

$$\overline{P} = \frac{\hat{q}}{4\pi a_0^2 r}\frac{d}{dt}\left[S_u(t)\cdot A_f(t)\right]$$

where $\overline{P}$ is the acoustic overpressure. Equation (3) is a significant relationship. It states that the acoustic overpressure generated by deflagrative combustion in three dimensions is given by the time rate of change of the rate of energy addition rather than the rate of energy addition as it is in strictly one-D flow.

The deflagrative combustion of a cloud generates a sound wave of very low frequency. The period of the sound source in this case is given by $\tau = 2\pi/\omega$ where τ is the burn time. Thus the requirement that the source region is acoustically simple becomes

$$[2\pi(r_m/\tau)/a_0]^2 << 1$$

where r_m is a major dimension of the cloud and approximates the distance traversed by the deflagration wave. Even with distortion and motion due to flame propagation $r/\tau \simeq S_u$ and the simple source requirement becomes $4\pi^2 M_{Su}^2 << 1$. If the quantity is to be less than 0.1 for no blast wave distortion S_u will be about 17.5 m/sec before deflagrative behavior produces significant distortion of the blast wave. Even for significantly higher values of S_u the first effect will be only to distort the shape of the blast wave. Higher overpressures than estimated here will not occur until S_u becomes quite large. This limitation of the theory has not been quantified as yet.

APPLICATION IN SPHERICAL GEOMETRY

Two classical problems have been treated in spherical geometry. These are the centrally-ignited, constant-velocity flame[3,4] and the Taylor piston problem.[5] For a constant-velocity flame Eq. (3) reduces to the relation

$$\overline{P} = \frac{\hat{q}S_u}{4\pi a_0^2 r}\frac{dA_f}{dt} \qquad (4)$$

where r is the distance from the source to the observer. For a growing spherical flame ball

$$A_f = 4\pi r_f^2$$

and

$$\frac{dr_f}{dt} = S_s = S_u\frac{V_b}{V_u}$$

Substituting yields

$$\overline{P} = \frac{2\hat{q}S_u^2 r_f}{a_0^2 r}\left(\frac{V_b}{V_u}\right)$$

where the r in the denominator must be equal to or greater than r_f. If we let it be equal to r_f, and if we substitute the definition of $\hat{q}$ given above, we obtain an expression for the acoustic pressure on the unburned side of the flame

$$\overline{P}_f = 2\gamma\ 1 - \left(\frac{V_u}{V_b}\right)\left(\frac{V_b}{V_u}\right)^2 M_{Su}^2 \qquad (6)$$

which is identical to Eq. (19) of Strehlow et al..[3]

If we write $r_f = S_s t$ for the flame propagation in Eq. (5) and recall that away from the flame t becomes

$$\left(t - \frac{r}{a_0}\right)$$

Eq. (5) may be written

$$\overline{P} = 2\gamma\left(1 - \frac{V_u}{V_b}\right)\left(\frac{V_b}{V_u}\right)^3 M_{Su}^3\left(\frac{a_0 t}{r} - 1\right) \qquad (7)$$

However, the transformation from a flame to a Taylor spherical piston is

$$M_p = \left(1 - \frac{V_u}{V_b}\right)^{1/3}\frac{V_u}{V_b}M_{Su}$$

Therefore, Eq. (7) reduces to

$$\overline{P} = 2\gamma_o M_p^3 \left\{ \frac{a_o t}{r} - 1 \right\} \qquad (8)$$

where the range on r is $S_p t < r < a_o t$. Equation (8) is identical to the Taylor solution for a spherical piston as $M_p \to 0$ (Eq. (13), Strehlow et al.(3)).

We note that in these equations the rate of flame area growth is proportional to r and the rate of decay of a spherical acoustic pulse is proportional to r^{-1}. Thus, for the spherical case the acoustic pressure remains constant at the flame front and the wave that is generated during flame motion is self-similar in r/t. This is a fundamental acoustic property of a constant-velocity flame that is completely surrounded by combustible material when it burns. It has been verified by numerous theoretical and numerical investigations.(3,4)

APPLICATION TO FREE CLOUDS

We restrict ourselves to clouds of arbitrary shape in free space. To simplify the treatment, we will consider free clouds which have a horizontal plane of symmetry for cloud shape, cloud distortion during deflagration and deflagration wave shape. Since we now wish to simply examine the implications of Eq. (3)for general geometries, we for the moment restrict ourselves to considering only constant velocity flames. This means that Eq. (3) reduces Eq. (4) where S_u is an effective normal burning velocity and r is the distance of an observer from the cloud center.

In general, the evaluation of the dA_f/dt term in Eq. (4) is complicated because during deflagrative combustion the cloud ahead of the flame will move and distort with time. This distortion will cause A_f to be a different function of time than that calculated from purely geometric considerations using a constant velocity flame propagating through a stationary cloud.

In order to determine how to best approximate the effect of such distortion for arbitrary ignition of an arbitrary cloud one must first examine flow behavior in two simple limit geometries; strictly one-dimensional flow caused by deflagration and volumetric heat addition in three dimensions.

The wave diagram for a strictly one-D flow driven by a constant-velocity deflagration wave is shown in Figure 1. If we model the combustion of a stoichiometric methane-air mixture as proposed in Strehlow et al.,(3) the gas to the right of the flame and to the left of the contact surface (regions 0, 1, & 3) can be assumed to have a heat capacity ratio $\gamma_0 = \gamma_1 = \gamma_3 = 1.4$, while the gas in region 2 has a $\gamma_2 = 1.202$. Additionally, the effective value of dimensionless energy addition is $\tilde{q} = 8.430$. This means (from Appendix A) that $\hat{q} = 9.242$ and that $V_b/V_u = 7.602$ for constant pressure combustion. Furthermore, for this system the Chapman-Jouguet deflagration Mach number is 0.16631.

The properties of the wave system of Figure 1 were calculated using the shock relationships from Liepman and Roshko(6) for the shocks S_1 and S_3, the deflagration relationship from Strehlow et al.(3) for the flame and the contact surface requirement that $U_{p2} = U_{p3}$ and $P_2 = P_3$. The results of this calculation that are important to this paper are summarized in Figure 2.

Figure 2 shows that particle velocities generated by flame propagation are relatively high and that the apparent flame speed relative to an observer is from 4.3 to 9.3 times the actual normal burning velocity. Also it is less than the space velocity of a spherical flame if the one-dimensional flame is traveling at less than about 0.9 of its maximum (CJ) velocity. Furthermore, the ratio $|U_{p3}/U_{p1}|$ lies between 1.0 and 1.4 in this same velocity range. This last observation simply means that even relatively high speed flame propagation from a free surface in one-dimensional space displaces the gas rather uniformly to the front and back. This is to be constrasted to the behavior of a one-dimensional CJ detonation, which displaces the gas primarily in the direction of detonation propagation.

We now wish to determine what happens if we relax the confinement normal to the flame propagation direction for the case of deflagrative combustion. To estimate this effect we assume that if energy is added relatively slowly to a gaseous volume, the volume will expand an equal <u>distance</u> in all directions. During flame propagation through a cloud the flame can be assumed to be a relatively thin sheet of energy addition when compared to the cloud dimensions. In other words the flame can be assumed to occupy a volume $A\Delta L$ when $\Delta L << \sqrt{A}$. Under these circumstances one would expect the major expansion to occur in a direction normal to the local orientation of the flame even though the pressure is propagating almost equally in all three dimensions.

This assumption has been qualitatively verified experimentally by observing flame propagation in a propane-air elongated pancake cloud above a flat plate of about 50 mm height and 0.6 m x 0.3 m, edge ignited at the center of the 0.3 m edge.(7) The flame's apparent propagation speed was about 3.5 times the normal burning velocity, and its luminous height was about equal to the initial cloud height. Furthermore the flame flashed to encompass an area considerably larger than the original lateral extent of the cloud. The assumption has also been reasonably well verified by numerical calculations made by Dr. Len Hazelman (8) at Lawrence Livermore Laboratories using a 2 D hydrocode. He calculated flow and flame propagation in an open 2 D channel. His calculations yielded a flame height almost twice the initial cloud height and an apparent flame speed about three times the normal burning velocity. However, his numerical flame had a thickness only one-third of the cloud height and this should lead to more vertical distortion.

These results indicate that for any arbitrary free cloud and igniter geometry in which the flame is not completely surrounded by a combustible mixture one may make the following conservative assumptions about flame-induced distortion.

1. The burning velocity that should be used is S_u. This is because, irrespective of the motion of the flame induced by the flow, the entire cloud will burn in a time, τ, given by $S_u\tau = r_m$ where r_m is the distance from the igniter to the farthest point in the initially quiescent cloud.

2. The increase of flame area in a free cloud due to distortion of the cloud by flame-induced flow will never exceed the value V_b/V_u calculated for isoboric combustion. This is because this is the value of distortion of the unburned material produced during combustion of a completely surrounded (spherical) flame, and relief due to product gas motion away from the flame can only reduce this factor.

Thus the general procedure for determining a first conservative approximation to the effect of cloud shape and igniter location on the overpressure-time behavior in the blast wave when the igniter is not completely surrounded by the cloud is as follows:

1. Assume a value of S_u.
2. Calculate dA_f/dt for a flame traveling at velocity S_u through the quiescent cloud.
4. Correct for distortion by multiplying by V_b/V_u.

Examples for Specific Geometries

a. Central (line) ignition of a pancake cloud.
The flame area is given by

$$A_f = 2\pi r_f H$$

where r_f is the flame radius and H is the height of the pancake. Therefore

$$\frac{dA_f}{dt} = 2\pi H \frac{dr_f}{dt} = 2\pi H S_u$$

and

$$\overline{P}_{max} = \frac{\hat{q} S_u^2 H}{2a_o^2 R_{obs}} \left[\frac{V_b}{V_u}\right] \qquad (9)$$

where R_{obs} is the distance of an observer from the center of the cloud.

b. Edge (line) ignition of a pancake cloud.

The area of a flame propagating from an edge of a cloud of radius R is given in terms of the distance propagated by the formula

$$A_f = 2Hr_f \cos^{-1} \frac{r_f}{2R}$$

Now

$$\frac{dA_f}{dt} = \frac{dA_f}{dr_f} \cdot \frac{dr_f}{dt} = \frac{dA_f}{dr_f} \cdot S_u$$

Therefore

$$\overline{P} = \frac{\hat{q}S_u^2H}{2\pi a_o^2 r}\left[\cos^{-1}\frac{r}{2R} - \frac{\frac{r}{2R}}{\sqrt{1-\left(\frac{r}{2R}\right)^2}}\right]\left(\frac{V_b}{V_u}\right)$$

or

$$\overline{P}_{max} = \frac{\hat{q}S_u^2H}{4a_o^2R_{obs}}\left(\frac{V_b}{V_u}\right) \qquad (10)$$

which is exactly half that predicted for a centrally-ignited pancake.

c. End (point) ignition of an ellipsoid (cigar-shaped) cloud.

Assume a flat flame of location r_f propagates from one end of a long cloud with a minor radius of rotation d and major radius b.

Flame area is therefore

$$A_f = \pi d^2\left[1 - \frac{r_f^2}{b^2}\right]$$

or

$$dA_f = -2\pi r_f\frac{d^2}{b^2}$$

Now let

$$-b = -L/2 \le r \le L/2 = b \quad \text{and } D = 2d$$

where L is the length of the ellipsoid. Therefore

$$\overline{P} = \frac{-\hat{q}S_u^2D^2}{2a_o^2rL^2}r_f\left(\frac{V_b}{V_u}\right)$$

and

$$\overline{P}_{max} = \frac{\hat{q}S_u^2}{4a_o^2}\cdot\frac{D}{R_{obs}}\cdot\frac{D}{L}\left(\frac{V_b}{V_u}\right) \qquad (11)$$

Now assume that $\hat{q}$, S_u, a_o, and (V_b/V_u) are independent of cloud shape and that the observer is at a fixed distance from the center of a cloud of fixed volume but of these different shapes. Call the volumes $V_s = V_p = V_c$, where the subscripts s, p, and c refer to spherical, pancake and cigar shaped clouds, respectively. Also define the aspect ratio of the cloud $\mathcal{R}$ as $\mathcal{R}_p = 2R_p/H_p$ and $\mathcal{R}_c = D_c/L_c$. Using this notation we obtain

$$H_p = D\sqrt[3]{\frac{2}{3\mathcal{R}^2}} \qquad (12)$$

and

$$D_c = \frac{D_s}{\sqrt[3]{\mathcal{R}}} \qquad (13)$$

Combining Eqs. (5), (9), (10) and (11) with Eqs. (12) and (13) yields

$$\left(\frac{\overline{P}_p}{\overline{P}_s}\right)_{max} = \frac{1}{\left[12\left(\frac{\hat{q}}{\gamma_o}+1\right)\mathcal{R}^2\right]^{1/3}} \qquad (14)$$

and

$$\left.\frac{\overline{P}_c}{\overline{P}_s}\right|_{max} = \left[\frac{2}{3\left(\frac{\hat{q}}{\gamma_o}+1\right)\mathcal{R}^4}\right]^{1/3} \qquad (15)$$

where the factor $\frac{\hat{q}}{\gamma_o} + 1 = V_b/V_u$ arises because we are considering the initial cloud diameter and Eq. (5) is written in terms of the final radius. Eq. (14) is for central-line ignition of a pancake: for edge-line ignition $(\overline{P}_c/\overline{P}_s)_{max}$ is one half of this value.

Figure 3 is a plot of Eqs. (13) and (14). It shows that the deflagration, after edge ignition of a large aspect-ratio cloud, produces much lower overpressures than central ignition of a spherical cloud. It should be pointed out that point ignition of the edge of a free spherical cloud $\mathcal{R}_c = 1$ would yield a higher overpressure than calculated with Eq. (14) because the derivation of Eq. (11) assumed a flat flame. In reality one would expect a flame of roughly spherical shape to propagate away from a point source and the initial rate of flame area increase would be larger in this case than for a flat flame.

EFFECT OF CLOUD SIZE

Equation (3) can be used to give an order of magnitude estimate of flame areas and accelerations that are necessary for a deflagration to produce a damaging blast wave. If we define the threshold of damage to be 0.1 Bar and assume that this level of overpressure is to be produced 100 m from the cloud center in an atmosphere with a velocity of sound of 350 m/s, Eq. (3) reduces to

$$1.6 \times 10^7 = A_f \frac{dS_u}{dt} + S_u \frac{dA_f}{dt} \qquad (16)$$

Table I was constructed from Eq. (16) under the two limit assumptions that (1) $dS_u/dt = 0$ and (2) $dA_f/dt = 0$.

Notice from Table I that even for very high flame velocities the rate of flame area increase must also be very high if even a weak blast wave is to be generated. On the other hand, only extremely large initial flame areas, exhibiting very large flame acceleration over their entire frontal area are necessary to produce a weak blast wave from deflagrative combustion. Both of these observations imply that extremely large clouds are required if one is to produce significant overpressure by deflagrative combustion alone.

DISCUSSION AND CONCLUSIONS

The application of simple source acoustic theory to deflagrative combustion of an unconfined cloud shows that:

1) It is very difficult to produce a damaging blast wave by deflagrative combustion once the flame is not completely surrounded by a combustible mixture.

2) The maximum overpressure produced, other circumstances being equal, is proportional to the ratio of the minor dimension of the cloud to the distance to the observer.

3) Blast pressure is rather uniformly distributed in all directions, i.e., the blast is roughly spherical.

4) The cloud must be very large if a damaging blast wave is to be produced.

5) Spherical flame propagation calculations such as those of Kuhl *et al.* and Strehlow *et al.* greatly overestimate blast pressures from deflagrative combustion following edge ignition of clouds with large aspect ratios.

From various accident accounts of the sequence of events that led to the production of a damaging blast wave after delayed ignition of a massive spill of combustible material it appears that:

1) There is a threshold spill size below which blast damage does not occur. Gugan's(9) documentation of incidents shows that blast damage has been observed for spills of less than 2000 Kg but more than 100 Kg only for the fuels H_2, H_2-CO mixture, CH_4 and C_2H_4. Blast damage has been recorded only for spills greater than 2000 Kg for all other fuels.

2) In the majority of cases where blast

TABLE I

(1) $dS_u/dt = 0$		(2) $dA_f/dt = 0$	
S_u, m/s	$\frac{dA_f}{dt}$, m²/s	A_f, m²	$\frac{dS_u}{dt}$, m/s²
1	1.7×10^7	100	1.7×10^5
10	1.7×10^6	10,000	1.7×10^3
100	1.7×10^5	1,000,000	17

damage occurred, fire was present for a considerable period before the blast occurred. 3) In many cases damage is highly directional.

These observations when coupled with the results of simple source acoustic theory for deflagrative combustion lead to the following conclusions.

1. There should be a size threshold below which blast will not occur as long as ignition is "soft", i.e., does not directly trigger detonation.

2. The fact that fire is present early after ignition indicates that massive flame accelerations are necessary to lead to blast wave formation. Since the flame must have burned through the cloud edge by the time the blast is produced, simple acoustic source theory must be operative for even high deflagration velocities. Thus, the blast must arise from some sort of effectively supersonic combustion process or from very rapid increases in effective surface area of the flame.

3. Simple acoustic source theory for deflagrative combustion shows that deflagrative processes per se cannot produce highly directional effects. However, it is well known that detonative combustion of a cloud does produce highly directional blast wave effects.

LITERATURE CITED

1. Stokes, Phil. Mag. XXXIV, Ser. 3, p. 52. (1849)
2. Lighthill, J., Waves in Fluids, Chapter 1, Cambridge Univ. Press, Cambridge (1978).
3. Strehlow, R.A., Luckritz, R.T., Adamczyk, A.A., and Shimpi, S.A., "The Blast Wave Generated by Spherical Flames", Combustion and Flame 35, 297-310 (1979).
4. Kuhl, A.L., Kamel, M.M., and Oppenheim, A.K., Fourteenth Symposium (International) on Combustion, The Combustion Institute, Pittsburgh, 1973, pp. 1201-1215.
5. Taylor, G.I., Proc. Roy. Soc. A 186, 273-292 (1946).
6. Liepman, H.W. and Roshko, A., Elements of Gas Dynamics, Chapter 3, p. 62, Wiley, New York, (1957).
7. Huseman, P., Private communication, Urbana, IL (1979).
8. Hazelman, L., Private communication, Livermore, CA (1979).
9. Gugan, K., Unconfined Vapour Cloud Explosions, The Institute of Chemical Engineers, London (1979), pp. 28-35, 39.

ACKNOWLEDGMENTS

The author wishes to thank Dr. Len Hazelman of Lawrence Livermore Laboratory, Livermore, California, for performing some 2 D hydrocode calculations. He also wishes to thank Mr. Peter Huseman for performing a few critical experiments and Dr. Harold O. Barthel for a number of helpful discussions.

This work was supported by Arthur D. Little, Inc. and by AFOSR Grant 77-3336, Dr. B. T. Wolfson, Technical Monitor.

APPENDIX A

The Generalized Heat Addition Hugoniot

Strehlow et al.(3) have shown that a generalized heat addition Hugoniot quite accurately fits the real Hugoniot for hydrocarbon-air combustion over the range of interest for vapor cloud explosions. Briefly they assume that

$$H_o = C_{p_o} T$$

and

$$h_1 = C_{p_1} T - \tilde{Q}$$

where $\tilde{Q}$ is the energy added at the deflagrative/detonative discontinuity and C_{p_o} and C_{p_1} are different heat capacities for the reactive and product gases respectively.

Using this approach they define a dimensionless energy addition as

$$\tilde{q} = (\gamma_1 - 1)\tilde{Q}/P_o V_o$$

Thus the volume change for constant pressure combustion is given by the expression

$$\frac{V_b-V_u}{V_u} = \frac{(\gamma_1-\gamma_o)}{\gamma_1(\gamma_o-1)} + \frac{\tilde{q}}{\gamma_1}$$

therefore

$$\frac{V_b-V_u}{V_u} = \frac{(\gamma_1-\gamma_o)}{\gamma_1 (\cdots}$$

Now define

$$\frac{V_b-V_u}{V_u} = \frac{\hat{q}}{\gamma_o}$$

for ease of nondimensionalizing the acoustic problem. This yields

$$\hat{q} = \frac{\gamma_o(\gamma_1-\gamma_o)}{\gamma_1(\gamma_o-1)} + \frac{\gamma_o}{\gamma_1}\tilde{q}$$

For stoichiometric methane-air mixtures(3)

$$\gamma_o = 1.4 \qquad \gamma_1 = 1.202 \quad \text{and} \quad \tilde{q} = 8.430$$

Therefore

$$\frac{V_b-V_u}{V_u} = \frac{\hat{q}}{\gamma_o} = 6.6015$$

Thus $\hat{q} = 9.242$

and $\frac{V_b}{V_u} = 7.6015$

for a stoichiometric methane cloud.

STREHLOW, R.A.

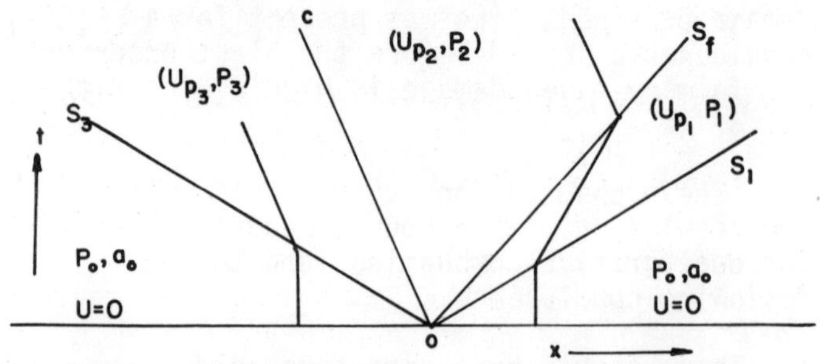

Figure 1. The wave diagram for strictly one-dimensional, constant-velocity deflagration of a free cloud after edge ignition (edge located at $x = 0$, S_f is apparent flame speed).

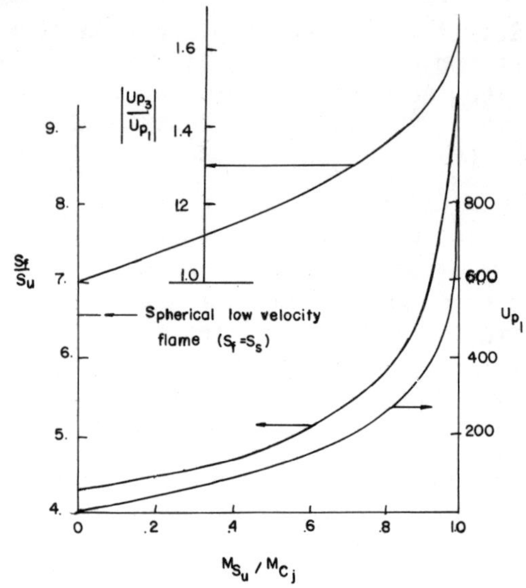

Figure 2. Flow velocities and apparent speeds associated with strictly one-dimensional deflagration combustion illustrated in Figure 1.

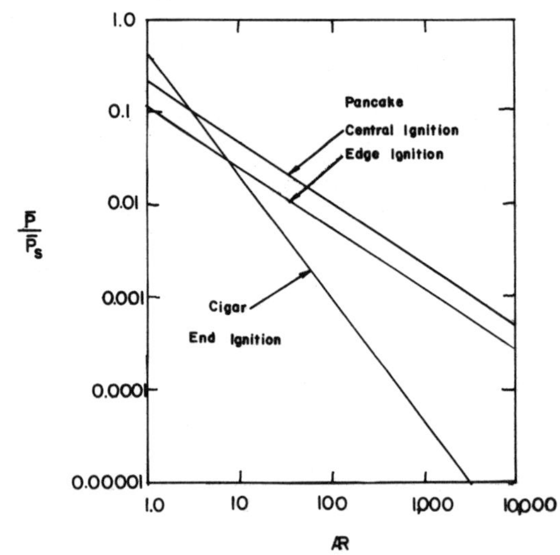

Figure 3. Effect of the aspect ratio, Æ, on the maximum blast wave pressure rise for the deflagrative combustion of pancake and cigar-shaped clouds. Cloud volume, normal burning velocity and observer distance from cloud center all assumed to be constant from cloud to cloud.

DISCUSSION

IAN SWIFT, Union Carbide: Bill, would you care to comment on the recent work that's been done both by John Lee and Rolf Eckoff in Norway and some work done by Derek Bradley and Tom Smith in Leeds, England? You can get these flame velocity acceleration mechanisms coming into play through the effect of turbulence you may produce by grids or, in the case that Bradley and Smith showed, you can get the turbulence produced by the atmospheric turbulence that you can get with typical wind speeds.

BRASIE: I've looked at some of John Lee's work; I'm less familiar with Bradley's. I think you have to consider this model as one end of a spectrum, because the basis here is that the derivative of SU (the basic flame velocity) is zero. In those processes that John Lee and other people have looked at, SU which is the local instantaneous burning rate, rather than the expansion and cloud movement, is being enhanced. If this theory were modified, I am sure it will move in the direction of predicting overpressures faster. You've got to take this model and integrate it. You have to keep determining the local delta SU and integrate your way along the cloud length. I presume you may then come up with the kind of results that John Lee and some other folks have shown.

SWIFT: I think that Flixboro gives us an example of one of the in-between situations. I think that Bradley and Smith looked at the meteorological data that was available for the Flixboro area at the time of the incident and they found that just the wind that was present could have produced accelerations of up to 40 times the initial burning velocity. Also, I would like to thank you on behalf of everyone for presenting Roger's paper Thank you.

BRASIE: I think the real world is somewhere in-between, and I think that Flixboro gives us an example of one of these in-between situations. I don't doubt that if any of you have been in fairly tightly packed plants like in Chicago streets, the wind velocities in-between different segments of buildings are several times the local ambient wind velocity. This effect is due to vortexing and venturi-type effects and that sort of thing. I'm convinced they do play a role. Unfortunately, how do you model these ahead of time? Do you take the big detonation instantaneous release model, or do you take some slow buildup which says the flame dies before it achieves any of these intermediate effects?

Fabrication and Use of a 20-L Spherical Dust Testing Apparatus

For dust explosivity evaluation, a 20-L spherical test device will provide reliable results which are scaleable to full-size equipment. The authors of this paper designed and fabricated a simple, low-cost 20-L unit.

R. E. Cocks and R. C. Meyer
Procter & Gamble Co., Cincinnati, OH.

As a result of entering the cake mix business in the late 1950's, we developed an awareness of the hazards involved in handling dry materials such as flour and sugar. To test the explosivity of these and other combustible dusts our Process Safety Group fabricated a demonstration explosimeter and a modified Hartmann device. Both units were patterned after Bureau of Mines equipment.

Our standard procedure was to initially determine if the dust was explosive using the demonstration explosimeter (Figure 1). Twenty consecutive negative results, testing a range of concentrations, indicated that the dust was non-explosive. Any positive result was followed by testing on the modified Hartmann, again through a range of concentrations.

Our modified Hartmann, now nineteen years old, provided an abundance of data for our many product areas. It should be obvious, however, that a cylindrical shape is not ideally suited for evaluating a spherical flame front.

In early 1978, we became quite concerned about the quality of our dust explosivity data due to a series of grain elevator explosions and a serious dust explosion and fire at a competitor's disposable diaper plant. During the next twelve months, we tested a great many of our combustible dusts, and we learned that many materials previously considered non-explosive could explode under the right conditions--paper dust as one example.

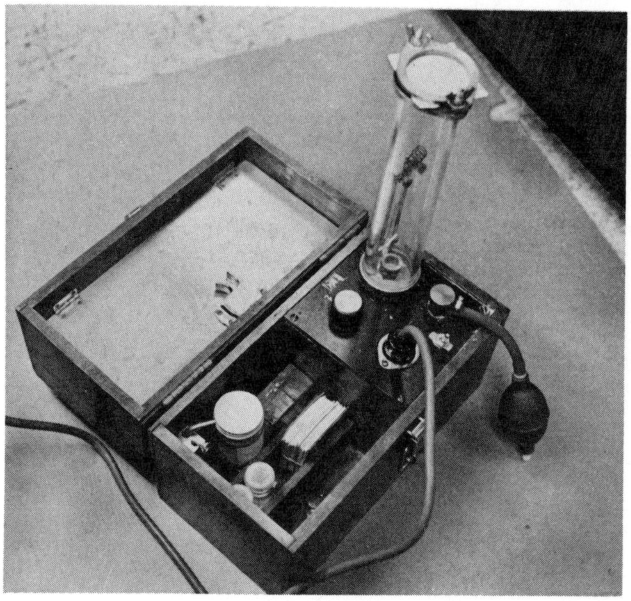

Figure 1

Coincident with our in-house testing work was a growing recognition of the outstanding work done by Dr. Bartknecht in this field and of his recommendation that 20 litre spherical devices be adopted as an

0149-3701/81/4007 $02.00 © 1981 AIChE

industry standard. In late 1978, we decided to obtain a 20 litre device. Since the supplier, Adolph Kühner in Basel, Switzerland, was not ready to begin shipping units, we decided to fabricate our own.

This paper will describe what we fabricated, its approximate cost, optimization of sample dispersion and ignition, and a comparison of results to Kühner's commercial unit.

FABRICATING A 20 LITRE BOMB

We discovered that two standard 14 inch (356 mm) hemispherical heads, with nominal 3/8 inch (9.5 mm) wall thickness, when joined together into a sphere equal a 20 litre volume (Figure 2). Two heads in 304 stainless steel cost about $200. By matching the volume of Kühner's commercial bomb, results should be comparable.

To provide access to the bomb, we used a top mounted Tube-turns hinged closure (Figure 3). Overpressure protection is provided by a union type rupture disc with vacuum support, rated at 1,720 kPa burst pressure. The burst pressure is determined by the hinged closure which Tube-turns rates at 1,850 kPa.

Figure 2

Figure 3

The bomb was fabricated in one of our shops and equipped with various penetrations for sample introduction, exhaust/vacuum valves, and ignitor holders. The unit was piped with the ability to introduce the sample and/or dispersion air from either the top or bottom of the bomb (Figure 4). Dispersion air from a compressed gas cylinder is stored in a small reservoir with a volume of 0.8 litre. The sample is placed in the cyclone shaped chamber, which is equipped with a quick-opening lid. Actuation of a solenoid valve permits dispersion air to sweep the sample into the bomb.

Instrumentation consists of a piezoelectric type transducer, an electronic signal conditioner, and an oscilliscope. Slope measurements are easily taken off the grid type screen. A permanent record is made by taking a photograph typically of a group of three consecutive runs.

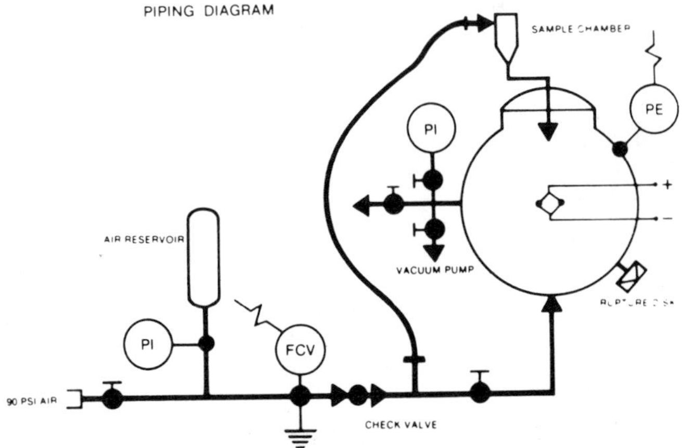

Figure 4

As our test setup is now assembled, the approximate costs are:

Purchase fittings and fabricate bomb	$2,000
Purchase piping components	250
Purchase ignition control components	100
Purchase instrumentation	3,000
Total	$5,350

This assumes in-lab piping and wiring--both relatively simple tasks.

We've spent another $3,000 in throw-away costs and obviously have invested considerable engineer effort in learning how to use our unit. The result, however, is a relatively simple, easy to use device which should be within the reach of most industrial firms.

SAMPLE IGNITION

Initial attempts to operate the unit centered around the use of either a hot wire or a high voltage spark as an ignition source. We didn't obtain consistent results using either the hot wire or the spark. This was probably due to the low ignition energy (about ten joules maximum) available from either source. This was further clouded by our parallel efforts to optimize sample dispersion. The control circuitry was complicated due to the need to preheat the hot wire prior to sample introduction, and to leave the hot wire or spark current on for a finite period of time after sample introduction.

Conversations with people at the Bureau of Mines, and later during a visit to Europe where we visited Kühner, Ciba-Geigy and the Bergbau-Versuchsstrecke Institute in West Germany, convinced us of the merits of using electric matches as an ignition source. We now use these exclusively and as a result of this decision, we were able to fabricate an extremely simple ignition control unit (Figure 5).

Figure 6 is a schematic of the control circuit. The design permits easy calibration of the time delay relay.

For the initial calibration runs, we used the same electric match as the Bureau of Mines, an ICI Atlas M103. We're pleased with electric matches as an ignition source. They're safe to handle, reliable, permit the use of simple control circuitry, and introduce significantly more energy than either a hot wire or spark.

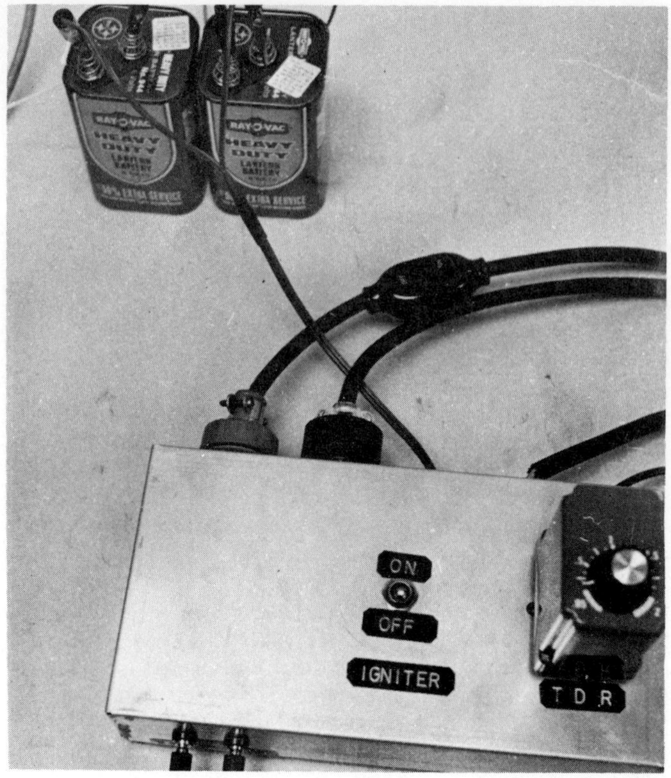

Figure 5

WIRING DIAGRAM

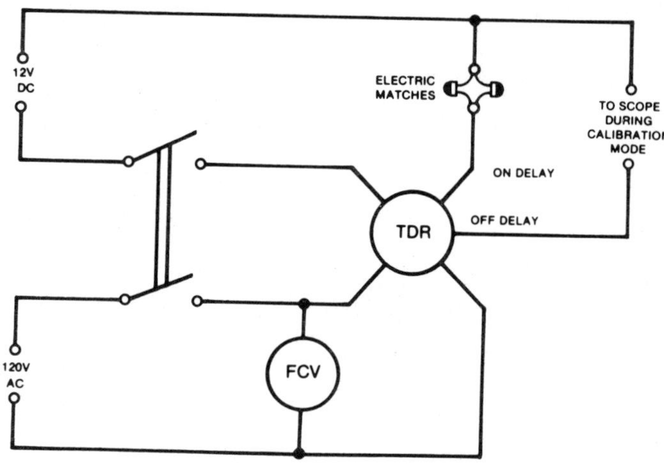

Figure 6

Two matches were used at a time, with a total measured energy release of 275 joules. Figure 7 shows dp/dt data for Tinuvin at 500 g/m^3 versus number of electric matches. As can be seen, the number of matches used in this case did not affect dp/dt results. However, we've tested other materials where we believe higher ignition energy definitely would have

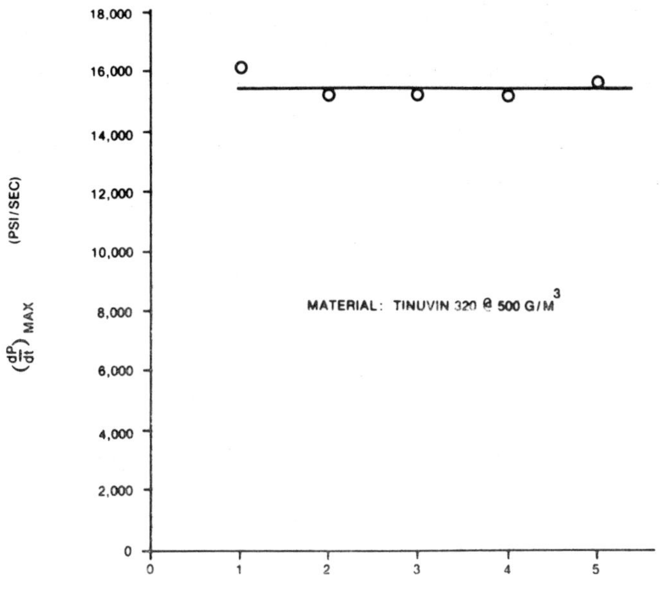

Figure 7

yielded higher dp/dt results, and are planning to begin using higher energy ignition sources.

Dr. Bartknecht and Kühner strongly recommend the use of two 5000 joule pyrotechnic ignitors as an ignition source; a 36 fold energy increase from the Atlas match. The use of two 5000 joule matches will provide a "worst case" event from a strong ignition source, which is what we always want to design for.

Our in-house procedure for sizing explosion vents utilizes dp/dt data from sample testing and will size vents for equipment rated at low MAP's, such as dust filters. At the time this paper was written, serious consideration was being given to converting to the Bartknecht nomographs for vent sizing. Our primary concern is that the nomographs only go down to a 0.2 bar maximum vessel pressure during venting. Unfortunately, many rectangular dust filters fail at about 0.2 bar. There's a need to include a 0.1 bar curve on the nomographs.

SAMPLE DISPERSION

The Kühner unit introduces the sample from the bottom through a horseshoe shaped ring containing numerous small diameter holes. Many of the samples we test are fibrous and would soon plug these holes. In fact, samples of our paper dust run by Kühner did quickly plug the horseshoe.

We introduce the sample from the top, using about 1/3 of the total air volume. The balance of the air enters simultaneously from the bottom. The resulting turbulence results in good sample dispersion. A number of options have been tried at the sample entry point, including straight pipes, the use of a static mixer element at the end of a pipe, and various configurations of holes at the end of a capped pipe. We currently use a pattern of five 3/16 inch (5 mm) holes at the end of a length of capped tubing. For fibrous samples an open pipe can be used.

Results for both open and capped pipes are about the same. We use capped tubing when possible because both intuitively and during visual observations dispersion is better than with an open pipe.

Ciba-Geigy is experimenting with other sample entry systems, and there are indications that the horseshoe shaped device might be replaced with something less apt to plug.

TEST PROCEDURES

If we have a process which uses or generates dusts which are suspected of being explosive, we obtain representative samples of the materials in air tight bags. We test samples of each raw material plus intermediate and final products.

Upon receipt of a dust sample, we initially determine if the dust is explosive by running a series of tests in a demonstration explosimeter. Figure 8 shows in decision tree format the entire testing procedure. The samples are checked for moisture content and, if necessary, dried to less than 5% moisture in a convection oven. A seive analysis is run on the sample using 250, 125, 63 and 45 micron screens. Twenty runs of the through 250 micron portion at various concentrations are then tested in the demonstration explosimeter. A positive result (i.e., burst paper diaphragm on the explosimeter) at any concentration indicates that the dust is explosive and that further tests are required to quantify the explosive characteristics of the material.

We have three demonstration explosimeters, two of which are regularly loaned to plants and technical centers for training, and occasionally, some preliminary testing work.

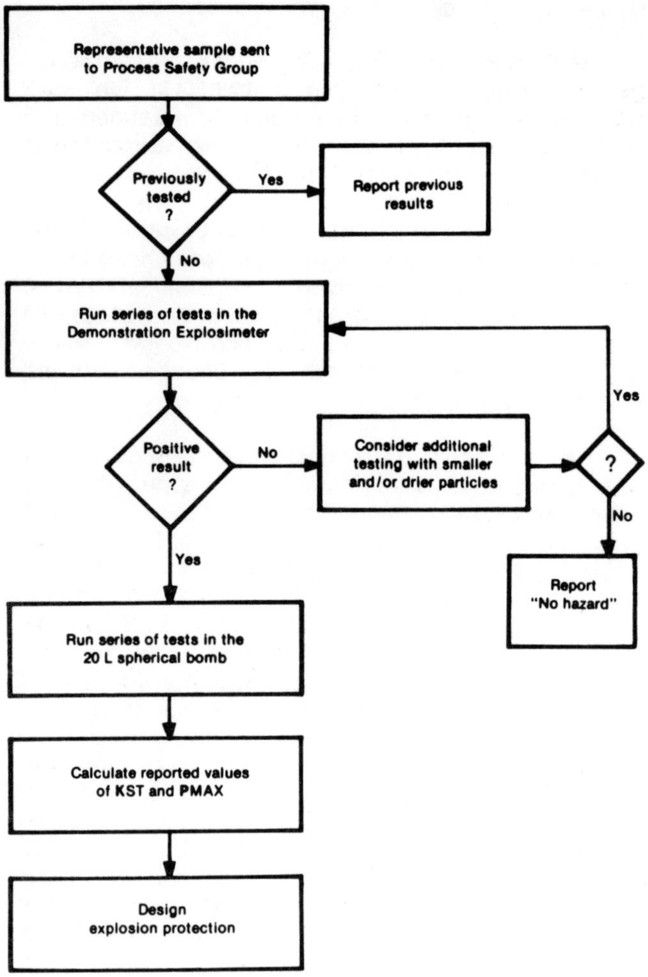

Figure 8

Samples are then evaluated in the 20 litre bomb as follows:

A. First, the optimum time delay between the opening of the solenoid valve and the firing of the electric matches is determined. Starting at 100 ms, test runs are repeated at a concentration of 500 g/m^3 with various timer settings until the highest value for maximum rate of pressure rise is determined (Figure 9). That time delay setting is used for the remaining test series. For this example, we would use 120 ms.

B. The test is continued by halving the concentration in each succeeding run until a negative result is obtained. This concentration can be interpreted as the lower explosive limit.

C. The sample concentration is then increased to 750 g/m^3 and beyond in increments of 250 g until the peak values of maximum pressure and rate of pressure rise have been established.

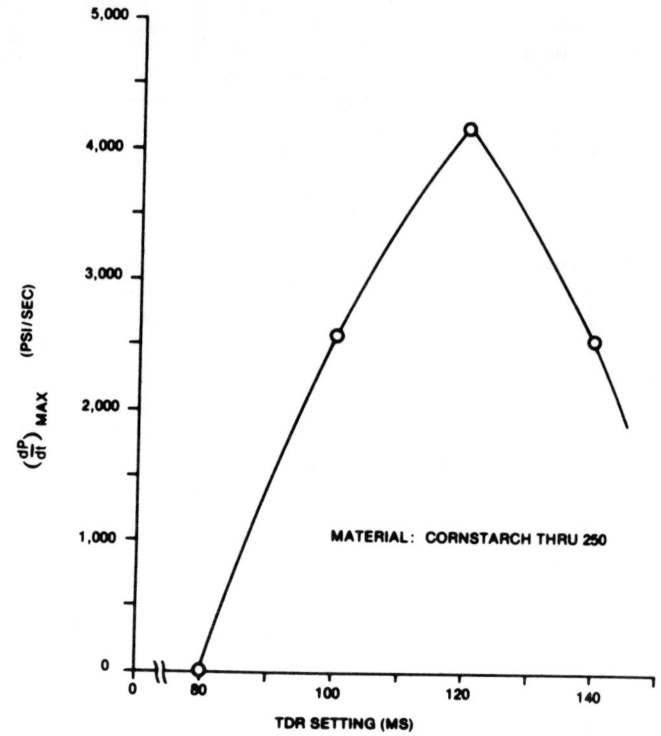

Figure 9

D. Finally, the maximum pressures and the maximum rates of pressure rise are plotted versus concentration for the sample (Figure 10). Two additional samples are then run at the concentration value which gave the maximum rate of pressure rise.

The sequence of an actual test run is as follows:

1. Calibrate the time delay relay using the oscilliscope.

2. Install the ignitors.

3. Weigh out the sample, place in sample chamber, and close bomb.

4. With the use of a vacuum pump, evacuate the bomb and pressurize the air reservoir. This is done so that the bomb chamber is at atmospheric pressure at the time of ignition.

5. Initiate the event by energizing the switch on the control panel.

6. The event is recorded on the oscilliscope in a pressure-time curve from which the maximum pressure obtained and the maximum rate of pressure rise can be read.

7. Vent and clean the bomb to prepare it for the next test run.

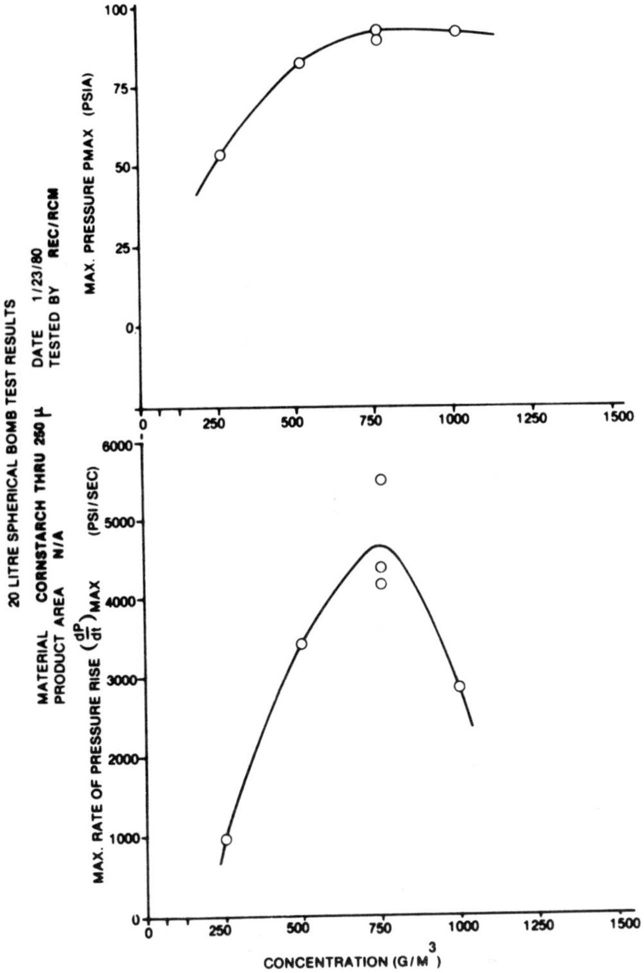

Figure 10

COMPARISON OF RESULTS

The apparatus was fabricated and ready for use in the spring of 1979. Startup was assigned as a summer project to an engineering student. During startup, we made about 100-200 runs using cornstarch, none of which provided any conclusive calibration data. We did, however, demonstrate the merits of using electric matches as an ignition source. We also satisifed ourselves that sample introduction from the top while introducing additional air from the bottom provided adequate dust dispersion. Any attempt to reduce bottom air significantly affected results.

During the October visit to Kühner, they were kind enough to provide us with two samples of material used as calibration standards for their commercial 20 litre apparatus. One material, Tinuvin 320, is an optical brightener. The second, Irganox 1010, is a plasticizer. Both are Ciba-Geigy products.

Our goal during the final calibration period was to match the results for the standards. During this period, we made 175 runs. The early runs used lycopodium powder, a dependable performer which was used during final tuning of the bomb. Tuning refers to final modifications to the sample dispersion and ignition systems. Once this was completed, we began testing the Tinuvin and Irganox samples. We also did some work using a sample of Pittsburgh Standard coal supplied by the Bureau of Mines.

Before we show the calibration results, let's talk about consistency. We're most impressed with the ability of our 20 litre spherical bomb to give consistent results. Figure 11 shows the oscilliscope screen containing 4 consecutive Tinuvin runs. You'll note that 3 of the 4 traces are virtually superimposed one on the other. Peak pressure for all runs was 7.6 bar, with slopes ranging from 950 to 1,190 bar/s.

Our work demonstrated that there are two factors which could significantly affect our results. First is having the optimum time delay before ignition. Figure 12 shows the calibration of our TDR before a test run. Optimum test results will occur over a band of about 20 ms; therefore, operating in the outer 5 ms of this band can give erratic results. This is one of the reasons we are changing our control system to a microcomputer as described later in this paper.

Secondly, we learned that obtaining accurate results is also influenced by regular system cleanout. Many materials

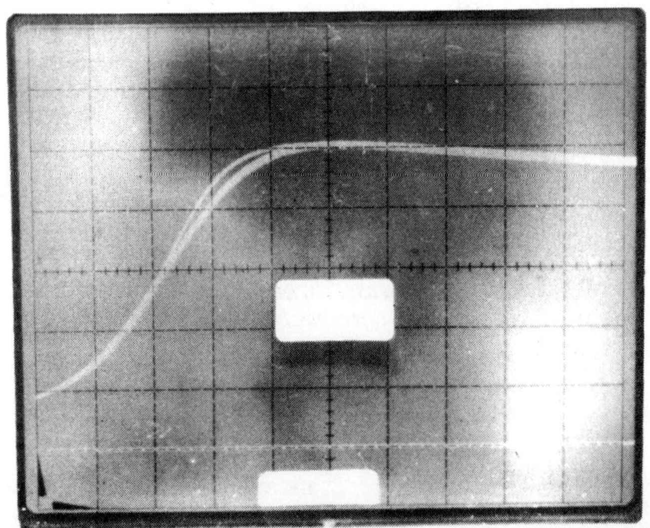

Figure 11

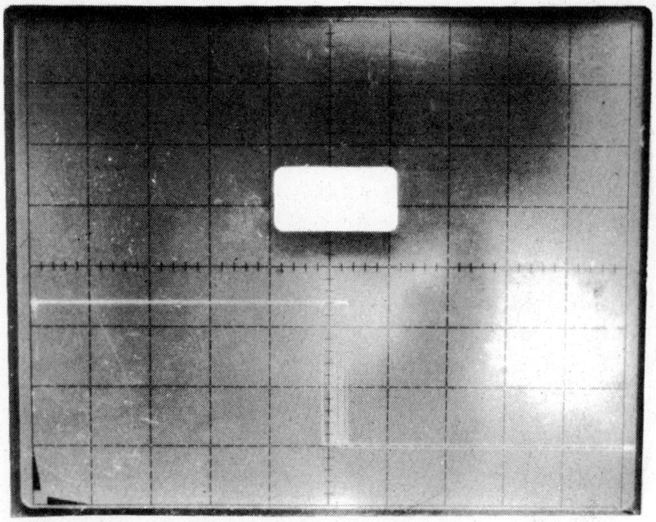

Figure 12

will back flash through the sample chamber and piping causing carbonization and/or line blockage. At concentrations of 750 g/m^3 and above, it's important to regularly break the piping and clean back to bare metal. The interior of the bomb is always vacuumed after each run, and the sample chamber is always visually inspected before introducing the sample.

Figure 13 shows the Kühner Tinuvin results for each concentration as a range of values in solid black with a line connecting the average of three reported values. The shaded lines are Procter & Gamble's results. You'll note, that our P_{max} values for Tinuvin typically are 0.2 to 0.3 bar higher than the Kühner results. This might indicate more complete combustion as a result of better sample dispersion. The dp/dt results match the standard reasonably well in the mid-range of concentration. We're not sure why our results are lower at 250 and 1000 g/m^3. We suspect that an increase in sample inlet piping diameter from 1/2 inch (12 mm) to 3/4 inch (19 mm) might improve results at the higher concentration.

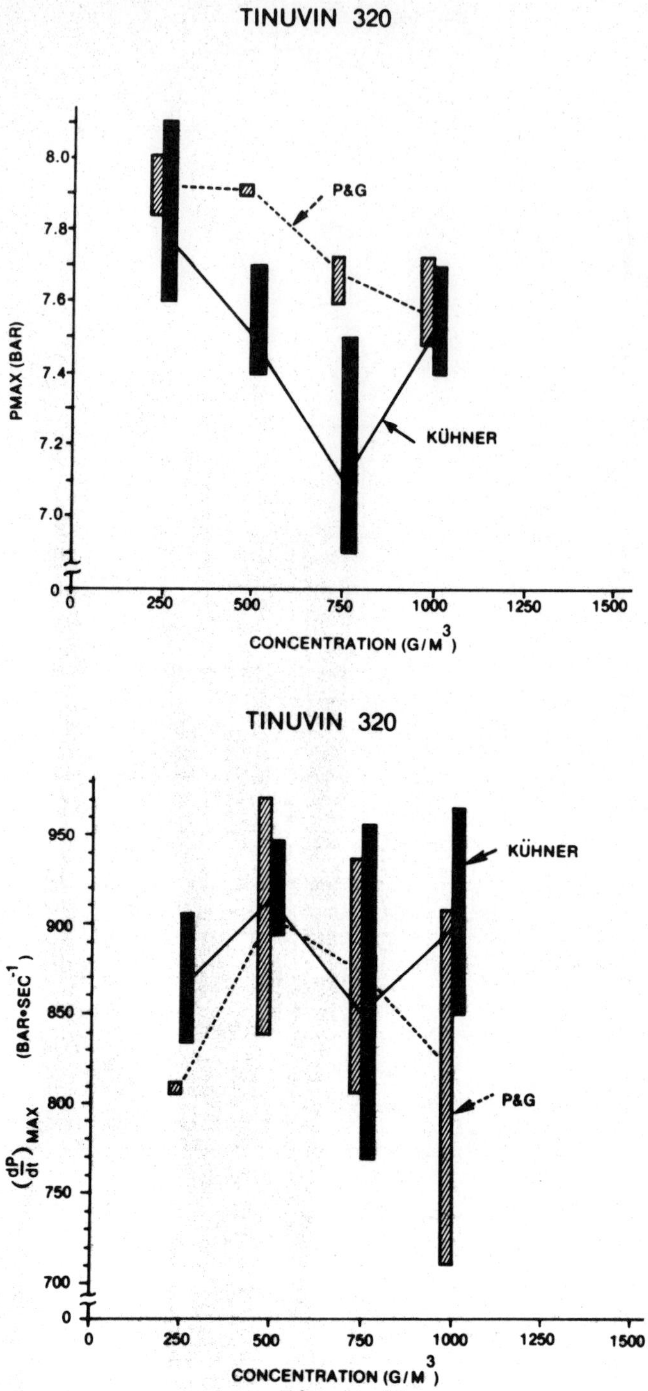

Figure 13

The Irganox results are shown in Figure 14. Irganox tended to quickly foul the inlet piping, and obtaining consistent results required considerable attention to system cleanout. For this material our results matched the standard for P_{max} and were slightly higher than Kühner's results for dp/dt.

The Pittsburgh Standard coal samples were run at a concentration of 500 g/m^3 and results were compared to data reported in the Bureau of Mines Report of Investigations 8360 "The Flammability of Coal Dust-Air Mixtures". For coal dust with a mean particle size of 22 microns, dp/dt for a 500 g/m^3 concentration is reported as 97 bar/s. This work was done in an 8 litre cylindrical vessel. The 97 bar/s. result can be scaled to a 20 litre volume using the cubic law:

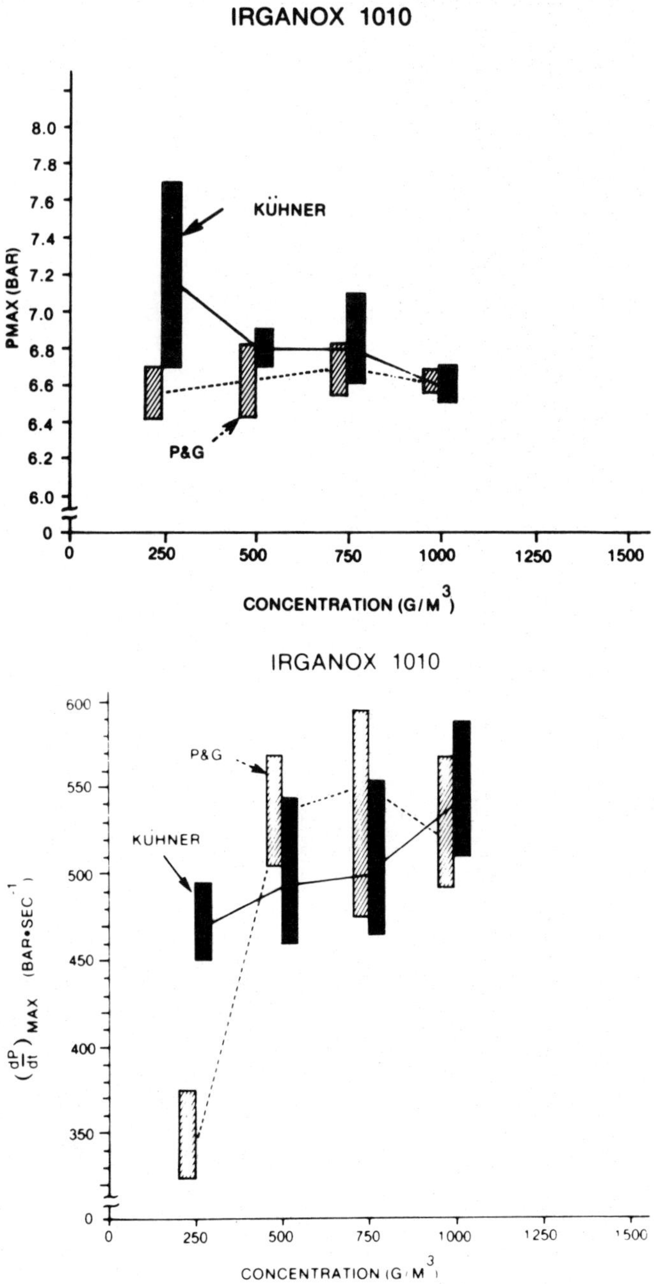

Figure 14

$$\frac{dp}{dt} \cdot V^{1/3} = \text{Constant}$$

The BOM result (97 bar/s) scaled to a 20 litre volume would indicate a result of 71.5 bar/s. Our results averaged about 200 bar/s for material through a 45 micron screen. A discrepancy of this magnitude is not unexpected when comparing results from a small cylindrical device to what Dr. Bartknecht has determined to be the minimum size spherical device.

UTILIZATION OF DATA

A 20 litre bomb is relatively simple to use. The results obtained clearly are more representative of full scale equipment than data formerly obtained on a Hartmann. Because of the simplicity of the test equipment, engineers involved in process safety work have been trained how to use the spherical bomb, and our lab is utilized as a walk-in facility. When a sample has to be tested, the engineers will usually do the work themselves. For extensive runs, technicians can be brought in and quickly trained.

We use the results to size explosion vents for items of equipment or process areas such as rooms or buildings. Whether the NFPA nomographs or some other procedure is used to size vents, only dp/dt data from testing is utilized.

Our current dp/dt measurement technique can result in up to a 6% error in reading the data. This data could be read more accurately by connecting the transducer output to a small microcomputer (Figure 15). We're now purchasing a system using an Apple II Plus computer, a TV monitor and a printer interface which, including programming and interfacing effort, will cost about $5,000. Once installed, this will accurately control time delay as well as provide an instantaneous readout of dp/dt and P_{max} data with an extremely high level of accuracy. With some additional programming effort, it could even handle determination of vent size for any inputted vessel or room configuration. However, if vent sizing is done using the NFPA nomographs, the need for more accurately

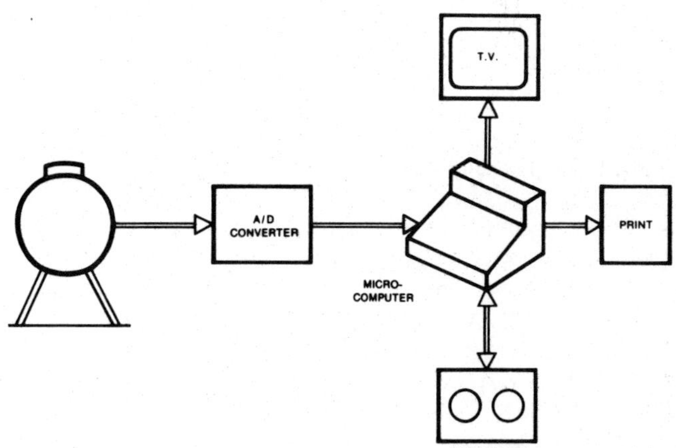

Figure 15

read results is questionable since the calculated K_{St} value is placed within one of three ranges and then used to enter a field of curves as follows:

$$1 \leq K_{St} \leq 100 \ : \ St\ 1$$

$$101 \leq K_{St} \leq 200 \ : \ St\ 2$$

$$K_{St} \geq 201 \ : \ St\ 3$$

These nomographs obviously don't require data readout with the degree of precision obtainable from a computer.

We're extremely pleased with the quality of results obtainable from our 20 litre spherical bomb. This test equipment for my company represents a valuable addition to our process safety technology. I sincerely hope that other organizations with interests in combustible dusts will consider investing in similar devices.

REFERENCES

1. National Fire Protection Association, "NFPA 68--Explosion Venting 1978".

2. Palmer, K. N., Dust Explosions and Fires, London, Chapman and Hall.

3. Bureau of Mines Report of Investigations; RI 5753, "Explosibility of Agricultural Dusts"; RI 5971, "Explosibility of Dusts Used in the Plastics Industry"; and RI 6516, "Explosibility of Metal Dusts".

4. Dorsett, Jacobson, Nagy, and Williams, Bureau of Mines Report of Investigations 5624, "Laboratory Equipment and Test Procedures for Evaluating Explosibility of Dusts".

5. Bartknecht, W., "Explosion Pressure Relief", 1977 AIChE Loss Prevention Symposium, pp. 93-105.

6. Bartknecht, W., "Gas, Vapour & Dust Explosions--Fundamentals, Prevention, Control", 1978 International Symposium on Grain Elevator Explosions, Washington, DC, pp. 55-159.

7. Bartknecht, W., "Determination of Coefficients of Explosions of Combustible Powders", December, 1977, Nuerenberg, W. Germany.

8. Hertzberg, Cashdollar & Opferman, U.S. Bureau of Mines Report of Investigations 8360, "The Flammability of Coal Dust-Air Mixtures".

9. Scholl, Reeh, Wiemann, and Faber, "Brenn-und Explosions--Kenngrössen von Stäuben". Bergbau--Versuchsstrecke Institut für Explosionsschutz und Sprengtechnik.

10. Adolf Kühner AG Apparatebau, Angensteinerstrasse 46, CH-4052 Basel, Switzerland.

COCKS, R.E.

MEYER, R.C.

DISCUSSION

BOB COFFEE, Eastman Kodak: A very interesting paper, Dick. You have been talking about using a computer to obtain the maximum slope and so forth. If you want to save a little money, you might consider Kistler equipment. Put a second pressure transducer in there, take the derivative directly, and then have peak reading meters read both. One will read P max. and the other one will read dp/dt max.; thus, you obtain the two values instantaneously. They can do it automatically for about $1700 a piece. It's a lot cheaper than paying many thousands for a computer.

IAN SWIFT, Union Carbide: I would like to congratulate Dick and Procter & Gamble in publicizing the work they've done on the spherical vessel. Perhaps, I could share with you some experiences we've had in these lines as well. Since 1977, Union Carbide has been using a 100-gal vessel to determine dust explosivity. About a year ago, we found that we're spending far too much time doing the task with the large vessel. Of course, tests always take much longer to do in large vessels, in fact the time is usually proportionate to the cube of the volume. So we decided to go to a smaller spherical vessel.

We built a 26-L spherical vessel and we got some quite interesting results from it in comparison to our 100-gal vessel and our Hartman Bomb. We found that we were getting Kst values approximately three times greater in our smaller vessel. We attributed this to the more intense turbulence that we obtained using our small vessel.

Incidentally, we use the horse shoe sparger that the Germans and the Swiss use. We found also that we would have to reclassify quite a number of dusts that we had tested previously in the Hartman. For example, one agricultural dust that we retested recently had a K value that would put it almost in dust class 3. In the Hartman apparatus, it was shown to be only a weak explosion hazard or fire hazard. So, it was quite an extreme change in classification using this new test. Just to carry on a little, we use a microcomputer both to control the time sequence of the inlet of the dust and the firing of the ignition circuit and to record the data. This gives us very reporducible results. We get a very accurate measure of the time delay and we can take data very rapidly and also plot out the curves, the slope and maximum pressure.

W.H. DOYLE, Consultant: I think Ian Swift has partly answered my question. Have you noticed in comparing your results with that of Hartman if there is any change in the relative ranking? Ian says they have found one where the change was drastic. Have you noticed anything of that nature? I understand we may have major changes. But, are there any in the ranking? Aluminum is worse than coal dust which is worse than flour, do you find changes in such rankings? Ian found one and I was wondering if you had. Thank you.

COCKS: The AFTM Committee on the C27 is about to start on round robbin tests with people with spherical vessels. So, hopefully, we'll be getting some comparative data to the Hartman tests that have been run already.

W.W. RUSSELL, Monsanto Co.: Dick, I very much appreciated your paper and was wondering if you could comment on what you think the possibilities might be of your test equipment for doing actual venting studies. As you know the data and methods that are available are limited as to pressure range. Do you think we might be able to get data from your equipment to fill in the gaps?

COCKS: Since our 20-L bomb has been in operation, we have run a series of detergent dust samples, run dust samples on a cellulosic product and on one other sample. We have yet to actually run a comparison with the Hartmann. The data, I've seen a great deal in the past three years from Dr. Bartknect and his associates, show some very dramatic differences between results run on a 20-L spherical device and results run on Hartmann and Hartmann-sized equipment.

JOHN ROGERSON, Procter & Gamble: Just to take up your question, Bill, the ASTM #27 Committee is about to start a round robin test using people with spherical vessels. Hopefully, we'll be getting some data to compare with the Hartmann tests that have been run already.

COCKS: You're suggesting, Bill, actually modifying a 20-L device to have the ability to try various size vent areas.

RUSSELL: Right.

COCKS: I see no reason why that wouldn't be feasible. One of the nice things about using this equipment, as I said earlier, is that the cost of the spherical bomb is quite low. Once you're set up with the basic instrumentation, I would imagine that we quite economically could experiment with exactly what you are proposing—run some basic venting studies for a fairly minimal cost.

We haven't considered this, but it certainly is a possibility.

I might add that Dr. Bartknecht published a textbook on combustible dusts in Europe about a year ago; there is an English version that's going to be published here in the States this fall. I've seen the German version and I'm sure it will be a valuable reference for many of you.

There is also a fairly sizable body of data available from the West German Bureau of Mines equivalent to that I mentioned earlier. This organization recently published a soft cover manual which is an extremely comprehensive listing of tests they have run on a wide range of agricultural, metal and chemical dusts; basically the type of work that the Bureau of

Mines was involved in 20 years ago but actually stopped doing in the mid 60's. Odds are good that many of the materials U.S. companies handle are contained in this West German reference edition. Their testing has all been on a 1-m size apparatus; I'm sure that the data is extremely solid.

COFFEE: There have been a number of questions raised about comparing data from the 20-L bomb or the 26-L bomb with the Hartmann data. One thing you should all keep in mind is that there is a great deal of Hartmann data in the literature (other than the Bureau of Mines') which are not correct. They are not correct because the equipment has never been standardized. ASTM committee E-27 in running round robbins and trying to develop a standardized test found that the major fault with the Hartmann apparatus involves the check valve. If the equipment is not checked and calibrated frequently, you can obtain answers that often vary by at least three or four to one.

I have seen some European data that report values about a third of those obtainable from the Hartmann apparatus when it is in good working order. So, I would say that until committee E-27 comes out with a standard (probably later this year), which shows how to check your equipment, do not depend on Hartmann data for comparisons. I think it is going to take another couple of years before we have some real good data for correlations or comparisons which will establish how good the Hartmann is compared to the 20-L or 1-m^3 vessel. It is going to take some time, unfortunately, but that's the way the situation is at the moment.

DAVE KIRBY, Factory Mutual: Have you or do you know of anyone who has run tests at higher than ambient temperatures with these dust-testing devices?

COCKS: The Bureau of Mines has some insulated test equipment, their West German counterpart has two 1-m bombs which are insulated, and Ciba-Geigy has insulated test equipment. All are capable of initiating events from elevated temperatures.

How Far Should We Go in Bringing Old Plants up to Modern Standards?

If the change on an old plant is as easy to make as on a new one it should be made, unless the plant is to be shut down within a few years or there is a cheaper and easier way of achieving the same objective of increased safety.

T.A. Kletz
Imperial Chemical Industries Limited
Wilton, Cleveland, UK

SUMMARY

There is no simple answer to the question asked in the title, but this paper considers some new features which have been or might be added to existing plants. Improvements in the layout are impossible but gas detectors, emergency isolation valves, better ventilation, steam and water curtains, better drainage, flanged or welded pipe joints and larger relief valves can, to some extent at least, be added to old plants. Often deficiencies in the hardware have to be compensated for by better software and this means that a high quality of operation and management is necessary. Old plants must not be starved of resources. There is no simple formula or rule but a number of examples are considered and these may help when other problems arise.

1 INTRODUCTION

Plant operators have always had to answer this question, but never more so than in the last decade when many changes have been made to the design of new plants to increase safety. This paper attempts to provide some information which may help in answering the question.

If we take extreme examples, as we do first, the question is easy to answer. Some features cannot possibly be added to an existing plant - we cannot alter the spacing between the various sections for example - while at the other extreme some features are as easy, or almost as easy, to install on an existing plant as on a new one. Problems arise when installation of the new safety features is possible, but more expensive than it would be on a new plant.

Also we have to consider the future life of the old plant - how long is it expected to run? - and the time taken to obtain the new equipment. Also major alterations to an existing plant are not without their risks (1), though they have been carried out successfully and safely many times, particularly in connection with debottlenecking projects.

2 AN IMPOSSIBLE MODIFICATION - BETTER LAYOUT AND LOCATION

Standards of plant layout have been under discussion for some years (2 & 3) but more so since Flixborough (4). Wide spacing between the sections of a plant is recommended to restrict the spread of fire. There is little or nothing that can be done to improve the layout of existing plants. All one can do is to see that fire-fighting facilities are sufficient to cope with any fire that might occur, possibly add water sprays or steam curtains (see 6 below) to confine and disperse gas leaks, provide water sprays to cool structures exposed to fire and check that fire insulation is adequate.

New plants should be located so that explosions do not cause such serious damage to neighbouring equipment that there is a domino effect or adjoining communities are affected. However, there is little that can be done to an existing plant, short of shutting it down. That extreme solution has been adopted in a few cases. After Flixborough a small plant operating a similar process and located inside an urban area was shut down.

It may be possible to prevent adverse changes in the location of an existing plant by preventing housing or other unsuitable plants being built too close to an existing plant. This can be done by buying the surrounding land or by coming to an agreement with the land-owners. In the UK the Health and Safety Executive advise planning authorities whether or not developments near a hazardous plant should be permitted.

3 AN EASY MODIFICATION - GAS DETECTORS

The installation of these detectors, so that leaks are detected quickly, has been recommended, particularly on plants or parts of plants, that are not continuously manned (5 and 6). Thus it is more important to install them in the areas underneath compressor houses than in compressor houses themselves. Combustible gas detectors, which detect any flammable gas, are the types most widely used, but other types may be particularly useful on particular plants.

It is as easy, or almost as easy, to install gas detectors on an existing plant as on a new one - and if there is a case for installing them on new plants - as I think there is - they should be added to old plants, unless they are expected to shut down soon. Even when they shut down the detectors can be recovered and re-used (though not the cabling which may be over half the cost).

Needless to say gas detectors do not remove the need for regular patrols by operators. Despite our gas detectors, half our leaks are detected by people.

4 A SOMEWHAT MORE DIFFICULT MODIFICATION - EMERGENCY ISOLATION VALVES.

These valves have been installed on many plants in recent years. They are located so that the lines leading to (or from) equipment which may leak can be isolated from a safe distance by pressing a button. They cannot, of course, be installed on the lines leading to all equipment which might leak but they can be installed on lines leading to equipment which experience shows are particularly liable to leak or where the consequences of a leak are particularly serious. (5) gives more information on the valves and on ways of deciding when they are necessary.

It is usually not difficult to install these valves on existing plants. Cutting and welding existing pipelines can usually be avoided by fixing pneumatic or electric actuators to existing valves. In Fig 1, while a new plant might have a single emergency valve on the common suction line to the two pumps, on an existing plant it would be easier to motorise the two existing suction valves. This has the added advantage that the valves will be easier to test. This example shows how a safety measure can often be modified to make it easier to install on existing plants.

I would urge all of you to consider whether more of these valves should be installed on your plants.

5 OPEN-SIDED BUILDINGS

A number of reports (for example, (7)) have described explosions in closed buildings and recommended that, whenever possible, equipment should be located in the open air, so as to obtain the full benefit of natural ventilation. While several tonnes of flammable gas or vapour are needed for an explosion in the open air, a few tens of kilograms are sufficient indoors. There is no objection to a roof over equipment such as compressors (provided there are ridge ventilators) but there should be no walls at equipment level.

During the last decade many compressor houses have been modified by removal of their walls. A particularly successful example of this treatment has been described by Monsanto (8). They paid particular attention to the provision of portable wind shields and heaters for the protection of maintenance workers.

Demolition of a wall is harder than not building it in the first place but nevertheless is usually practicable, not unduly expensive and should be considered again by those companies who have not yet done so. One of the biggest practical difficulties is often relocation of electrical and other equipment fastened to existing walls and some compromise may be possible, odd sections of wall being left. At the very least, if you are not prepared to knock down your walls, windows and frames can be removed.

6 STEAM AND WATER CURTAINS

Steam curtains (9) have been successfully added to several naphtha crackers and have been used 'in anger' to confine and disperse leaks of flammable gas. Water curtains (10) can be used as an alternative and have been used on a few polythene plants. Whether or not there is a need for them varies from plant to plant and a procedure for deciding is given in (11). On many plants they are not necessary - leaks will disperse naturally before they reach known sources of ignition. In others there may be no practicable location for the curtains, but there may be some where installation is both practicable and desirable. All I suggest is that you ask yourself if there is a need for them on your plant. In many cases effective water curtains can be provided by the use of portable monitors.

Water curtains require no extra water supply as fire-fighting water can be used. If the plant catches fire the curtain has failed and can be switched off. On the other hand, water curtains may require extra drainage to get rid of the water. Steam curtains are therefore probably the most economic where adequate steam is available, but not when an extra boiler would have to be installed. The choice between the two should be based on costs.

7 DRAINAGE

On older plants with inadequate drains, fire-water is often unable to get away and collects under the plant, unburnt oil floating on top (12). Modern plants have adequate drains (though not necessarily big enough to take the maximum quantity of fire-water that might be used) and the ground underneath them is usually sloped so that any spillage runs away from the plant ((5) and Fig 2).

While modern plants have closed drains, some old plants use open drain channels which provide a means of spreading burning liquid round the plant. Equipment for covering the channels with 'light water' should be considered.

Altering drains and the slope of the ground is difficult but nevertheless, in Petrochemicals Division, we have tested the capacity of the drains on all our major units and, where necessary, made some improvements. Their nature varies. In one or two cases a drainage sump has been built next to the plants, in others additional pumps have been provided for pumping fire-water away, in a few cases extra drains have been installed or concrete laid to alter the slope of the ground. The action taken depends on the size of the hazard and the adequacy of the existing arrangements. On some old plants, areas left unconcreted have absorbed spillages which then re-appear whenever there is heavy rain. These should be dug out and concreted.

8 PIPE-JOINTS - TWO CASE HISTORIES

(a) A liquid-phase hydrocarbon oxidation plant was constructed using caf gaskets in the flanged joints. Nowadays our specifications call for spiral wound gaskets. A number of flange leaks occurred during the early years of operation and a change to spiral-wound gaskets was suggested. As there are many thousands of flanged joints on the plants, the magnitude of the task appeared daunting but further consideration led to a practicable solution.

It was decided to replace all the gaskets in lines carrying liquid but not those in vapour lines, as leaks of vapour are less serious than leaks of liquid. Three thousand gaskets were changed over 2 years, during the normal 6-monthly shutdowns, the gaskets in lines carrying liquids above their boiling points being changed first. Completion of this programme has been accompanied by a big reduction in flange leaks, though other factors may also have contributed.

This example provides a good illustration of the way in which an improvement to an existing plant which, at first sight, appears too large to contemplate, becomes possible when it is pruned a little and spread over several years.

(b) Our current specifications permit the use of only flanged or welded joints (or special high-pressure joints); screwed joints are not permitted except on certain small bore instrument lines (after the first isolation valve) and on domestic water lines. Screwed joints have failed on a number of occasions, causing injury, and the wisdom of the specification is not disputed. However, the hazard is not considered sufficiently great to justify replacement of the numerous screwed joints on some older plants and replacements have been selective. On one plant, for example, all joints on main process lines were replaced up to the first branch isolation valve, but joints downstream of these valves were left unchanged.

9 CONTROL BUILDINGS

Since Flixborough, where most of the men killed were in the control room, there has been a move towards the construction of blast-resistant control buildings (13 & 14). For a new building the increase in cost is not great, about 30% of the initial building cost. However, strengthening existing buildings is very expensive - several times the cost of a new building - and would cause considerable disruption to plant operation, and this disruption could be hazardous. In Petrochemicals Division therefore, we have not felt that strengthening existing control buildings was justified but that the hazard to the occupants could be reduced by several measures that do not cost very much: Reducing the size of large windows (or removing them altogether), covering windows with plastic film and removing heavy lighting fittings.

An attempt has been made to quantify the size of the risk and this shows that, while there is a case for strengthening new control buildings (or moving them to a less hazardous area), it is not an overwhelming case (15). The large sums that would be required to strengthen existing buildings would probably be better spent on ways of reducing the chance that an explosion will occur.

10 MATERIALS OF CONSTRUCTION

Older naphtha crackers contain mild steel pipes and equipment which can become too cold and therefore brittle under abnormal operating conditions. This has caused a number of serious leaks and explosions (16) and newer plants make much more extensive use of special steels.

What should we do with the older plants? Complete replacement of pipes and equipment is obviously impracticable. We have carried out detailed studies of the pipework on older plants in order to assess its 'fitness for purpose'. Some pipes have been replaced; in other cases existing welded pipes have been radiographed to check that they contain no gross defects and have been stress relieved, thus making them suitable for service down to -50°C or below. In other cases additional alarms or trips have been installed to detect low temperatures and operator training has emphasised the importance of watching temperatures during start-up and reacting quickly to any abnormality.

11 RELIEF VALVES

Methods for the calculation of relief valve size have been revised during the last decade and many older plants therefore have undersized relief valves by modern standards. In Petrochemicals Divison we have recalculated relief valve sizes on our older plants and many relief valves have been replaced by larger ones. This is straightforward, but the flare headers may now be too small and their replacement would be expensive. Trip systems can be used (17) to avoid or reduce the need for larger relief valves or, if larger valves are fitted, to avoid a corresponding increase in the size of the flare headers.

12 SOFTWARE

Although we cannot always bring the hardware on our plants up to modern standards, we can ensure that our software - that is our methods of operation, training, instructions, testing and inspection, auditing and monitoring - are fully up to standard. This is at least half the problem. Accident investigations show that in at least half the accidents that occur no reasonable change can be made to the hardware to prevent a recurrence; the answer must lie in changes to the software. While 'safety by design' is a laudable objective, it is often impossible.

Thus, if we cannot bring the hardware of our old plants up to modern standards, we can at least bring the software up to standard and we shall not be as far below standard as might appear from an examination of the hardware alone.

In some cases (see 10 above) we can make up for deficiencies in the hardware by paying extra attention to the software.

It is, of course, not only necessary to bring the software up to standard, but it has to be maintained at the standard required. Without monitoring and support, without, that is, constant management interest, the software suffers from a 'social corrosion' every bit as damaging as the conventional sort.

Maintaining the necessary level of expertise in maintenance, operation and management becomes more difficult as more and more software is added. This often coincides with the extra maintenance and operating problems that are produced by the ageing of the plant.

Regular surveys are necessary on older plants to look for detailed features which are below standard, for example, drain and sample points and instrument connections. These items pass unnoticed unless a systematic survey is made.

Some modification of the software is sometimes necessary on old plants. Petrochemicals Division, some years ago, following some serious incidents, reviewed its standards of isolation for maintenance and decided that isolation by slip-plates should be the normal method (18). It was soon realised that on old plants there was often insufficient spring in the pipelines to allow full-thickness slip-plates to be inserted and that somewhat thinner ones might have to be used. If over-pressured they are unlikely to burst but have been known to become convex and are then difficult to remove. Newer plants are, of course, designed to take full thickness slip-plates.

13 CONCLUSIONS

It would be helpful to have a simple rule of thumb or numerical technique to help us decide whether or not a particular feature should be installed on an existing plant. Unfortunately, I do not see how one can be devised. Each case must be considered on its merits and the examples discussed above may help us in dealing with other cases. A few general comments may be made.

(i) If the change is as easy, or almost as easy, to make on an old plant as on a new one, then it should be made, unless the plant is to be shut-down within a few years (see item 3) or there is a cheaper and easier way of achieving the same objective of increased safety (see, for example, the 'fitness for purpose' evaluation in item 10).

(ii) We should not accept a higher level of risk to the operators (or to the public living nearby) on an old plant than we would be prepared to accept on a new plant, but we may have to use other methods to achieve an acceptable level of safety. We may have to modify the hardware design we would use on a new plant (see item 4) or we may have to rely on software procedures (see items 10 and 12).

If we do rely on the software, this must not be mere lip-service. Monitoring is essential to make sure that the procedures adopted in place of hardware changes are, in fact, being carried out. The results of these checks should be recorded and reported to more senior levels of management.

(iii) Where new standards are designed to prevent damage to plant and loss of production rather than injury to people, then it is legitimate to leave an older plant unchanged. The decision depends on a balance between probablities and costs and particularly if the plant is small or used only when peak output is required, change may not be justified.

(iv) Do we know how far our plants fall short of modern standards? Hazard and operability studies (19-21) and relief and blowdown reviews may be required.

Resources tend to be concentrated on expansion and development rather than upon running existing units which have operated reasonably well. There is a danger that the quantity and quality of our effort, hardware and software, on our older, less glamorous units can become second best, from operators through to professional staff. An insurer has pointed out that the worst losses occur on medium sized plants and has suggested that this may be due to the diversion of resources onto new large plants (22). Older units, on which hardware deficiencies are covered by increased attention to the software, require a high standard of operation and management, and we must see that our older plants are not starved of these resources.

ACKNOWLEDGEMENTS

Thanks are due to the many colleagues who helped in the preparation of this paper, particularly Messrs E H Frank, K E Scanlon and E A Turrill.

LITERATURE CITED

1 Kletz, T.A., Chem. Eng. Prog., 72 (11), 48 (1976).

2 Simpson, H.G., I. Chem. E. Symposium Series No. 34, 105 (1971).

3 Mecklenburg, J. C., (Editor), "Plant Layout - A Guide to the Layout of Process Plants on Sites", Leonard Hill in association with I. Chem. E., London (1973).

4 Kletz, T. A., Loss Prevention, 13, 147 (1980).

5 Kletz, T. A., Chem. Eng. Prog., 71 (9), 63 (1975).

6 Barker, G.F., T.A. Kletz and H. A. Knight, Chem. Eng. Prog., 73 (9), 4 (1977).

7 Howard, W. B., Loss Prevention, 6, 68 (1972).

8 Morris, D. H. A., Proceedings of the 1st International Symposium on Loss Prevention and Safety Promotion in the Process Industries, Elsevier, Amsterdam, 369 (1974).

9 Simpson H. G., Power and Works Engineering, 8 (8 May 1974).

10 McQuaid, J., Proceedings of the 2nd International Symposium on Loss Prevention and Safety Promotion in the Process Industries, Dechema, Frankfurt, 511 (1977).

11 McQuaid, J., presented at the course on "Process Safety - Theory and Practice", Teesside Polytechnic (July 1978).

12 Kletz, T. A., Hydrocarbon Processing, 58 (1), 243 (1979).

13 Advisory Committee on Major Hazards, Second Report, Her Majesty's Stationery Office, London (1979).

14 "Process Plant Hazard and Control Building Design", Chemical Industries Association, London (1979).

15 Kletz, T. A., Chem. Eng. Prog., 74 (10), 47 (1978).

16 van Eijnatten, A. L. M., Chem. Eng. Prog., 73 (9), 69 (1977).

17 Kletz, T. A., Chemical Processing, 77 (1974).

18 Kletz, T. A., as Ref 8, 1.

19 Lawley, H. G., Chem. Eng. Prog., 70 (4), 45 (1974).

20 Lawley, H. G., Hydrocarbon Processing, 55 (4), 247 (1976).

21 Hazard and Operability Studies, Chemical Industries Association, London (1978).

22 Neil, D. G., I. Chem. E. Symposium Series No. 34, 13 (1971).

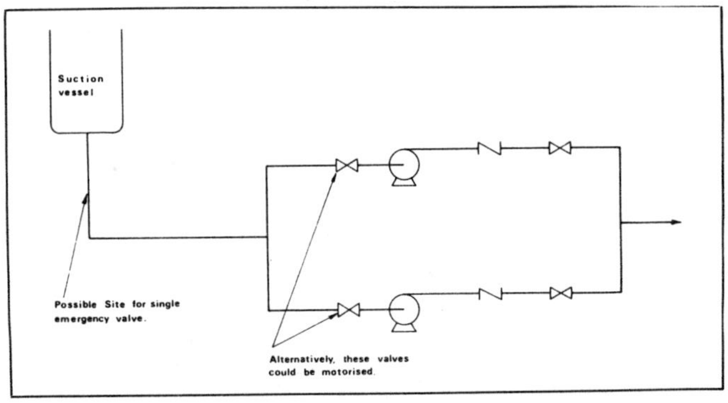

Figure 1. Possible approaches to installing emergency isolation valves (EIV) on an existing plant.

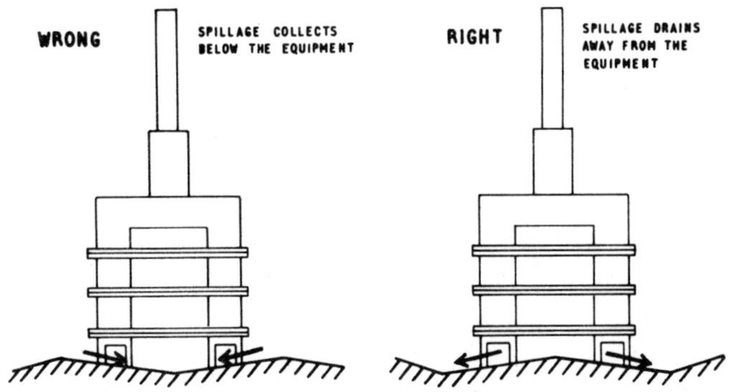

Figure 2. Drainage.

Figure 3. Gas detectors.

KLETZ, T.A.

DISCUSSION

STAN SKINNER, Monsanto: I have two questions, Trevor. First, in relation to the gasket placement, how much refacing did you have to do to the surfaces on which the gaskets mate? I was just wondering as the spiral wound gaskets would need a better standard of surface, this might contribute to the reduction in leaks.

TREVOR KLETZ, ICI: No resurfacing had to be done. The finish on the flanges was good enough for the operating pressure, about 8 bars.

SKINNER: The other point was with respect to relief valves and using a trip system to prevent use of relief valves. I assume this was in parallel to the relief valve and not in series, because if it is in series it might contravene a number of codes of practice.

KLETZ: I wouldn't say it was in series, but it was used as an alternative to a relief valve. We have a number of vessels on which there are no relief valves or the relief valves are not big enough. We have other groups of vessels where, if all the relief valves lift at once, say after a cooling water failure, the flare header is too small and will be overpressured. We are relying on instruments to make sure that no more than, say, one relief valve in five lifts. We are putting our faith on instruments.

SKINNER: Could you isolate the line up to the relief valve?

KLETZ: No, we wouldn't put a block valve underneath a relief valve. But, we have cases where the relief valve is either not there or not big enough and we are relying on trips. A number of papers including *Chemical Processing* (Sept., 1974, p.77) and *Chemical Engineering* (May 12, 1975, p.81) describe this philosophy which has been slow to catch on because mechanical engineers are very conservative and very devoted to relief valves and tend to say: "Put a high pressure trip on the vessel if you like, but if you don't mind, I'll put a relief valve on as well."

It is not the relief valves that we really want to avoid, but large flare systems. One way to keep flare system size down is to rely on instruments. There is a very interesting difference between the American and the U.K. codes on this. According to BS 5500, the British code on pressure vessels, a relief valve must be fitted to every closed vessel "in which the pressure is liable to rise to a dangerous degree." If a reliable trip system is installed and properly maintained, the pressure is not liable to rise to a dangerous degree. The U.S. code does not contain these words.

SKINNER: I'm a chemical engineer and not a mechanical engineer and I still have somewhat of a hangup on this point. I think that the relief valve as a device is inherently more reliable in a software sense than the trip system, in our experience.

KLETZ: If you use instruments instead of relief valves, you must have a management that is committed to the philosophy of testing the instruments. If you don't have that, you had better stick to relief valves. While you must test a relief valve every two years, you must test an instrument system every month. But, if you test an instrument system every month, you can get the same reliability as you would with a relief valve. In fact, you can get 10 times better, if the instruments are duplicated.

W.H. DOYLE, Consultant: I have a peripheral question, Trevor. When these remote actuated control valves are put in, do you fireproof the actuator and the power supply as a general rule?

KLETZ: We didn't originally. But, we're now starting to fireproof the actuator, if it is close to the source of fire. Fireproofing the power supply is easy; you use mineral-insulated, copper-covered cable. Fireproofing the actuator is more difficult; you've got to build a fireproof box around it or you can use a fireproof sack over it. It shouldn't take more than a few minutes to press the button and get the valve to close, so, 15-minute fire protection should be ample.

JOHN ELLINGTON, Mobil Chemical: I just wanted to comment on the first of your talk. You mentioned the problem with housing and so forth, closing in on our old plants. We in this country are dealing with urban sprawl all the time, but as chemical engineers we have to watch some of the things that are happening in our cities. I know of a large southern city where there has been a chemical plant for over 50 years. Houses are bad enough, but the zoning commission has approved and construction is under way on a multistory hospital within a few hundred yards of a large chlorination plant. The plant, you're right, might not be there. But this sort of gross, poor planning is very bad. All of you should be aware of this kind of thing, because it's something that we're going to have to face.

KLETZ: The same thing used to happen in the United Kingdom. About five years ago, a major sports complex was put up within a few hundred yards of a wharf where explosives are loaded. There's a lot of debate going on at the moment as to what should be done. But, now, there are in the U.K. procedures to prevent this happening. Before you an develop any land within about 2 km of a major hazard plant, that is one that contains more than 20 tons of LPG or 5 tons of chlorine, the Health and Safety Executive (our equivalent of OSHA) has to be consulted and for all practical purposes have the power of veto.

Sizing Safety Valve Inlet Lines

Limiting inlet pressure drop to 3% is not always the optimum method of preventing chatter and it can limit the designer's flexibility in certain instances.

O. J. Cox, Jr. and M. L. Weirick, Union Carbide Corp., South Charleston, W. Va. 25303

The American Petroleum Institute's Recommended Practice 520 (API-RP-520, Part II, Paragraph 2.2) (*1*) specifies that inlet pressure drop be less than 3% of a valve's set pressure for gas, two phase, or flashing liquid service. The 3% criterion prevents reduction of the valve's capacity, either by decreased driving pressure or by "chattering." Chattering is the rapid opening and closing of a valve, which can cause mechanical damage, capacity reduction of 50% or more, and valve sticking (either open or closed). Each percent increase in inlet loss causes approximately a 1% decrease in valve capacity. Figure 1 shows the proper operation of a safety valve having negligible inlet frictional pressure drop.

The underlying premise of the 3% inlet pressure drop criterion is that chattering will not occur, if the inlet pressure drop is less than the valve's blowdown. The 3% limit is well below valve manufacturer's recommended blowdown setting of 5 to 7%. (Blowdown is the difference between the safety valve's opening pressure and its closing pressure, expressed as a percent of the set pressure.)

Inlet line sizing method

Based on 3% inlet loss, this method will eliminate errors and reduce costs. Designers using it can size inlet lines without lengthy calculations or erroneous rules-of-thumb. A similar method can be developed for pressure drops that differ from the 3% limit. (Figures 2 to 4 illustrate inlet piping configurations.)

The method is outlined step-by-step below. By following this logic, the number of calculations will be minimized. The list of safety valves in the steps below do not include all valve sizes available, but do include the common sizes.

Step 1. Estimate inlet piping frictional loss velocity heads (K). See Table 1 for typical fitting values.

Step 2. If the inlet velocity heads (K) are greater than 1.5, skip to *Step 7,* otherwise continue to *Step 3.*

Step 3. If any one of the following items are true, continue to *Step 4.* Otherwise, skip to *Step 5.*

- A rupture disk is in the inlet line.
- The valve size is 4P6, 6R8, 6R10, or 8T10.
- The inlet line friction loss is greater than 0.7 velocity heads (or more than a full port shutoff valve, five diameters of pipe length, and a flush, square-edged connection). Any of the following valve sizes are used: 1½H3, 2J3, 2½J4, 3L4, or 6Q8.

Step 4. The inlet line should be one size larger than the valve inlet connection.

Step 5. If the valve size is 1D2, 1½D2, 1½D2½, 1E2, 1½E2, 1½E2½, 1½F2, 1½F2½, 1½G2½, 2G3, 1½H3, 2H3, 2J3, 2½J4, 3J4, 3K4, 3K6, 3L4, 4L6, 4M6, or 4N6, continue to *Step 6.* Otherwise, skip to *Step 7.*

Step 6. The inlet line should be the same size as the valve inlet connection.

Step 7. Calculations of the inlet pressure drop are required. As a first iteration, assume the inlet line to be the same size as the valve inlet connection. Continue to *Step 8.*

Step 8. Calculate inlet line pressure drop as shown in the "calculations" section. If the result is less than 3%, the line size assumed is adequate. Otherwise, continue to *Step 9.*

Step 9. The line size assumed is too small. Increase all or part of the inlet line as judgement indicates. The vessel connection should be at least as large as the largest inlet line component. Repeat *Step 8.*

Calculations

Inlet pressure drop includes the frictional (nonrecoverable) losses for all components between the protected vessel and the valve inlet flange. Losses are calculated at the actual flow rate (not the rated flow rate) at the set pressure. Percent inlet pressure drop is defined as:

Figure 1. Proper operation of a safety valve having negligible inlet frictional pressure drop is shown on the face of this pressure gauge.

0149-3701/81/4258 $02.00 © 1981 AIChE

Figure 2. Inlet pipings.

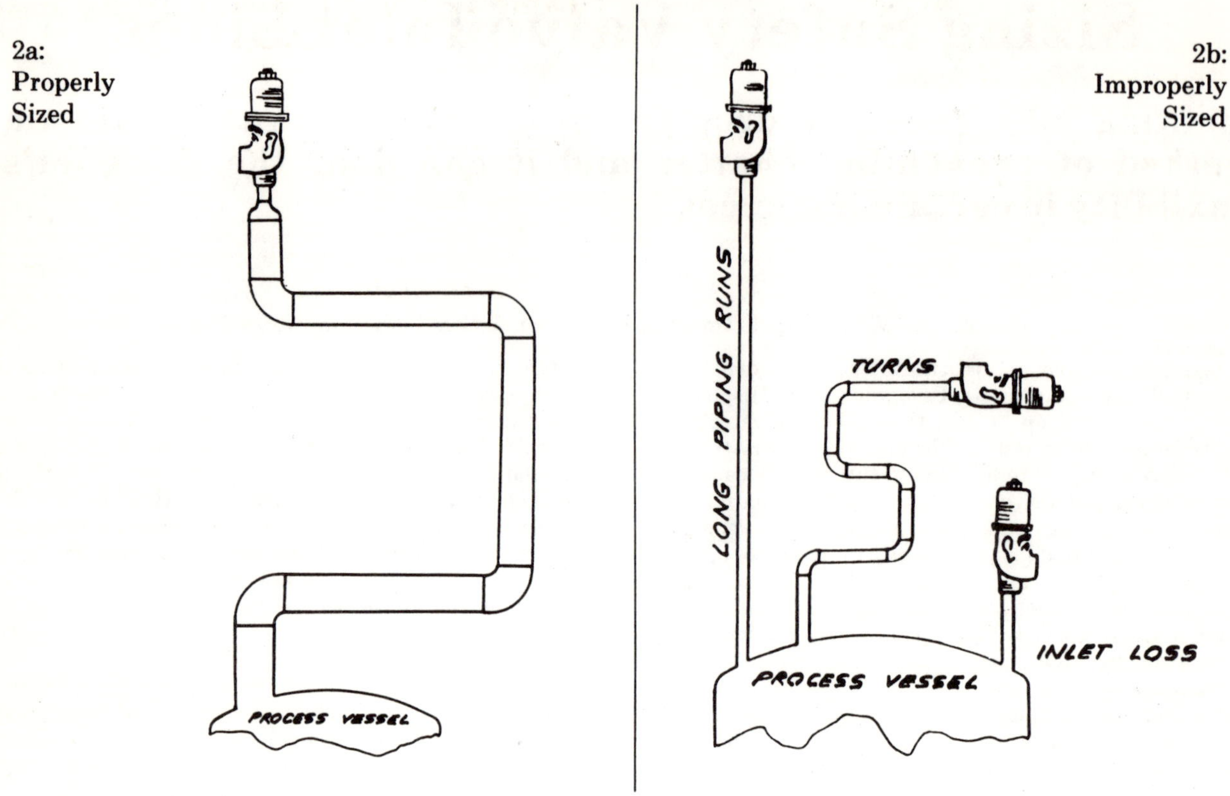

Illustrations in Figures 2, 3, and 4 courtesy Anderson, Greenwood & Co.

$$\%\Delta P = \frac{\text{inlet frictional pres. loss at actual flow, kPa}}{\text{safety valve set pres., kPa}}$$

Percent inlet frictional pressure is calculated by either Eq. 1 or 2. Eq. 1 is sounder but requires knowing the rated valve capacity. Eq. 2 assumes a valve orifice coefficient of 0.975 and an isentropic coefficient of 1.4.

$$\%\Delta P = 3.86 \times 10^{-11} \frac{K(W/0.9)^2}{A^2 \rho P_s} \qquad (1)$$

$$\%\Delta P = 28 \frac{K A_n^2}{A^2} \frac{P_s + 1}{P_s} \qquad (2)$$

where

K = Inlet piping frictional loss in velocity heads. (See Table 1 for values.)

ρ = Fluid density at valve set pressure and temperature, kg/m^3.

W = Rated safety valve flow rate at set pressure, kg/h.

P_s = Safety valve gauge set pressure, Bars (gauge).

A = Inner cross-section area of safety valve inlet connection, m^2.

A_n = Historical standard orifice areas, m^2.

In units consistent with manufacturer's literature, Eq. 1 is:

$$\%\Delta P = 1.73 \times 10^{-5} \frac{K[(W, \text{1b/h})/0.9]^2}{(A, \text{in}^2)^2 (\rho, \text{1b/ft}^3)(P_s, \text{psig})} \qquad (3)$$

Inlet frictional drop is the same as the change in stagnation pressure. The velocity head associated with acceleration of the fluid is ignored, because valve dynamics depend on stagnation pressure and not on static pressure. The standard velocity head approach is used to analyze fric-

Figure 3.

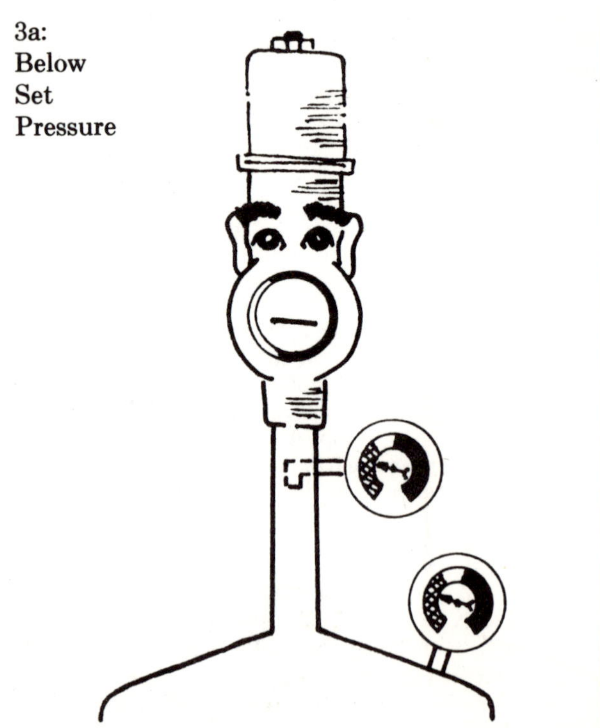

tional loss (*4*) with incompressible flow assumed.

$$\Delta P = \frac{K \rho V^2}{2gc}$$

where

V = Valve inlet connection velocity.

gc = Gravitational constant.

The pressure drop is calculated at the actual flow rate at the valve's set pressure. The ASME Code (*5*) requires the rated (nameplate) valve flowrate to be 90% of the actual (measured) flowrate. Inlet pressure drop calculations need the actual flowrate at the set pressure, because chattering occurs at those conditions, not at sizing conditions.

The short-cut method of Eq. 2 has the following additional assumptions:

- Sonic flow at the valve orifice
- Constant isentropic coefficient of 1.4
- Orifice coefficient of 0.975

3% limit: its merits and drawbacks

API-RP-520's criterion to prevent chattering is straightforward. Inlet pressure drop must always be less than 3%. Therefore, the blowdown must be greater than 3% (although API-RI-520 does not set limits for blowdown). There are no inherent design limitations that prevent valve blowdown below 3% (although it may be difficult). Nor does API-RP-520 specifically prohibit blowdowns below 3%. Valve inlet lines can be easily sized to meet the criterion if the guidelines presented here are used. The 3% limit, however, does have drawbacks.

Limiting inlet pressure drop to 3% is not always the optimum method of preventing chattering and it can limit

Table 1. Velocity heads fitting losses (*2, 3*).

Fitting	*k*
Square-Edged Inlet Nozzle	0.5
Protruding Inlet Nozzle	0.8
90° Long Radius Elbow	0.3
Concentric Reducer	0.1
Scored Reverse Buckling Rupture Disc	1.5
Full Area Stop Valve	0.1
Three-Way Full Area Value	0.1
Ball Valve	0.1

1. The piping loss is calculated as:

$$K = 4f\frac{L}{D}$$

where, in consistent units: L = pipe length, D = inlet ID, and f = Fanning friction factor.

2. Fitting losses of inlet line whose diameter differs from the valve inlet flange diameter are corrected by multiplying the fitting losses by:

$$\left(\frac{\text{valve inlet connection inner diameter}}{\text{fitting inner diameter}}\right)^4$$

High inlet frictional pressure drop.

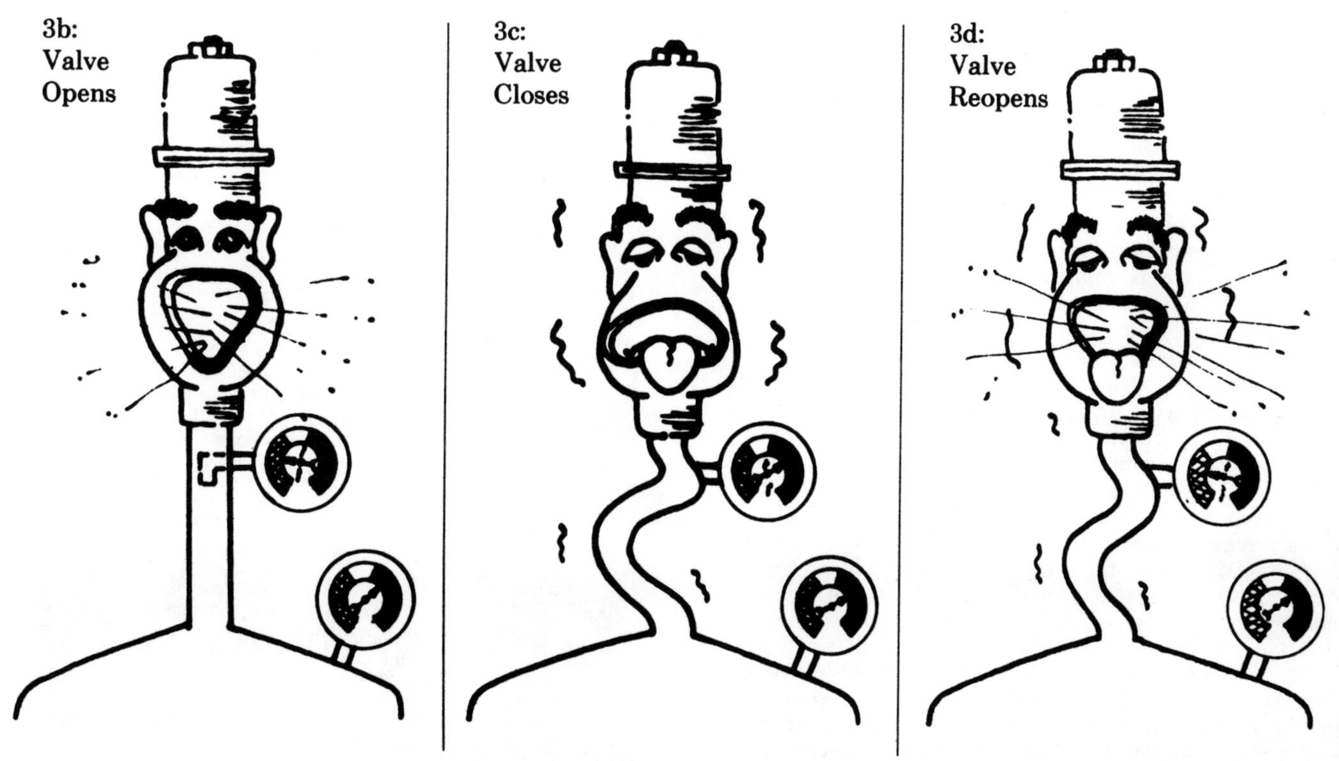

Figure 4. Well-designed inlet line.

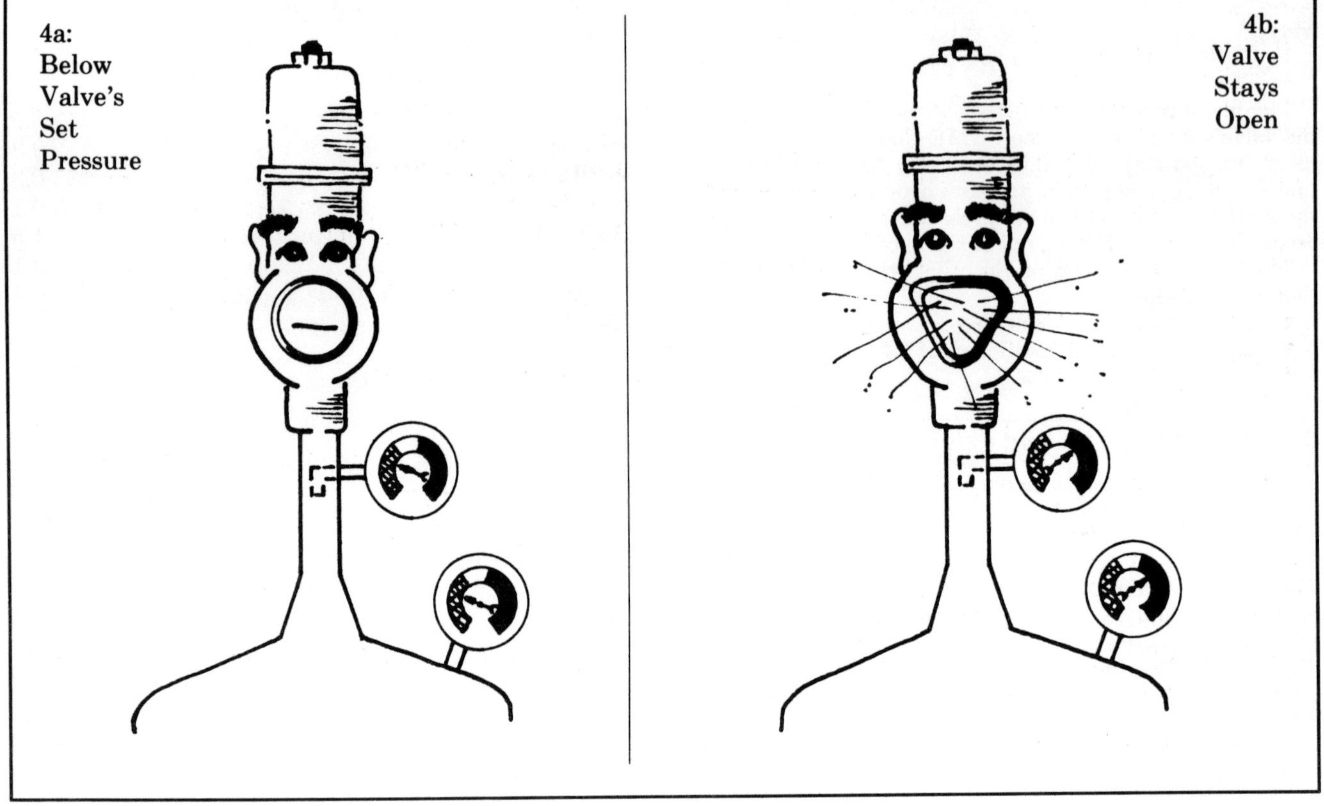

the designer's flexibility in certain instances. Valves with relatively large ratios of orifice area to inlet connection area (>0.4) require enlarged inlet lines to stay below 3% drop. The additional cost of enlarged inlet components can be considerable.

High costs can be incurred when process changes require increases in relief capacity of existing equipment. Replacement of the vessel nozzle and even vessel heat treating may be needed to stay below the 3% limit. A lower cost alternative not allowed by the 3% criterion would be to adjust the valve blowdown to be greater than inlet pressure drop.

The weakness of API-RP-520's criterion is that the blowdown is frequently unknown, and cannot be easily measured. The blowdown can be adjusted and manufacturers do recommend how to set the blowdown, but the value of the blowdown is seldom determined. While the 3% criterion greatly reduces the possibility of chattering, chattering can be eliminated only by proper testing and adjustment of field installations.

An additional deficiency of the 3% criterion is its failure to recognize the combined effect of inlet pressure drop and backpressure on chattering. Backpressure alone can cause chattering; when combined with high inlet losses, it can be particularly troublesome—a situation not covered by API-RP-520 recommendations.

For example, if an unbalanced valve were installed with a 2.5% inlet drop and an 8% backpressure, would it chatter? What blowdown setting would prevent chattering? The answers could be determined only by testing the individual model valve. The potential cost savings to the user of a more flexible criterion (that would require a knowledge of blowdown settings for each valve) are apparent. Testing expense to obtain this information, although substantial, could be prorated over future installations.

Obviously, there is a need for information on every marketed safety valve. Until that information is available, the designer is left to follow criteria such as API-RP-520 and such methods as presented here. #

Literature cited

1. *"American Petroleum Institute Recommended Practice 520,"* 4th ed., Part II, Para. 2.2, Washington, D.C.
2. Crane, "Flow of Fluids," Technical Paper No. 410, Crane Co., New York (1976).
3. Simpson, L. L., and M. L. Weirick, *Chem. Eng.* (April 3, 1978).
4. McCabe, W. L., and J. C. Smith, "Unit Operations of Chemical Engineering," McGraw-Hill, New York (1967).
5. ASME Boiler and Pressure Vessel Code, Section 8, American Society of Mechanical Engineers, New York (1979).

O. Cox, Jr. is a senior engineer in the Engineering and Hydrocarbons Div. of Union Carbide Corp. For the past 15 years, he has worked extensively in the field of venting technology on both new projects and plant improvement projects.

M. L. Weirick is an advanced engineer in the Engineering and Hydrocarbons Div. of Union Carbide Corp. A fluid dynamics specialist in the Heat Transfer and Fluid Dynamics Technology Group, he earned a B.S. degree at Virginia Polytechnic Institute. He is a registered professional engineer in West Virginia.

DISCUSSION

CHARLIE SOLOMON, Essochem Europe, Inc.: Perhaps you can answer a question which has been bothering me for some time. Which flow rate should be used for checking that the pressure drop in the inlet lines to safety valves do not exceed 3% of the set pressure?

API 520 states that you should take the "maximum" rated flow through the valve, but many designers use the design flow rate. You mentioned in your paper that when a safety valve opens it pops open or almost fully open so that, at that instant, the flow through the valve must be close to the maximum capacity of the valve and quite independent of the design contingency flow rate. Which flow rate do you recommend be used for checking the 3% rule?

MARTIN L. WEIRICK, Union Carbide: Actual Capacity depends a little bit on the definitions passed in the ASME Codes in the 1960s that requires a safety valve to vent 111% of the design. In other words, the safety valve has to be tested and a coefficient put on that valve so that it is rated at 90% of what it will vent. The actual conditions we recommend for calculating the chatter is at the set pressure; the actual flowrate, not the rated 90% of what's tested but the 111% of the design flow rate calculated using the set pressure and the valve sizing equation. That is our opinion of the correct flow rate. Chattering occurs at the opening pressure, not at 110% of the opening pressure. If you rise some percent above the opening pressure, the chattering may cease. Does that answer your question?

SOLOMON: Does that mean you don't take the design capacity, what you are designing for?

WEIRICK: No, because you always overrate the safety valve. You want to use the capacity of the valve not the process needs. Let's say you come up with venting requirements of 50,000 kg/h; the valve you have to buy passes 55. Well, it is rated to pass 55; you divide by 0.9 and will come out something like 63. That flow rate would be used to calculate the pressure drop. Also, if you're calculating back pressure you use that number, because that's what is really going through the valve; you want to use a number that is really happening.

SOLOMON: Right, thank you very much.

WEIRICK: Well, I'm aware of this problem. I'm not sure if the sum of the inlet and back pressure should be 10% or if it should be 9%. But, the actual force balance on the piston is such that I don't believe it would be exactly at 10%. I think you would have to take a more detailed analysis than we have to determine what happens to the valve when it opens. I mentioned a similar case in the paper as an example. Say, you had a valve at 8% back pressure and 2% inlet loss, well that valve obviously meets the published criterion but I don't think you can tell me if it will chatter. We think the only way to determine whether the valve will chatter is to test it. As Otto said, there is no where in the United States we know of where you can test the blowdown on a valve larger than 3 in. in diameter.

OTTO J. COX, JR., Union Carbide: I'd like to add something to what Marty just said. The built-up back pressure tests that I have seen done have been under ideal conditions for the inlet pressure drop and in fact the inlet pressure drop for all practical purposes was nil. Then, they begin to pinch down on the outlet piping to the point where the valve chatters and this 10% rule really comes into effect under those conditions. Any other questions?

S.S. GROSSEL, Hoffmann-LaRoche: I was discussing this problem last week with one of our instrument engineers who actually sizes our relief valve and rupture discs from the process information which we supply. The question came up about chattering and how to avoid it. The rule of thumb for unbalanced safety valves is that the back pressure should not exceed 10% of the set pressure. What he told me, and I don't know this for a fact, and I would like your opinion about this, is that this value really should also include the inlet pressure drop so that really it's 3% on the inlet and 7% on the outlet, for a total of 10%.

Explosive Evolution of Gas in the Manufacture of Ethyl Polysilicate

This incident proves that a very high energy release may not be necessary if the reaction is fast and can result in the production of large quantities of gas.

S.J. Skinner
Monsanto Ltd., London, U.K.

This incident, which occurred in late 1978, resulted in considerable property and consequential losses. The most probable cause was considered to be formation of separate layers between the two reactants. When this occurs the rate of reaction is limited by diffusion to the interface and may be very slow. However, the gas evolved from the reaction can cause some agitation which in turn leads to further gas evolutions and agitation, and thus the reaction once started can proceed at a rapidly accelerating (and uncontrolled) rate. Although the present incident relates to a specific reaction it is considered that there must be many other different reactions where a similar situation could occur.

PROCESS

The process (see Figure 1) involves a reaction between silicon tetrachloride and industrial methylated spirits (I.M.S.) containing some water. The reactants are fed simultaneously to the reactor, a stirred steam-jacketed, vessel. The initial reaction is normally rapid and endothermic and is followed by further hydrolysis and condensation reactions. Hydrogen chloride is evolved during the reaction and piped to an adjacent plant for recovery.

The vessel was fitted with a rupture disc.

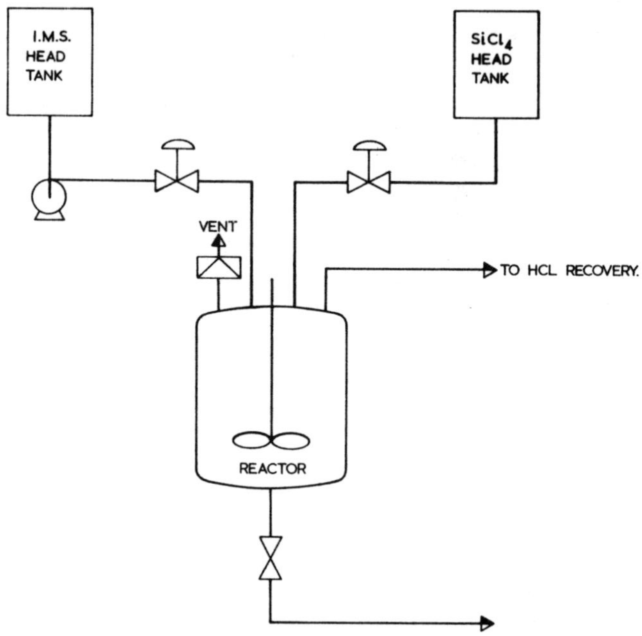

Figure 1.

0149-3701/81/4395 $02.00 © 1981 AIChE

INCIDENT

The operator was running two similar reactors. The process was nearing completion in the first reactor and the operator was between this vessel and the nearby foreman's office (see Figure 2) when the top cover of the No. 2 reactor was ejected upwardss and he was blown a distance of one metre against nearby handrailing. The plant was enveloped in hydrogen chloride fumes. The operator escaped by a stairway and fortunately was not seriously injured.

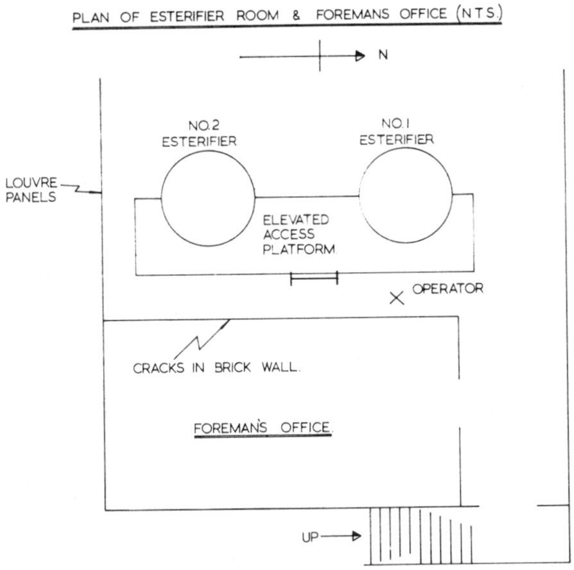

Figure 2.

DAMAGE

The reactor, which was a glass-lined vessel, was extensively damaged. The clamps around the lid were deformed and the lid was blown off and distorted. The agitator motor lodged in steelwork above the reactor and the gearbox was blown 13 metres.

Extensive damage was done to the concrete floor and to steelwork above the agitator.

Figure 3 shows a general view of the damage to the reactor.

The investigations included a careful assessment of the damage done and the possible energy release which was estimated as less than $\frac{1}{2}$lb (0.226 kg.) TNT equivalent. The reactor was an old one, most of the clamps being 25 years old and of an outdated design. It seems likely that the clamps failed prematurely at around 200 psi (1380 kPa). This may have reduced the effect of the explosion and there was no damage to the rest of the vessel. Other damage was consistent with the calculated energy release.

The pressures needed to distort the head to the degree observed would have been about 650 psi (4480 kPa) but it is not thought that this pressure was approached and that the deformation was the result of a number of other factors, resulting in assymetrical stressing of the head.

Figure 3.

RESULTS OF INVESTIGATION

A full investigation of the incident was carried out. The following possible causes were considered:

1. Combustion of hydrogen evolved by a reaction between exposed mild steel and hydrogen chloride.

2. Combustion of hydrogen evolved in a reaction between trichlorosilane and hydrogen chloride.

3. Combustion of ethanol-air mixture.

4. Reaction between $SiCl_4$ and water.

5. Reaction between $SiCl_4$ and alcohol.

Of these possible causes only the last was consistent with the calculated energy release (i.e. the TNT equivalent), the pressures needed to effect the observed damage and the other observations and measurements carried out. Subsequent laboratory work confirmed this as the most likely cause.

It was noted that the entry pipes into the vessel were situated so that the denser reactant $SiCl_4$ was fed to the bottom and layering was likely to occur. The difference in density between $SiCl_4$ and industrial methylated spirits is high and increased the probability.

The cause implied that agitation was not present, and that the reactants had formed two layers. This presupposed that the reactor agitator was not switched on due to an oversight, or that the electrics failed, or the belts slipped. All were possible. The plant was poorly instrumented, it was not easy to see whether the agitator was rotating and the operators relied on sound. In retrospect, the fact that the plant had run without a major incident for over 20 years indicated a remarkable combination of operator liability and good fortune.

IMPROVEMENTS IMPLEMENTED

The accident investigation committee drew up a list of recommendations, many of which were effected before the plant was restarted. The major recommendations related to better instrumentation. This included a positive means of indicating and alarming the agitator drive shaft rotation. Agitator rotation was also interlocked to the charging valves so as to ensure that the agitator was running prior to charging reactants. This did not allow for the possibility of loss of agitator blades which was considered very unlikely to occur without prior warning.

The dip pipes were modified so that they were the same length as the agitator thus reducing any tendency to formation of layers.

CONCLUSIONS

To summarise, the reactants were of very different densities and formed distinct layers which reacted very slowly initially. Once the reaction had started, however, the agitation, promoted by rapid evolution of hydrogen chloride, caused rapid acceleration of the reaction with sufficiently fast evolution of gas to overpressurise the reactor and lift the top. The improvements made were to ensure agitation during addition of reactants. Although this particular case involved $SiCl_4$ and industrial methylated spirits, a similar situation may be possible with many other reactions. As in this case, a very high energy release may not be necessary if the reaction is fast and can result in the production of large quantities of gas.

It should be noted that the rate of reaction from such an incident can be virtually instantaneous so that the evolution of gas is extremely fast. It is therefore difficult to design a satisfactory venting system as calculations show this to be extremely large if based on the worst case. In the present instance the disc had been shown to be inadequate and the largest practicable size was fitted, based on existing manhole covers. This decision was taken in the light of the relatively high design pressure of the standard reactor. However, it is considered that the first emphasis in such a situation should be to reduce the probability of such an incident occurring.

SKINNER, S.J.

CASE HISTORIES ON DUCT FIRE INCIDENT

W.H. Doyle, Consultant

I do have a couple of brief case histories. One accident occurred at a refinery operation where there was a change in management and the engineering group had been going over the safety situation with a fine tooth comb. They found that butane was being stored in spheres where the metallurgy was such that in the winter the metal was below its null-ductility temperature.

Of course, in the winter, the pressure of butane is low but they didn't like the idea. So, they increased the safety by installing heaters; I believe they were steam, in the spheres, thermostatically controlled to maintain the temperatures of the liquid above the null-ductility temperature of the metal; a step forward in overall safety. They used a thermostat for temperature control, it turned out to be unreliable. One of the spheres overheated in the nighttime (a January nighttime), butane came out and collected in a cloud along the ground; the gaugers drove into the cloud—two of them—their vehicle stalled. They spent about 10 to 15 minutes trying to start the vehicle and one of the efforts resulted in igniting the cloud and they both died of their burns.

I was reminded of the other one by Damon Snow's paper on duct fires. This is a duct fire with a difference. It was an operation where a viscose solution is made up using carbon disulfide. The solution is forced through spinerettes and run through a sulfuric acid bath, rinsed, and the fiber wound in a cake. The fumes are collected in a hood and sucked out and dissipated through a high stack so it is possible to survive in the neighborhood without being overcome with mecaptans, etc., etc., etc. The duct work was fiberglas-reinforced plastic and led down to a fan at the ground level, the spinning tanks were one floor up above grade level.

The fan discharged into a basement duct that went out into the brick stack. Because of corrosion problems, the fan was lined with plastic and the fan blades were constructed of plastic. The fan bearings were inboard. The composition of the mixture being discharged through the fan was at about 5 to 8% of the lower explosive limit; no problem. However, one of the inboard bearings did heat and the auto ignition temperature of carbon disulfide is on the very low end of the scale on the order of 212°F (100°C): Bob Coffee can do it lower; other people can do it higher.

But, the temperature of the bearing got hot enough so the vapor was incinerated on the bearing and incineration of carbon disulfide even in a very dilute stream produces heat. The heat aggravated the bearing problem and the fan blades began to soften and the fan lost efficiency. Therefore, the concentration in the duct gradually went up. Eventually, it got high enough, so fire flashed back to the spinning bath; the spinning baths were metal filled with sulfuric acid; the ducts were overhead. This was a reinforced concrete roof deck on steel supports; obviously not an area that needs sprinkler protection; nothing to burn. You people may have heard that phrase before.

The carbon disulfide coming off the bath started to burn and after the baths were installed, they had corrosion problems so they were covered about 1/2 in. thick with a flammable resin, which caught fire and there was enough fire from the resin to drop the roof. And, they had themselves a real nice accident; no people were killed, so it didn't get beyond the local newspapers. That was a really unusual accident and I am reminded of it because the first reports we got were that this was essentially a duct fire and that started research on protection of ducts against fire and some of the developments that Damon showed for protecting ducts were a direct outgrowth of the misinterpretation of what caused the problem there. Thank you.

INCIDENT INVOLVING A LOADED ANHYDROUS AMMONIA ROAD TANKER

E.A. George
ICI
England

In August, 1978, an articulated road tanker loaded with 17 Tes. anhydrous ammonia overturned at the Four Went Ways roundabout at the junction of the A.604 and A.11 trunk roads near Cambridge, England. The steps taken to deal with the problem and some of the lessons learned are discussed here.

Probable reasons for roll-over

So called "slow speed rollovers" during certain types of road vehicle maneuver have, regrettably, been on the increase in the U.K. over the past 3-4 years. The increase cannot be accounted for purely by an increase in the number of articulated vehicles in use.

The mishap usually occurs during a maneuver involving a reverse curve or "S" motion. The term "slow speed rollover" is, perhaps unfortunate, implying to some drivers that increasing speed helps. It does not. Experiments show that the typical U.K. 32-ton gross vehicle weight articulated petroleum tanker has a critical speed as low as 18 mph. Some of the factors which appear to need consideration are:

1. Roll resistance of the suspension
2. Height of center of gravity
3. Power assisted steering
4. The high torque diesel engine
5. Driver awareness
6. Roll resistance of the kingpin/5th wheel coupling
7. Road design
8. Effect of 'slosh' of liquid loads

It is my view that before factors 3 and 4 became standard equipment on U.K. heavy goods vehicles, the problem was much less severe. Without power steering, a driver cannot speed through a roundabout—the effort of putting a left-hand lock, right hard and left again, is too great. Having been forced to slow to a crawl, without the high torque diesel, acceleration to cruising speed takes time.

With the advent of power steering and the high torque diesel, a driver can apply lock fast and accelerate quickly. If he accelerates after his cab has approached to straighten up on leaving a roundabout but before his trailer has left the curve, he is in trouble. It is thought the rollover of this ammonia tanker is the "slow speed" type. With no conflicting traffic on a quiet Sunday, it is possible the driver, experienced though he was, took the roundabout at too high a speed.

Figure 1. Overturned road tanker with anhydrous ammonia.

The road tanker overturned, ending on its side with the barrel resting on a kerb. Markings on the road and witness evidence, suggest "slow speed rollover" and that the vehicle did not slide far after turning.

Apart from a wisp of ammonia gas from a small amount of liquid trapped in the base of the pump, there was no leakage of product, nor was any cracking found in the pressure shell after the incident. Indeed, the only spillage needing attention was diesel fuel.

Action taken and lesson learned

The driver was able to take the first steps by "phoning (from a nearby Transport Drivers' Cafe)" the emergency center at his base and the Police. The nearest Company emergency center was alerted and set off for the scene.

When the emergency team arrived, they could see two impact dents on the barrel, but could not tell if other damage had occurred. They decided to adopt the Standard Pressure Difference method transfer. This involves connecting a receiving ammonia road tanker to the cripple, liquid-to-liquid. The vapor space of the receiving tanker is connected to the liquid line of a tanker of water. As ammonia vapor is absorbed in the water the pressure drops in the receivng ammonia tanker.

Product transfer was adopted to minimize the risk of there being a major release of ammonia, if by lifting the vehicle full the shell failed as a result of damage not visible beforehand.

The position of the vehicle on its side, and the consequent relation of shell nozzles to liquid level, made the operation of transfer a slow one. In the event, part of the load was transferred, radio isotope detection being used to monitor contents of the cripple.

After three hours, the level had been reduced to a point where it was decided safe to roll the cripple from its side onto its back. By this time, cranes and ICI lifting experts were on the scene. A second transfer started and continued until pressures balanced off. The cripple was then righted and the remaining ammonia transfered.

The whole operation from accident to reopening the road took 24 hours.

The main lessons from a well-handled incident, leaving aside the vehicle design, were:

- The emergency services in a predominantly rural area did not fully appreciate the problems of the recovery team. A course on chemical emergencies was arranged by the U.K. Chemical Industry and attended by the area's emergency services. This course was extremely well received.
- Better communications, on-site temporary facilities, and transport arrangements for team members were seen as important. Partly as a result of this incident, the Division decided to augment its emergency team equipment with back-up trailer. This trailer can be hauled by a standard motive unit and provides workshop, office and amenity facilities.
- A barrel constructed from a high-strength steel might well have cracked in an incident of this kind. The stress-relieved carbon steel shell of the vehicle in the accident was tough enough not to crack.

CASE HISTORIES ON TRANSPORATION

Trevor Kletz
ICI
England

Most of these incidents didn't occur in ICI, and that I'm grateful to all the companies concerned for allowing me to describe them.

Case 1. A tall crane, about 60-ft tall, was left standing overnight. A breeze came up and it blew over. By the greatest good fortune, it fell along the length of the roadway; if it had deviated 20 degrees to the left or to the right, it would have fallen on equipment and there would have been a serious incident.

Case 2. A long crane was slightly overloaded and fell over so that it balanced on two of the outriggers, with the end of the jib resting on the plant. The crane was not lifting a heavy weight, just a valve weighing about l00 lb, but it was doing so at the maximum extension of the jib. The driver set the jib in the right direction and then he extended it to its maximum length. When he lifted the relief valve, it was just enough to tip it over. The alarms on this crane were in full working order and they satisfied all the codes of practice that we had. But, those codes of practice were written for the mechanical strut crane. A hydraulic crane has an additional degree of freedom; as well as being able to turn the jib and raise and lower it, you can lengthen the jib. As you've got an extra degree of freedom, you really need an extra alarm but nobody thought of that. The message of this incident is even if you are following all the rules you may not be safe, because the rules may be out of date.

Case 3. A tank during the construction was damaged by a high wind. Tanks are designed to withstand winds, but we didn't realize until this happened during the construction period, before the roof is fitted, a tank will not withstand a strong wind.

Case 4. This is a tank that has been sucked in. We all do this regularly in a variety of ways. This tank had been cleaned for a special grade of product and somebody tied on a sheet of polyethylene over the vent to keep it nice and clean. When the tank cooled down, it was sucked in!

Case 5. Two air cooler fans, side by side, are called J345A and J345B. B had to be overhauled, so it was defused and a clearance issued for tho ovor haul. By mistake A was overhauled instead. The motor was disconnected and later rewired while the power was isolated only by the starter.

While this was taking place, neither fan was in use. When the overhaul was nearly complete, one of the fans was needed and as the clearance book showed that B was being overhauled, A was started. Fortunately, by this time, the overhaul of A was complete except that the guard had not been put back on the fan. The mistake occurred because three things were wrong:

(i) The fans were arranged, Figure 1. The B label was on the side of B cooler furthest away from B fan and near A fan. Not surprisingly, the men doing the overhaul assumed that B fan was the one nearest the B label. (The labels have now been moved.)

(ii) The fans were on the structure, 20 m above the ground. The start/stop buttons were on the ground, out-of-sight. When the operator pressed the A button, he could not see that the fan which started had no cover on it. The start/stop buttons have now been placed near the fans and the emergency stop buttons—previously near the fans—are now at ground level.

(iii) The process and maintenance supervisors should have visited the job, while it was in progress. One of them might then have noticed that men were working on the wrong fan.

Case 6. The roof of a tank was blown off. By a piece of extraordinary good luck, it missed all the equipment and landed in the drum park; it even landed on a bit of ground that was not covered with drums!

There were two methanol tanks, an old one in service and a new one that was still being built. Unknown to the operating team, the construction team connected up the nitrogen supply to the new tank. The construction team said afterwards that they wouldn't have connected up the methanol line, but they thought it would be alright to connect up the nitrogen line. They didn't realize that the two tanks were designed to be on balance with each other and that the nitrogen system contained methanol vapor and blew the roof off the tank.

Case 7. A reactor got too hot; a runaway reaction occurred; the bursting disc failed to lift. The joints on the reactor failed and the contents came out and set solid; the whole plant covered in a brittle solid, which looked exactly like molasses candy (treacle toffee).

The bursting disc had a vacuum and the vacuum support instead of being put below the disc was put on top. Fortunately, the reactor didn't burst; it was a fairly low pressure rating, 150 psi: at that pressure rating the flanges will give and relieve the pressure.

Case 8. Figure 2 shows a batch reaction system. A batch of glycerol is placed in the reactor and

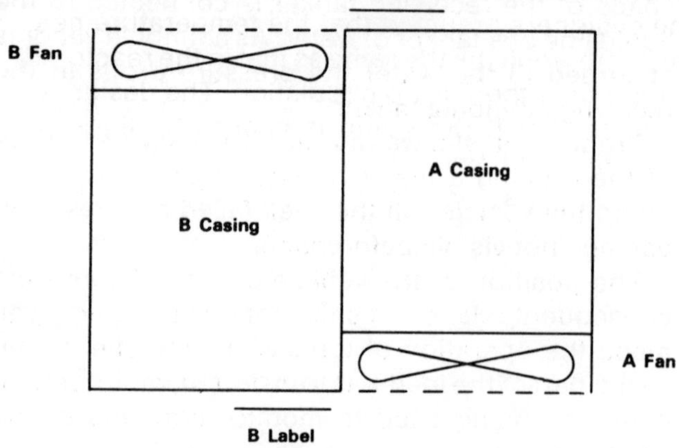

Figure 1. Arrangement of the fans.

circulated through a heat-exchanger which can act as both a heater and a cooler. Initially, it is used as a heater and when the temperature reaches 115° C, addition of ethylene oxide is started. The reaction is exothermic and the exchanger is now used as a cooler.

(a) The circulation pump is running.

(b) The temperature is above 115° C. Otherwise, the ethylene oxide will not react.

(c) The temperature is below 125° C. Otherwise, the reaction is too fast.

Despite these precautions an explosion occurred. One day, when ethylene oxide addition was started, the pressure in the reactor rose. This showed that the ethylene oxide was not reacting. The operator decided that perhaps the temperature point was reading low, perhaps a bit more heat was required to start the reaction, so he adjusted the trip setting and allowed the indicated temperature to rise to 200° C. Still, the pressure did not fall.

Always try to measure the property we wish to know directly, rather than another property from

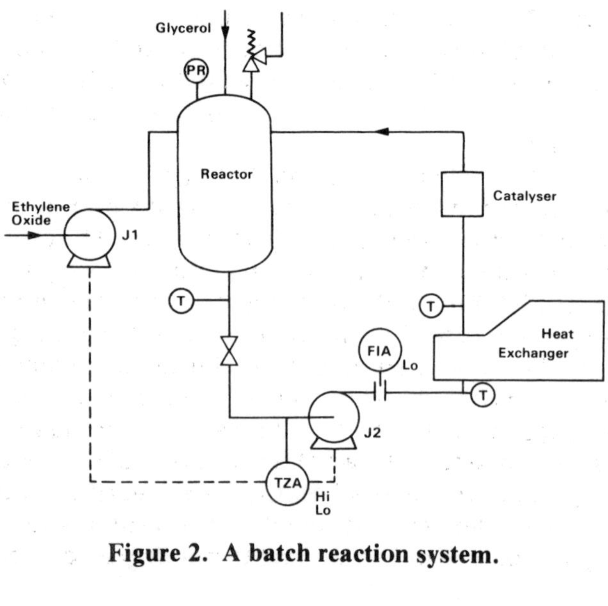

Figure 2. A batch reaction system.

which the property we wish to know can be inferred. The designers assumed that the temperature near the pump would be the same as that in the reactor. It will not be if there is no circulation. The designers assumed that if the pump is energized, liquid is circulating. But, this not always the case. If you were taking part in an operability study, would you have picked up these points?

- More effort might have been made to keep the flow indicator alarm in working order.
- A high-pressure trip should be installed on the reactor.
- Operators should be trained to "look before they leap," when they find valves wrongly set.

DISCUSSION

ANONYMOUS: Was the reactor fitted with a relief valve?

KLETZ: No relief can cope with a runaway reaction of this sort. The vessel was fitted with a relief device, but if you get a violent runaway reaction due to an excess of ethylene oxide, no relief device can help you. You've got to put your trust in a good control system. This brings me back to the comment I was making to Stan Skinner earlier. Whatever the arguments for or against instruments in place of relief valves, in cases like this no relief valve is going to help you and you have got to rely on instruments.

GENE DEHAVEN, Dow Chemical I have missed the function of the flowmeter and the discharge of the pump on the diagram you just showed. Could you review that, please?

KLETZ: If the low flow alarm had been working, the operator would have known that there was no circulation. This flow indicator had never worked since the startup of the plant and after reporting it a few times, the operators gave up worrying about it and ran the plant on temperature alone. They found they did not need the flow indicator.

CASE HISTORY ON TRANSPORTATION

E.A. George
ICI,
England

Referring to what was said by Trevor Kletz on the lock, we found that it is possible that it contravenes the United Kingdom Construction & Use Regulations which, in general, take no real notice at all of anything above the wheels. The regulations are primarily concerned with the automotive scene and interfering with the truck braking system may not be allowed. We are, therefore, developing a new system using security locks approved by the Ministry of Defence. We think these are probably the only locks of a quality likely to stop drivers finding ways around them.

I would like to explain about the scenes of our rollover accident and the arrangmenet underneath the vehicle. There is a hydraulically driven cargo pump and there are various openings and valves. Because these tankers have never been subjected to operability studies, modifications have been made when drivers said "wouldn't it be a good idea if ...." So, there were a large number of purge valves, cross-overs, additional points, etc.

At Billingham, we produce over 4,000 tons per day of ammonia from four ammonia plants and distribute it to downstream plants, to road filling, sea filling, rail filling and storage tanks. We also have a low-pressure ammonia gas system which takes vapor return from a number of plants including road filling and distributes it to various plants. Consider a new road filling facility, a tanker at the filling point, an articulated arm connected to the liquid fill point on the station, and a hose connecting the vapor return to the plants. Now, under these conditions, the road tanker barrel is an ammonia plant vessel and not a transport vessel, being connected right through the main system to the ammonia plants, albeit indirectly. Consider now an operability study and in particular four points.

First, *Flow.* When we applied the techniques of low flow, back flow, too much flow, we came up with the possibility of failure of the articulated arm or of leakage. We found that the sophisticated pneumatic operated trip system on the plant could not operate the internal valve in the tanker. Unless the leak was so large that the tanker excess flow valve operated, we would have had to wait for that tanker to discharge itself. So, all future designs (and we will retrofit the existing ones) will have air cylinders operating the internal valves, these cylinders connected by a tether line to the plant trip system so that operating the plant trip automatically closes off the tanker.

Now, incidentally, this provides one with a bit of "drive-away" protection, if the tether line is correctly adjusted. The first thing to break is the tether line; I don't mean you want to drive it away, but at least a valve will close if someone does. Also, use of air means that the vehicle, if parked up overnight while the driver is taking a rest, cannot have the internal valve opened by children unless they obtain an air cylinder!

Secondly, *Pressure*. We don't have relief valves on our ammonia tankers in the United Kingdom. It is not allowed, and we discovered that the whole of the Billingham site ammonia system was designed to a higher pressure than the tankers. There was, thus, no protection for the tankers under certain circumstances. We had to modify the filling point to protect the tanker.

Thirdly, *Temperature*. On the Billingham site, we produce ammonia at -33° C, at -6° C and above 0° C. The road tankers are not allowed to be moved if the steel is below 0° C, because they are not built from impact tested steel. We have had to introduce low-temperature alarms and trips to the filling point. But, if a leak develops as the pressure drops, the temperature drops and this situation can be a problem.

With liquid CO_2 it is much worse, we discovered that some customers actually want liquid CO_2 at below normal pressure. We now have to have selected tankers to deal with those customers, which is a headache for the traffic planners. But, again, all future vehicles will be built, in the case of ammonia, for -33° C; for liquid CO_2 for -50° C, to get over this problem.

Fourthly, *Contaminants*. You might say "what on earth can we contaminate in an ammonia road tanker." But, if you remember that the tanker should be inspected internally at regular intervals, we are going to need to get rid of the ammonia and to introduce air. We found, on investigation, that because there was no facility at Billingham at the time to empty tankers (it was only there to fill them), the road transport people were so fed up with delays, they took it upon themselves to prepare tankers for entry.

This meant, of course, they were handing things over to themselves which is against our normal safety rules. But, fortunately, our road transport depot, although adjacent to the factory, is in a wilderness, and they were discharging ammonia to atmosphere when no one was looking, filling the tanker up with water and emptying the water down the drain, then certifying that it was safe to go inside. Also, they returned the tanker full of air but didn't bother to tell the filling point. We then realized why it was that some of the downstream plants had mysterious tripouts due to oxygen alarms.

It was simply that one of these tankers had been returned to the filling point and a slug of oxygen had got into the ammonia return line. So, we have now taken the opportunity of converting the old road filling point, rendered redundant by the new ones, into a point for preparing vehicles for maintenance and ammoniating them after return. There is a more serious point and that is that by having oxygen in the tank and then letting ammonia in, there is a remote risk we might induce stress corrosion cracking of the tanker. So, we cleared that one up as well.

Operability study techniques, treating a tanker as a piece of production plant, produced some very interesting results. In fact, it is now a rule of the Division that all liquified gas tankers for which we propose changes in design must be treated as pieces of process plant. We have also, in fact, improved some of the customer-end installations and made changes on liquid CO_2 tanks which have saved weight. Unfortunately, on ammonia tankers we have to add weight to make them safer, so we are losing a bit there on payload.